Advanced Oxidation Processes for Water and Wastewater Treatment

Contents

About the Editors vii

Marco S. Lucas, José A. Peres and Gianluca Li Puma
Advanced Oxidation Processes for Water and Wastewater Treatment
Reprinted from: *Water* **2021**, *13*, 1309, doi:10.3390/w13091309 1

Khanh Chau Dao, Chih-Chi Yang, Ku-Fan Chen and Yung-Pin Tsai
Recent Trends in Removal Pharmaceuticals and Personal Care Products by Electrochemical Oxidation and Combined Systems
Reprinted from: *Water* **2020**, *12*, 1043, doi:10.3390/w12041043 5

Jin Ni, Huimin Shi, Yuansheng Xu and Qunhui Wang
A Comparison of the Mechanism of TOC and COD Degradation in Rhodamine B Wastewater by a Recycling-Flow Two- and Three-dimensional Electro-Reactor System
Reprinted from: *Water* **2020**, *12*, 1853, doi:10.3390/w12071853 35

Yu-Jung Liu, Yung-Ling Huang, Shang-Lien Lo and Ching-Yao Hu
Comparing the Effects of Types of Electrode on the Removal of Multiple Pharmaceuticals from Water by Electrochemical Methods
Reprinted from: *Water* **2020**, *12*, 2332, doi:10.3390/w12092332 51

Jian Wang, Yonghui Song, Feng Qian, Cong Du, Huibin Yu and Liancheng Xiang
Removal Characteristics of Effluent Organic Matter (EfOM) in Pharmaceutical Tailwater by a Combined Coagulation and UV/O$_3$ Process
Reprinted from: *Water* **2020**, *12*, 2773, doi:10.3390/w12102773 61

Xiaofan Lv, Yiyang Ma, Yangyang Li and Qi Yang
Heterogeneous Fenton-Like Catalytic Degradation of 2,4-Dichlorophenoxyacetic Acid by Nano-Scale Zero-Valent Iron Assembled on Magnetite Nanoparticles
Reprinted from: *Water* **2020**, *12*, 2909, doi:10.3390/w12102909 73

De-Ming Gu, Chang-Sheng Guo, Qi-Yan Feng, Heng Zhang and Jian Xu
Degradation of Ketamine and Methamphetamine by the UV/H$_2$O$_2$ System: Kinetics, Mechanisms and Comparison
Reprinted from: *Water* **2020**, *12*, 2999, doi:10.3390/w12112999 93

Philipp Otter, Katharina Mette, Robert Wesch, Tobias Gerhardt, Frank-Marc Krüger, Alexander Goldmaier, Florian Benz, Pradyut Malakar and Thomas Grischek
Oxidation of Selected Trace Organic Compounds through the Combination of Inline Electro-Chlorination with UV Radiation (UV/ECl$_2$) as Alternative AOP for Decentralized Drinking Water Treatment
Reprinted from: *Water* **2020**, *12*, 3275, doi:10.3390/w12113275 107

Vanessa Guimarães, Ana R. Teixeira, Marco S. Lucas and José A. Peres
Effect of Zr Impregnation on Clay-Based Materials for H$_2$O$_2$-Assisted Photocatalytic Wet Oxidation of Winery Wastewater
Reprinted from: *Water* **2020**, *12*, 3387, doi:10.3390/w12123387 127

Cristian Ferreiro, Natalia Villota, José Ignacio Lombraña and María J. Rivero
Heterogeneous Catalytic Ozonation of Aniline-Contaminated Waters: A Three-Phase Modelling Approach Using TiO$_2$/GAC
Reprinted from: *Water* **2020**, *12*, 3448, doi:10.3390/w12123448 143

Anna Serra-Clusellas, Luca Sbardella, Pol Herrero, Antoni Delpino-Rius, Marc Riu, María de Lourdes Correa, Anna Casadellà, Núria Canela and Xavier Martínez-Lladó
Erythromycin Abatement from Water by Electro-Fenton and Peroxyelectrocoagulation Treatments
Reprinted from: *Water* **2021**, *13*, 1129, doi:10.3390/w13081129 . . . . **169**

About the Editors

Marco S. Lucas

Marco S. Lucas is Assistant Researcher and Deputy Head of the Chemistry Department at the University of Trás-os-Montes and Alto Douro (UTAD). He holds a PhD in Chemistry, a Masters in Environmental Technology and a Degree in Environmental Engineering and Natural Resources. His main research interests are i) water and wastewater treatment, ii) advanced oxidation processes, iii) environmental chemistry, iv) biosorption and v) biologic processes. To date he has published 43 papers in top-ranked peer-reviewed journals, 1 patent, 2 book chapters and more than 100 conference proceedings and communications. He has an h-index of 24 and >2350 citations. He has been awarded with three prizes: i) the Dr. Francisco da Fonseca Henriques Prize, Portugal; ii) the APRH 2010/11 Prize, awarded by the Associação Portuguesa dos Recursos Hídricos, Portugal and iii) the Young Researcher 2013 - UTAD Prize.

José A. Peres

José Alcides Peres graduated in Chemical Engineering at the University of Coimbra, Portugal. He completed his PhD degree in Chemistry at University of Trás-os-Montes and Alto Douro (UTAD) in 2002 and obtained a Habilitation degree in 2011 at UTAD. He has been Associate Professor of Habilitation at UTAD since June 2013. He is the author and co-author of more than 60 full-papers published SCI journals. He has more than 200 publications in national and international conferences, 15 papers in national scientific periodicals, 1 book chapter and is co-inventor of 1 patent. Has been involved in 15 research projects and 7 enterprise projects. He has been working for 20 years in the development of Advanced Oxidation Processes (AOPs) such as Fenton reagent, homogeneous and heterogeneous photo-Fenton, ozonation, ferrioxalate/solar radiation and heterogeneous photocatalysis for industrial wastewater treatment, water disinfection and application on chemical contaminants degradation.

Gianluca Li Puma

Gianluca Li Puma is Professor of Chemical and Environmental Engineering at Loughborough University. His academic qualifications include: Laurea, Chemical Engineering (University of Palermo, Italy, 1992) MPhil, Chemical Engineering (University of Bath, 1993) PhD, Chemical Engineering (Hong Kong University of Science and Technology, 1996). He has more than 370 scientific publications (h-index 46, >6500 citations, >122 journal papers, 2 books for the RSC, 5 book chapters) and 2 patents. Professor Li Puma has achieved sustained evidence of research excellence and significant international standing and impact, research income, publications, editorial and committee membership presence. He is Editor of *"Applied Catalysis B: Environmental"*, one of the most prestigious catalysis journals focusing on catalytic elimination of environmental contaminants, environmental catalysts, catalyst for clean energy, clean catalytic combustions and photocatalysis.

Editorial

Advanced Oxidation Processes for Water and Wastewater Treatment

Marco S. Lucas [1,*], José A. Peres [1] and Gianluca Li Puma [2]

1. Centro de Química de Vila Real, Departamento de Química, UTAD–Universidade de Trás-os-Montes e Alto Douro, Quinta de Prados, 5000-801 Vila Real, Portugal; jperes@utad.pt
2. Environmental Nanocatalysis and Photoreaction Engineering, Department of Chemical Engineering, Loughborough University, Loughborough LE11 3TU, UK; g.lipuma@lboro.ac.uk
* Correspondence: mlucas@utad.pt

Citation: Lucas, M.S.; Peres, J.A.; Li Puma, G. Advanced Oxidation Processes for Water and Wastewater Treatment. *Water* **2021**, *13*, 1309. https://doi.org/10.3390/w13091309

Received: 26 April 2021
Accepted: 5 May 2021
Published: 7 May 2021

Publisher's Note: MDPI stays neutral with regard to jurisdictional claims in published maps and institutional affiliations.

Copyright: © 2021 by the authors. Licensee MDPI, Basel, Switzerland. This article is an open access article distributed under the terms and conditions of the Creative Commons Attribution (CC BY) license (https://creativecommons.org/licenses/by/4.0/).

Technical and scientific developments have facilitated an increase in human life expectancy and quality, which is reflected in a large growth of global population. This process has been accompanied by economic development and an immense urbanisation level. This growth has been followed by over-exploitation of water resources and an increase in environmental pollution. Therefore, the rising demand for clean and safe water is a major challenge. Water scarcity, elimination of emerging contaminants and escalating need for water reuse allied to limited treatment capacity of conventional water and wastewater plants are major barriers that need to be appropriately addressed.

Water and wastewater treatment with Advanced oxidation processes (AOPs) have received a great deal of attention in recent years, since these processes are known for their capacity to mineralize a wide range of pollutants into CO_2, H_2O and mineral acids.

In this Special Issue, the application of AOPs as an effective solution for the treatment of drinking water and urban/industrial wastewater was addressed. The removal of pharmaceuticals and organic pollutants through electrochemical oxidation processes ([1–5]), combined coagulation and UV/O_3 (Wang et al., 2020), UV/H_2O_2 system (Gu et al., 2020) and heterogeneous catalysis ([6,7]) are the main topics covered in this issue. In addition, the degradation of an agro-industrial wastewater through heterogeneous photocatalysis using Zr on clay-based materials ([8]) is also part of this special issue.

Since the call for papers was announced in March 2020 and after a rigorous peer-review process, ten articles were accepted for publication (52% of total submitted papers). This Special Issue of Water comprises contributions from over fifty authors and from five countries (China, Germany, India, Portugal and Spain). To gain a better insight into the essence of the Special Issue, we offer brief highlights of the published papers below.

1. Recent Trends in Removal Pharmaceuticals and Personal Care Products by Electrochemical Oxidation and Combined Systems

This review paper [1] focuses on the removal efficiency of pharmaceuticals and personal care products through electrochemical oxidation, alone or in combination with other treatment processes, in the last 10 years. The influence of reactor designs and configurations as well as electrode materials and operational conditions (e.g., initial concentration, electrolytes, current density, temperature, pH, stirring rate, electrode spacing, and fluid velocity) are also explored.

2. A Comparison of the Mechanism of TOC and COD Degradation in Rhodamine B Wastewater by a Recycling-Flow Two- and Three-dimensional Electro-Reactor System

This research article [3] designs a recycling-flow 3D electro-reactor system for the degradation of a dye wastewater. The mechanism of TOC and COD degradation of Rhodamine B wastewater was analysed in terms of mass transfer, electrochemically active surface area of electrodes, instant concentration of hydroxyl radicals, current efficiency and energy consumption in comparison to the recycling-flow 2D electro-reactor system.

3. Comparing the Effects of Types of Electrode on the Removal of Multiple Pharmaceuticals from Water by Electrochemical Methods

This work [2] investigates the effect of laboratory-scale electrochemical treatments on removing multiple pharmaceuticals (diclofenac, sulfamethoxazole and atenolol) from synthetic water and spiked wastewater by electro-coagulation and electro-oxidation using aluminium and graphite electrodes. The pharmaceutical removal with electro-oxidation was higher than those with the electro-coagulation process, which was obtained from a five-cell graphite electrode system, while the removal of pharmaceuticals with aluminium electrodes was about 20% (20 µM).

4. Removal Characteristics of Effluent Organic Matter (EfOM) in Pharmaceutical Tailwater by a Combined Coagulation and UV/O_3 Process

Coagulation and UV/O_3 processes are combined to remove the effluent organic matter from a biotreated pharmaceutical wastewater [9]. The removal behaviour of effluent organic matter by UV/O_3 process was characterized by synchronous fluorescence spectroscopy (SFS) integrating two-dimensional correlation (2D-COS) and principal component analysis technology. Five main components of pharmaceutical tail wastewater were identified by SFS. Spectral analysis revealed that UV/O_3 was selective for the removal of different fluorescent components, especially fulvic acid-like fluorescent (FLF) and humus-like fluorescent (HLF) components. The coagulation-UV/O_3 processes proven to be an attractive way to reduce the environmental risks of pharmaceutical tail wastewater.

5. Heterogeneous Fenton-Like Catalytic Degradation of 2,4-Dichlorophenoxyacetic Acid by Nanoscale Zero-Valent Iron Assembled on Magnetite Nanoparticles

Fe^0@Fe_3O_4 nanoparticles were synthesized in this study [7] and combined with hydrogen peroxide to obtain a heterogeneous Fenton-like system, which was further applied into the degradation of the 2,4-dichlorophenoxyacetic acid. The effect of different experimental conditions was assessed on the removal of 2,4-D, such as pH, hydrogen peroxide concentration, temperature and catalyst dosage. In addition, the possible mechanism of the Fe^0@Fe_3O_4 activated Fenton-like system was proposed.

6. Degradation of Ketamine and Methamphetamine by the UV/H_2O_2 System: Kinetics, Mechanisms and Comparison

The reaction kinetics and degradation mechanisms of ketamine (KET) and methamphetamine (METH) by UV/H_2O_2 are studied in this paper [10]. The influence of various operational parameters on KET and METH removal was evaluated: initial H_2O_2 dosage, pH and water background components. KET and METH degradation is inhibited by the presence of some anions (HCO_3^-, Cl^-, NO_3^-) and humic acid. The degradation products were analysed by UPLC-MS/MS, and potential transformation paths are proposed.

7. Oxidation of Selected Trace Organic Compounds through the Combination of Inline Electro-Chlorination with UV Radiation (UV/ECl_2) as Alternative AOP for Decentralized Drinking Water Treatment

In this study [4], a UV/ECl_2 system is tested in lab-scale conditions to assess chlorine production and radical formation changing chloride concentrations, pH, cell currents and UV energy input applied. Afterwards, a UV/ECl_2 pilot setting was tested under real conditions treating Elbe river water. The combination of ECl_2 with UV can be a feasible alternative to reduce Trace Organic Compounds without the need for any external chemicals and electricity supply.

8. Effect of Zr Impregnation on Clay-Based Materials for H_2O_2-Assisted Photocatalytic Wet Oxidation of Winery Wastewater

This work [8] successfully produced UV-activated Zr-doped composites through the impregnation of Zr on the crystal lattice of different clay materials (Zr-MT, (Zr)Al-PILC and (Zr)AlCu-PILC). The ((Zr)Al-Cu-PILC) photocatalyst was tested in the photodegradation of a real winery wastewater under UV-C irradiation, and the influence of Zr immobilization on the properties and photoactivity of the heterogeneous catalysts was assessed.

9. Heterogeneous Catalytic Ozonation of Aniline-Contaminated Waters: A Three-Phase Modelling Approach Using TiO_2/GAC

In this publication [6], the catalytic ozonation of aniline promoted by granular active carbon doped with TiO_2 is modelled. A three-phase modelling approach is proposed including ozone mass transfer parameters and rate constants (pseudo-first order kinetics) from both surface and liquid bulk reactions involving the adsorption process, oxidation in the liquid and the solid catalyst.

10. Erythromycin Abatement from Water by Electro-Fenton and Peroxyelectrocoagulation Treatments

In this final work [5], electro-Fenton and peroxyelectrocoagulation processes were investigated to mineralize erythromycin from ultrapure water, assessing the influence of some operational conditions, such as the anode material (BDD and Fe), current density (5 mA cm^{-2} and 10 mA cm^{-2}), oxygen flowrate injected to the cathode (0.8 L min^{-1} O_2 and 2.0 L min^{-1} O_2), pH (2.8, 5.0 and 7.0) and electricity costs.

Author Contributions: The three authors made equal contributions to this editorial. All authors have read and agreed to the published version of the manuscript.

Funding: This research received no external funding.

Institutional Review Board Statement: Not applicable.

Informed Consent Statement: Not applicable.

Data Availability Statement: No new data were created or analyzed in this study. Data sharing is not applicable to this article.

Acknowledgments: The authors appreciate the efforts of the Water editors and publication team at MDPI and the anonymous reviewers for their invaluable comments.

Conflicts of Interest: The authors declare no conflict of interest.

References

1. Dao, K.C.; Yang, C.-C.; Chen, K.-F.; Tsai, Y.-P. Recent Trends in Removal Pharmaceuticals and Personal Care Products by Electrochemical Oxidation and Combined Systems. *Water* **2020**, *12*, 1043. [CrossRef]
2. Liu, Y.-J.; Huang, Y.-L.; Lo, S.-L.; Hu, C.-Y. Comparing the Effects of Types of Electrode on the Removal of Multiple Pharmaceuticals from Water by Electrochemical Methods. *Water* **2020**, *12*, 2332. [CrossRef]
3. Ni, J.; Shi, H.; Xu, Y.; Wang, Q. A Comparison of the Mechanism of TOC and COD Degradation in Rhodamine B Wastewater by a Recycling-Flow Two- and Three-dimensional Electro-Reactor System. *Water* **2020**, *12*, 1853. [CrossRef]
4. Otter, P.; Mette, K.; Wesch, R.; Gerhardt, T.; Krüger, F.-M.; Goldmaier, A.; Benz, F.; Malakar, P.; Grischek, T. Oxidation of Selected Trace Organic Compounds through the Combination of Inline Electro-Chlorination with UV Radiation (UV/ECl_2) as Alternative AOP for Decentralized Drinking Water Treatment. *Water* **2020**, *12*, 3275. [CrossRef]
5. Serra-Clusellas, A.; Sbardella, L.; Herrero, P.; Delpino-Rius, A.; Riu, M.; de Lourdes Correa, M.; Casadellà, A.; Canela, N.; Martínez-Lladó, X. Erythromycin Abatement from Water by Electro-Fenton and Peroxyelectrocoagulation Treatments. *Water* **2021**, *13*, 1129. [CrossRef]
6. Ferreiro, C.; Villota, N.; Lombraña, J.I.; Rivero, M.J. Heterogeneous Catalytic Ozonation of Aniline-Contaminated Waters: A Three-Phase Modelling Approach Using TiO_2/GAC. *Water* **2020**, *12*, 3448. [CrossRef]
7. Lv, X.; Ma, Y.; Li, Y.; Yang, Q. Heterogeneous Fenton-Like Catalytic Degradation of 2,4-Dichlorophenoxyacetic Acid by Nano-Scale Zero-Valent Iron Assembled on Magnetite Nanoparticles. *Water* **2020**, *12*, 2909. [CrossRef]
8. Guimarães, V.; Teixeira, A.R.; Lucas, M.S.; Peres, J.A. Effect of Zr Impregnation on Clay-Based Materials for H_2O_2-Assisted Photocatalytic Wet Oxidation of Winery Wastewater. *Water* **2020**, *12*, 3387. [CrossRef]
9. Wang, J.; Song, Y.; Qian, F.; Du, C.; Yu, H.; Xiang, L. Removal Characteristics of Effluent Organic Matter (EfOM) in Pharmaceutical Tailwater by a Combined Coagulation and UV/O_3 Process. *Water* **2020**, *12*, 2773. [CrossRef]
10. Gu, D.-M.; Guo, C.-S.; Feng, Q.-Y.; Zhang, H.; Xu, J. Degradation of Ketamine and Methamphetamine by the UV/H_2O_2 System: Kinetics, Mechanisms and Comparison. *Water* **2020**, *12*, 2999. [CrossRef]

Review

Recent Trends in Removal Pharmaceuticals and Personal Care Products by Electrochemical Oxidation and Combined Systems

Khanh Chau Dao [1,2], Chih-Chi Yang [2], Ku-Fan Chen [2] and Yung-Pin Tsai [2,*]

[1] Department of Health and Applied Sciences, Dong Nai Technology University, Bien Hoa, Dong Nai 810000, Vietnam; daokhanhchau07@gmail.com
[2] Department of Civil Engineering, National Chi Nan University, Nantou Hsien 54561, Taiwan; chi813@gmail.com (C.-C.Y.); kfchen@ncnu.edu.tw (K.-F.C.)
* Correspondence: yptsai@ncnu.edu.tw; Tel.: +886-49-2910960 (ext. 4121)

Received: 4 March 2020; Accepted: 30 March 2020; Published: 7 April 2020

Abstract: Due to various potential toxicological threats to living organisms even at low concentrations, pharmaceuticals and personal care products in natural water are seen as an emerging environmental issue. The low efficiency of removal of pharmaceuticals and personal care products by conventional wastewater treatment plants calls for more efficient technology. Research on advanced oxidation processes has recently become a hot topic as it has been shown that these technologies can effectively oxidize most organic contaminants to inorganic carbon through mineralization. Among the advanced oxidation processes, the electrochemical advanced oxidation processes and, in general, electrochemical oxidation or anodic oxidation have shown good prospects at the lab-scale for the elimination of contamination caused by the presence of residual pharmaceuticals and personal care products in aqueous systems. This paper reviewed the effectiveness of electrochemical oxidation in removing pharmaceuticals and personal care products from liquid solutions, alone or in combination with other treatment processes, in the last 10 years. Reactor designs and configurations, electrode materials, operational factors (initial concentration, supporting electrolytes, current density, temperature, pH, stirring rate, electrode spacing, and fluid velocity) were also investigated.

Keywords: advanced oxidation processes; electrochemical advanced oxidation processes; pharmaceuticals and personal care products; electrochemical oxidation; anodic oxidation

---

1. Introduction

The concern for pharmaceuticals and personal care products (PPCPs) as toxic substances in the environment and the essential to assess their environmental risks have significantly increased recently. PPCPs are defined as a group of compounds that is including pharmaceutical drugs, cosmetic ingredients, food supplements, and ingredients in other consumer products (e.g., shampoos, lotions) [1]. Pharmaceuticals are used to prevent or treat diseases on humans and animals, whereas personal care products (PCPs) are used mostly to improve the quality of daily life [2]. They are considered as emerging pollutants (new products or chemicals without regulatory status) and whose effects on the environment and human health are unidentified [3]. Due to the widespread occurrence in water bodies, regardless of the low concentrations (normally ranging from ng/L to µg/L), residues of PPCP can harm human and animal health when it enters and accumulates in the food chain, causing unknown long-term effects [2,4].

During wastewater treatment (WWT) processes, many PPCPs experience microbial mediated reactions [5] in the environment. Thus, transformation products are formed. The transformation of PPCPs can occur during WWT, depending on the compound's physicochemical properties and conditions, where PPCPs can be destroyed or partially transformed or remained unchanged [6]. In this review, it can be seen that the effect of PPCPs in the environment does not only depend on concentration but also persistence, bioaccumulation, biotransformation, and elimination. Some PPCPs produce metabolites or by-products more harmful than the parent compounds. Toxicity evaluation is an important environmental pollution control factor since the degradation by-products from the initial structure can be more toxic.

Biodegradation, photodegradation, and other processes of abiotic transformation, such as hydrolysis [7], can reduce environmental concentrations of PPCPs and result in partial loss and mineralization of these compounds. Chiron et al. [8] revealed that acridine is a photodegradation product of carbamazepine under artificial estuarine water conditions, whereas tetracycline could not be photodegraded due to its sediment adsorption [9].

The electrochemical oxidation process (EOP) can be described as an electrochemical technology capable of achieving oxidation of contaminants from water or wastewater, either by direct or mediated oxidation processes originating on the anode surface of the electrochemical cell. This means that these oxidative processes should not actually be carried out on the anode, but only on its surface. As a consequence, this technique incorporates two main types of processes [10]: heterogeneous and homogeneous oxidation. Direct anodic oxidation or electrolysis occurs directly on the anode (M) with direct charge transfer reactions between the surface of the anode and the organic contaminants involved. The mechanism requires only the mediation of electrons that are capable of oxidizing such organic compounds at defined potentials more negative the oxygen evolution potential [11]. The indirect electrochemical oxidation by reactive oxygen species is based on the electro-generation of adsorbed *OH ($E°$ = 2.8 V/SHE) onto the anode surface as an intermediate of the OEP [10,12].

This paper intends to be a powerful tool for researchers in the pursuit of comprehensive information on the removal of PPCPs from liquid solutions by EOP, alone or in combination with other treatment processes. The remediation of aqueous or real wastewater was assessed, regarding many features like the configuration of the electrochemical reactor, anode and cathode characteristics, and operational parameters such as initial PPCPs concentration, supporting electrolytes, current density (j), temperature, pH, temperature, stirring rate, electrode spacing, and fluid velocity.

2. Origins and Classification of PPCPs

Direct and indirect pathways can introduce PPCPs into the environment. PPCPs may enter surface water by direct discharge into surface water from factories, hospitals, households, and WWTPs, as well as through land runoff in the case of biosolids distributed over agricultural land that may touch groundwater by leaching or bank filtration. Sediment can adsorb PPCPs within the surface water compartment because of various binding sites [13]. Soil may also be one of the PPCPs sinks. PPCPs can pass through irrigation into the soil with PPCPs containing treated and untreated wastewater. These can also be moved to the soil through an atmospheric wet deposition for some PPCPs [14].

Wastewater, including domestic, municipal, and hospital wastewater, are the primary sources that bring pharmaceuticals into the environment (both point- and nonpoint-sources) from various activities such as wastes (human and animal), landfill leachate, biosolid, and direct disposal of pharmaceuticals. Such pharmaceuticals then can not be biodegradable ultimately in WWTPs and enter the receiving waters [15–17]. In WWTPs, activated sludge is the main process for secondary treatment which can remove various kinds of PPCPs from wastewater. However, the removal rate depends greatly on physiochemical characteristics, reactors applied, and operational conditions (hydraulic retention time, sludge retention time, and pH) as well [18]. Table 1 summarizes the target PPCPs selected for this study and their structures, Table 2 updates the removal efficiency of PPCPs by combining biological treatment with other processes.

Table 1. Structures, chemical abstracts service registry number (CAS), and classification for the target pharmaceuticals and personal care products (PPCPs) selected for this study.

Compounds (CAS) Classification	Structure	Compounds (CAS) Classification	Structure
Aspirin (50-78-2) Nonsteroidal anti-inflammatory drugs (NSAIDs)		Lamivudine (134678-17-4) Antivirals	
Atenolol (29122-68-7) Beta-blockers		Levodopa (59-92-7) Antiparkinson Agents	
Berberine (2086-83-1) Antibiotics		Methotrexate (59-05-2) Antineoplastics	
Caffeine (58-08-2) Stimulant		Metronidazole (443-48-1) Antibiotics	
Carbamazepine (298-46-4) Anticonvulsants		Musk ketone (81-14-1) Fragrances	

Table 1. *Cont.*

Compounds (CAS) Classification	Structure	Compounds (CAS) Classification	Structure
Carboplatin (41575-94-4) Antineoplastics		Naproxen (22204-53-1) NSAIDs	
Ceftazidime (78439-06-2) Antibiotics		N,N-diethyl-m Toluamide (134-62-3) Insect repellents	
Ceftriaxone sodium (104376-79-6) Antibiotics		Norfloxacin (70458-96-7) Antibiotics	
Cephalexin (15686-71-2) Antibiotics		Ofloxacin (82419-36-1) Antibiotics	
Chloramphenicol (56-75-7) Antibiotics		Omeprazole (73590-58-6) Antibiotics	

Table 1. *Cont.*

Compounds (CAS) Classification	Structure	Compounds (CAS) Classification	Structure
Ciprofloxacin (85721-33-1) Antibiotics		Methyl Paraben (99-76-3) Preservatives	
Clofibric acid (882-09-7) Blood lipid regulators		Paracetamol (103-90-2) NSAIDs	
Diclofenac (15307-86-5) NSAIDs		Rifampicin (13292-46-1) Antibiotics	
Enrofloxacin (93106-60-6) Antibiotics		Salicylic acid (69-72-7) NSAIDs	
Estrone (53-16-7) Hormones		Sulfamethoxazole (723-46-6) Antibiotics	

Table 1. *Cont.*

Compounds (CAS) Classification	Structure	Compounds (CAS) Classification	Structure
Ibuprofen (15687-27-1) NSAIDs		Sulfachloropyrida-zine (80-32-0) Antibiotics	
Iohexol (66108-95-0) Radiological Non-Ionic Contrast Media		Sulfadiazine (68-35-9) Antibiotics	
2-methyl-4-isothiazolin-3-one (2682-20-4) Preservatives		Tetracycline (60-54-8) Antibiotics	
Ketoprofen (22071-15-4) NSAIDs			

Table 2. The removal efficiency of PPCPs by combining biological treatment with other processes.

Compounds	Initial Concentration	Treatment Processes	Removal Efficiency (%)	Ref.
Aspirin	930 ng/L	Modified Bardenpho process	92	[19]
Atenolol	255 ng/L	Grit tanks|primary sedimentation|bioreactor|clarifiers	47.1	[19]
	1197 ng/L	Pretreatment|primary (settling)|secondary activated sludge (AS)	14.4	[20]
Berberine	2.3 ± 2.0	Grit removal|primary clarifier|denitrification|nitrification|second clarifier	84	[21]
	75.0–375.0 mg/L	Upflow anaerobic sludge blanket (UASB)–membrane bioreactor (MBR)	99	[22]
Caffeine	82 ± 36 μg/L	Grit removal|primary clarifier|denitrification|nitrification|second clarifier	99.7	[21]
	22,849 ng/L	Anaerobic/Anoxic/Oxic (A2O)	94.9	[23]
Carbamazepine	208–416 ng/L	A series of different waste stabilization ponds	73	[24]
	129 ng/L	Pretreatment|primary (settling)|secondary AS	9.5	[20]
	2.0 ± 1.3 μg/L	Grit removal|primary clarifier|denitrification|nitrification|second clarifier	0	[21]
Carboplatin	4.7 to 145 μg/L	Adsorption to AS	70%	[25]
Ceftazidime	40 mg/L	Coupling ultraviolet (UV)|algae-algae treatment	97.26	[26]
Ceftriaxone	14 μg/L	AS process	<1	[27]
Cephalexin	4.6 mg/L	Grit channels|primary clarifies|conventional AS|Final settling	87	[28]
Chloramphenicol	206 ± 56 ng/L	Preliminary screening|primary sedimentation|conventional AS treatment	>70	[29]
	31 ± 16 ng/L	Screen|primary clarifier|AS system for denitrification and nitrification	50	[30]
Ciprofloxacin	2200 ng/L	Grit channels|primary clarifies|conventional AS	−88.6	[31]
	5524 ng/L	Pretreatment|primary (settling)|secondary AS	57	[20]
	2 mg/L	Aerobic sequencing batch reactors (SBRs) with mixed microbial cultures	51	[32]
Clofibric acid	0.25 ± 0.09 μg/L	Grit removal|primary clarifier|denitrification|nitrification|second clarifier	52	[21]
	26 ng/L	Pretreatment|primary (settling)|secondary AS	54.2	[20]
	20–70 mg/L	Primary treatment|Orbal oxidation ditch|UV disinfection	10–60	[33]
Diclofenac	2.0 ± 1.5 μg/L	Grit removal|primary clarifier|denitrification|nitrification|second clarifier	96	[21]
	232 ng/L	Pretreatment|primary (settling)|secondary AS	5	[20]
Enrofloxacin	9–170 ng/L	Conventional AS|UV disinfection	65	[34]
Estrone	57 ng/L	Grit channels|primary clarifies|conventional AS	93.7	[31]
	4500 ng/L	Grit channels|primary clarifies|conventional AS	99.7	[31]
Ibuprofen	3.4 ± 1.7 μg/L	Grit removal|primary clarifier|denitrification|nitrification|second clarifier	96	[21]
	2687 ng/L	Pretreatment|primary (settling)|secondary AS	95	[20]
Iohexol	9.0 ± 2.0 μg/L	Grit removal|primary clarifier|denitrification|nitrification|second clarifier	89	[21]
2-methyl-4-isothiazolin-3-one	1–3 mg/L	Aerobic process	80–100	[35]
Ketoprofen	441 ng/L	Anaerobic/Anoxic/Oxic (A2O)	11.2	[23]
Lamivudine	210 ± 13 ng/L	Screen|aerated grit-removal|primary clarifier|nitrification/denitrification	>76	[36]

Table 2. Cont.

Compounds	Initial Concentration	Treatment Processes	Removal Efficiency (%)	Ref.
Methotrexate	7.30–55.8 ng/L	Pretreatment|primary (settling)|secondary AS	100	[20]
Metronidazole	90 ng/L	Anaerobic/Anoxic/Oxic (A2O)	38.7	[23]
Musk ketone	0.640 ± 0.395 µg/L	Primary gravitational settling|AS	91.0 ± 5.2	[37]
Naproxen	3000 ng/L	Grit channels|primary clarifies|conventional AS	96.2	[31]
	2363 ng/L	Pretreatment|primary (settling)|secondary AS	60.9	[20]
DEET	503 ng/L	Primary|secondary treatment with AS	19.2–46.2	[38]
Norfloxacin	229 ± 42 ng/L	Screen|primary clarifier|AS system for denitrification and nitrification	66	[30]
	2100 ng/L	Grit channels|primary clarifies|conventional AS	124.2	[31]
Ofloxacin	2275 ng/L	Pretreatment|primary (settling)|secondary AS	64.1	[20]
Omeprazole	365 ng/L	Pretreatment|primary (settling)|secondary AS	8.5	[20]
Methyl Paraben	801 ng/L	Conventional biological treatment with P and N removal	100	[39]
Paracetamol	218,000 ng/L	Modified Bardenpho process	99	[19]
	23,202 ng/L	Pretreatment|primary (settling)|secondary AS	100	[20]
Rifampicin	0–31 ng/L	Secondary treatment process: AS, biological filtration oxygenated reactor, anoxic/oxic (A/O), cyclic AS technology (CAST), and A2O	0–100	[40]
Salicylic acid	5.866 µg/L	Primary|secondary treatment: trickling filter beds|final clarification.	>98	[41]
	7400 ng/L	Grit channels|primary clarifies|conventional AS	−35.8	[31]
Sulfamethoxazole	0.82 ± 0.23 µg/L	Grit removal|primary clarifier|denitrification|nitrification|second clarifier	24	[21]
	524 ng/L	Pretreatment|primary (settling)|secondary AS	31.2	[20]
	118 ± 17 ng/L	Screen|primary clarifier|AS system for denitrification and nitrification	64	[30]
Sulfachloropyridazine	0.19 µg/L	Conventional AS	62	[42]
Sulfadiazine	72 ± 22 ng/L	Screen|primary clarifier|AS system for denitrification and nitrification	50	[30]
Tetracycline	257 ± 176 ng/L	Preliminary screening|primary sedimentation|conventional AS treatment	69	[29]

3. Analytical Methods of PPCPs

Figure 1 shows the analytical method that is essential to investigate the occurrence of PPCPs in the environment, whichs consists of several main steps. This includes selecting appropriate analytical instruments (Table 3), which depend on the characteristics of PPCPs; extracting and purifying the samples by using techniques such as solid-phase extraction (SPE), liquid-liquid extraction (LLE), liquid-liquid micro-extraction (LLME), and solid-phase micro-extraction (SPME) that was introduced in various studies [43,44]; and optimizing of measurement parameters.

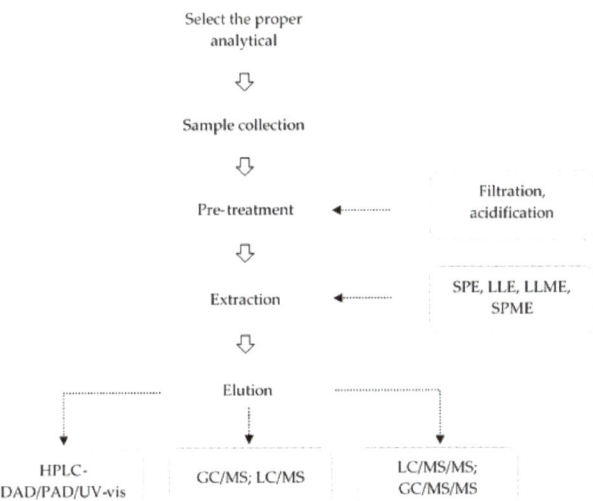

Figure 1. PPCPs analytical method procedure. Solid-phase extraction: SPE; liquid-liquid extraction: LLE; liquid-liquid micro-extraction: LLME; solid-phase micro-extraction: SPME; HPLC: High-performance liquid chromatography; DAD: Diode array detector; PAD: photodiode detector; UV-vis: ultraviolet-visible detector; GC/MS: Gas chromatography–mass spectrometry; LC/MS: Liquid chromatography–mass spectrometry.

Table 3. The analytical methods of PPCPs in the literature.

Analytical Methods	PPCPs
GC-MS	Ciprofloxacin, Chloramphenicol, Methyl paraben
HPLC	Lamivudine, Ceftazidime, Carboplatin, Aspirin, Cephalexin, Musk ketone, Norfloxacin, Ceftriaxone sodium, Levodopa, N,N-diethyl-m-Toluamide (DEET)
HPLC-DAD	Acetaminophen, Diclofenac, Sulfamethoxazole, Chloramphenicol, Ofloxacin, Berberine, Tetracycline
HPLC-UV/HPLC-UV vis/UV-vis	Ciprofloxacin, Rifampicin, Carbamazepine, Caffeine, Enrofloxacin, Sulfamethoxazole, Diclofenac, Isothdiazolin-3-ones, Metronidazole, Estrone, Paracetamol, Diclofenac, Methyl paraben, Clofibric acid, Sulfonamides
HPLC-HR-MS/HPLC-MS/HPLC-MS-MS	Carbamazepine, Iohexol, Ceftazidime, Methotrexate, Ibuprofen, Clofibric acid
HPLC-PDA	Atenolol, Paracetamol, Salicylic acid, Parabens, Sulfachloropyridazine, Omeprazole, Ibuprofen, Naproxen, Carbamazepine

4. Removal of PPCPs from Liquid Solutions by EOP

4.1. Electrochemical Reactor Designs and Configurations

There are two types of electrodes: two-dimensional and three-dimensional. Compared to two-dimensional, three-dimensional electrodes ensures a high electrode surface-to-cell volume ratio value. Due to the ease of scale up to a larger electrode size, more electrode pairs, or an increased number of cell stacks, cell designs using the parallel plate geometry in a filter press arrangement are widely used [45].

In the configuration of the reactor, the cell arrangement (divided and undivided cells) must be considered. The anolyte and catholyte are separated into divided cells by a porous diaphragm or ion-conducting membrane. Choosing the separating diaphragm or membrane is as critical for divided cells as choosing the correct electrode materials for proper electrolyte system functioning. Generally, the use of divided cells should be avoided wherever possible regarding the cost of separators, the complexity of reducing the electrode gap and the problems of the mechanic, and corrosion [46]. Undivided cells working in batch mode are often under magnetic stirring for mixing at a thermostatically controlled temperature (Figure 2). The number of electrodes can increase the active area per volume unit.

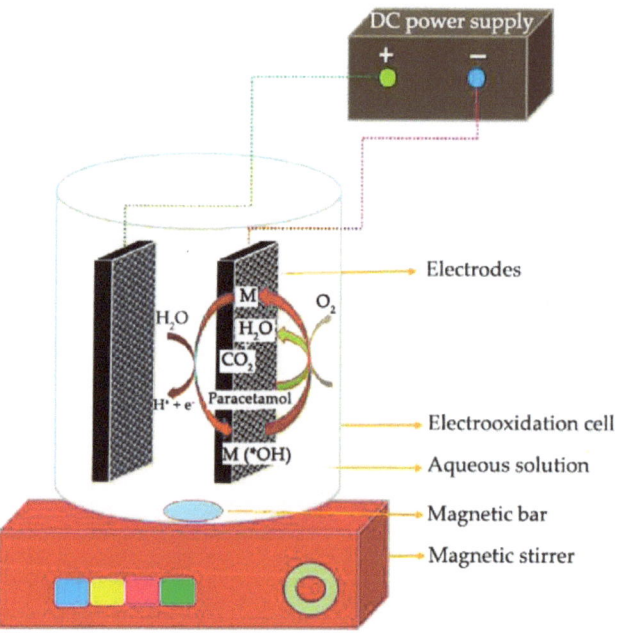

Figure 2. Diagram of the electrochemical reactor, using a glass beaker. The solution was stirred continuously throughout the process with a magnetic bar on a magnetic stirrer. The graphite anode was used as a working anode and a distance of 2 mm. Reprinted from Periyasamy and Muthuchamy [47], copyright © (2018) with permission from Elsevier.

Most of the studies were conducted in undivided electrochemical reactors, usually using solution volumes ranging from 100 to 500 mL, although 1 L or larger volumes were sometimes used [48–50]. Divided cells use a separator between anolyte and catholyte, which makes the treatment process more costly and challenging due to the penalty overvoltage of the separator. The investigation of norfloxacin degradation in an electrochemical reactor with the presence and absence of an ion-exchange membrane proved the use of the membrane is highly advantageous as it enhances the anodic reaction kinetics and improves the current efficiency. This leads to an improvement in the degradation of norfloxacin, mineralization, and the consequent mineralization current efficiency [51]. Moreover, Chen et al. [52] used successfully divided and thermostated cells and a Nafion 212 ion-exchange membrane separator to perform electrodegradation of DEET with total removal.

Since the metal deposition occurs on the surface of the cathode to boost the space-time yield, it is required to increase the surface area. Therefore, the fluidized bed electrode was developed, with granular graphite and glass beads for filling the gap between the main electrodes and used as the third electrode [32].

Filter-press cells have been used by coupling to a pump and a reservoir (Figure 3). One module including an anode, a cathode, and a membrane (if necessary) makes it relatively easy to operate and maintain the reactor.

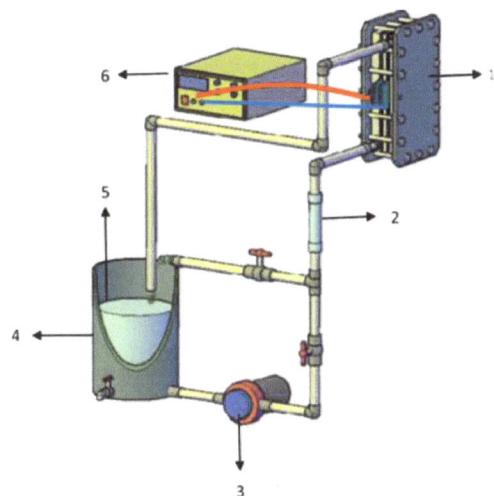

Figure 3. Experimental setup of 4 L undivided filter flow press reactor used for the treatment of paracetamol and diclofenac. 1. flow electrolytic cell, 2. flow meter, 3. peristaltic pump, 4. reservoir, 5. sampling, and 6. power supply. Reprinted from García-Montoya et al. [50], copyright © (2015), with permission from Elsevier.

4.2. Electrode Materials

It has also been shown that the anodes with high over-potential O_2 yield better electrochemical oxidation results [53–56]. Consequently, the electrode material (M) has a significant impact on the performance of PPCPs in oxidative degradation. Accordingly, an interesting issue is a systematic research on the comparative performance of electrode materials.

Sopaj et al. [57] tested on different electrode materials such as carbon felt, carbon fiber, carbon graphite, Platinum (Pt), lead dioxide, dimensionally stable anode (DSA) [58], (Ti/RuO_2–IrO_2), and boron-doped diamond (BDD) for removing of amoxicillin in aqueous media. BDD anode was more effective in oxidizing and mineralizing amoxicillin in water than the DSA. Moreover, it can be obtained very high electrolysis efficiency for the BDD electrode during the initial stage, even for high current densities.

Barışçı et al. [59] showed the performance of electrodes was significantly different for the anti-cancer drug carboplatin degradation with various mixed metal oxide (MMO) electrodes and BDD electrode (Figure 4). CV voltammograms unveiled that BDD, Ti/IrO_2-RuO_2, Ti/RuO_2, and Ti/IrO_2-Ta_2O_5 anodes had the highest levels of oxygen evolution and the poorest anodes were SnO_2/Pt, Ti/Pt and Ti/Ta_2O_5-SnO_2-IrO_2. Besides, higher oxygen evolution overpotential explained the formation of OH* on the surface of anode instead of molecular oxygen, which improved the efficiency.

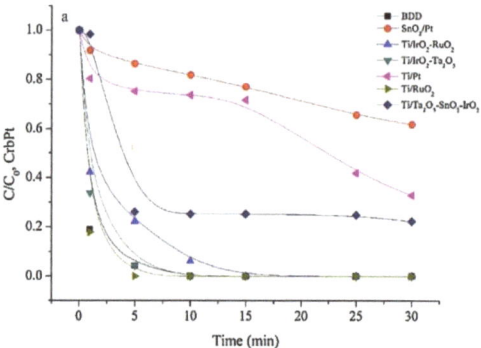

Figure 4. The effect of electrode material on anti-cancer drug carboplatin degradation under conditions: supporting electrolyte, 200 mg/L Na_2SO_4; pH 7; current density, 30 mA cm^{-2}. Reprinted from Barışçı et al. [59], copyright © (2018), with permission from Elsevier.

4.2.1. Lead and Lead Dioxide

Because of the stability, low cost, and high oxygen evolution potential, lead and lead dioxide have been used as anode materials [60] (Table 4). Recent studies have paid considerable efforts to improve the performance, including the addition of a new intermediate layer between the substrate and the oxidation layer, doping metal, or non-metallic ions and the adoption of new preparation methods [61,62].

Dai et al. [55] found the catalytic effect of La–Gd–PbO_2 showed the highest performance followed by that of La–PbO_2, Gd–PbO_2, PbO_2, respectively, in levodopa degradation. Moreover, compared to the pure PbO_2 electrode, the PbO_2 electrode with 1% Mo had a higher oxygen evolution potential and higher current of reduction and oxidation peaks, which led to increasing in electrochemical activity and decreasing of energy consumption [63].

Porous Ti plays an essential role in improving lead dioxide electrode performance compared to the traditional planar Ti substrate. Zhao et al. [64] found that compared to the traditional PbO_2 electrode, Ti/SnO_2-Sb_2O_3/PbO_2 had higher stability, safety, and removal performance of musk ketone. Xie et al. [65] developed a TiO_2-based SnO_2-Sb/polytetrafluoroethylene resin-PbO_2 electrode based on TiO_2 nanotubes and demonstrated the growing of TiO_2 nanotubes on Ti material led to an increase in current efficiency. Before electrons flow, the electrode needs a large overpotential that minimizes the oxygen evolution, decreases the production of hydrogen peroxide and ozone, and favors the creation of *OH, with the electron efficiency of 88.45%. The degradation of ibuprofen demonstrated the degradation rate constant over Ti/SnO_2-Sb/Ce-PbO_2 was two times of the value over Ti/Ce-PbO_2 [66].

4.2.2. DSA

In recent decades, MMO electrodes, known as DSA, have been made commercially available (Table 5). These consist of the corrosion-resistant base material, such as titanium or tantalum, coated with a metal oxide layer. DSA is catalytic oxide electrodes that, due to their low Cl_2 overpotential, can effectively produce active chlorine species [67].

Studies verify the performance of three-dimensional (3D) was much better, more cost-effective, and saved more energy than traditional two-dimensional (2D). The highest efficiency was recorded in the 3D process for removing carbamazepine compared to a 2D electrochemical process [68]. Furthermore, using a 3D electrode reactor to treat estriol, in batch mode, exhibited reaction rate per unit area was significantly higher and lower energy consumption than conventional 2D electrode reactor with indirect oxidation as the main contributor to the degradation in the batch 3D electrode reactor at all electrode distances [69]. Over 80% of the removal efficiency was attributed to indirect oxidation at an electrode distance of 2 cm (Figure 5).

Table 4. Selected results reported for PPCPs removal by electrochemical oxidation process (EOP) with lead and lead dioxide anodes.

PPCPs	Initial C	Electrolyte	j/mA cm^{-2}	Reactors/Operational Parameters	Electrodes Anode	Electrodes Cathode	pH	Reaction Time (min)	Removal (%)	Ref.
Lamivudine	5 mg/L	20 mM Na$_2$SO$_4$	≥10	Undivided cell, V 450 mL, current density (j) (6–14 mA cm^{-2})	Ti/SnO$_2$-Sb/Ce-PbO$_2$; 7 cm × 10 cm × 1 mm	Stainless steel (SS); 7 cm × 10 cm × 1 mm, gap 2 cm	3–11	240	70 (TOC)	[70]
Ciprofloxacin	50 mg/L	0.1 mol/L Na$_2$SO$_4$	30	Filter-press flow reactor; pH (3, 7, and 10), flow rate (qV = 2.5, 4.5, and 6.5 L min^{-1}), j (6.6, 20, and 30 mA cm^{-2}), and T = 10, 25, and 40 °C	Ti-Pt/β- PbO$_2$; 3.1 cm × 2.0 cm, 3.1 cm × 2.7 cm	AISI 304 SS plate	10	120	100	[71]
Ofloxacin	20 mg/L	Na$_2$SO$_4$	30	Differential column batch reactor, fluid velocity: 0.003 and 0.048 m/s, detention time: 10.3–0.54 min.	TiO$_2$-based SnO$_2$-Sb/FR-PbO$_2$; 2 cm × 5 cm	SS foil; Same shape and size, gap 0.5 and 3 cm	6.25	90	99.00	[65]
Enrofloxacin	10 mg/L	20 mM Na$_2$SO$_4$	8	Undivided electrolytic cell, V 30 mL, j (2–10 mA/cm^2), pH (~3–11)	Ti/SnO$_2$-Sb/La-PbO$_2$; 25 cm^2	Ti; Same area; gap 5 mm	3–11	30	95.1 (TOC)	[72]
Musk ketone	50 mg/L	0.06 mol/L Na$_2$SO$_4$	40	Cylindrical single compartment cell, V 100 mL, stirring rate 800 rmin^{-1}, j (10–50 mA cm^{-2}), pH (3–11)	Ti/SnO$_2$-Sb$_2$O$_3$/PbO$_2$; 1 cm × 1 cm	Stainless copper foil; (2 cm × 2 cm), gap 1.5 cm	7	120	99.93	[64]
Levodopa	100 mg/L	0.1 mol/L Na$_2$SO$_4$	50	Electrochemical system, V 250 mL, j (15–70 mA cm^{-2})	La-Gd–PbO$_2$; 12 cm × 2 cm, thickness: 1 mm, 14 cm^2	Ti; The same area; gap 4 cm	5.9	120	100.00	[55]

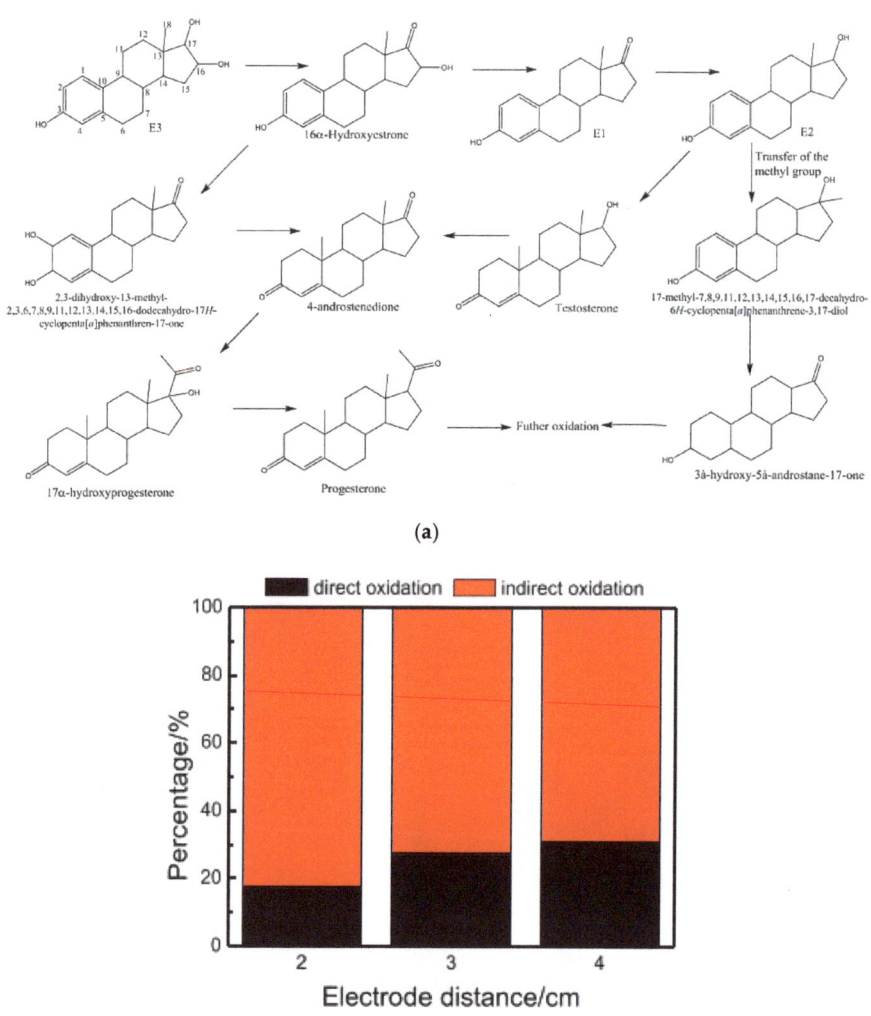

Figure 5. Proposed simplified pathways for estriol (E3) degradation in batch 3D electrolysis (**a**) and (**b**) the contribution of direct and indirect oxidation at various distances under operating conditions: C_0 = approximately 1000 µg/L, j = 2.2 mAcm^{-2}, constant current, Q = 150 mL/min. Reprinted from Shen et al. [69], copyright © (2017), with permission from Elsevier.

By adding powder activated carbon (PAC) or metal particles, the conductivity, mass transfer, or adsorption may also be increased in the 3D process [73]. The possibility of catalytic reaction and more reactive sites for adsorption are advantages of the 3D process that lead to better removal performance [74].

The SnO_2 electrode has been widely used in wastewater treatment because of its high oxidation activity, it lower toxicity than PbO_2, and it being more cost-effective than BDD. Sadly, it also contains limitations due to high energy consumption and instability. Adding TiO_2 could reduce the electrode's internal passivation and charge transfer resistance, improving its stability and efficiency in oxidation when Cu limits the growth of crack morphology and offer more effective active sites. They accelerated electronic transfer and decreased SnO_2 surface potential, improved the OEP, and increased the response current peak, which increased the electrode's oxidative degradation capacity [75]. Ti/SnO_2-Cu showed better stability and higher corrosion resistance than the conventional Ti/SnO_2-Sb electrode [76].

RuO_2/IrO_2-coated titanium anode with improved electrocatalytic behavior and stability are readily available in practical mesh geometries and have extended the lifetime and lower costs compared to BDD electrodes. Various DSA such as Ti/RuO_2, Ti/Pt, Ti/IrO_2-RuO_2, Ti/IrO_2-Ta_2O_5, Ti/Ta_2O_5-SnO_2-IrO_2, and Pt/SnO_2 used for removing X-ray contrast iohexol demonstrated that Ti/RuO_2 provided the highest degradation efficiency [77]. Barışçı et al. [59] found that Ti/RuO_2 could reach complete degradation of carboplatin anti-cancer drug in just 5 min and obtained zero toxicity at the end of the process. However, the use of IrO_2, RuO_2 on large scale is restricted by low abundance, high cost, and difficulty in their separation. Ir/IrO_2 nanoparticles could be immobilized on Fe_3O_4 core/ SiO_2 shell via surface-modified NH_2 functional groups resulted in high catalytic activity, high stability, and efficient recyclability.

4.2.3. Boron-Doped Diamond

The BDD anode showed high performance on various kinds of PPCPs, as seen in Table 6. The low-pressure conversion of carbon to diamond crystals has allowed a thin layer of diamond film to develop on suitable substrates like silicon, niobium, tungsten, molybdenum, and titanium [78]. He et al. [79] examined aspirin degradation with PbO_2, BDD, and porous Ti/BDD as the anode. On BDD electrodes, the electrochemical process involves direct and indirect electrochemical oxidation, whereas, on the PbO_2 electrode, only indirect oxidation. The kinetic results can be explained by the mechanism of aspirin degradation, which may take place in two distinct forms: direct oxidation at the electrode surface and indirect oxidation mediated by *OH. In indirect oxidation, the initial step involves the formation of *OH from water molecule discharge. The oxidation is indirectly mediated by *OH contributing to the mineralization of organic pollutants. Aspirin mineralization is mainly performed by reaction with *OH. Porous Ti/BDD is the highest excellent potential for aspirin relative to flat BDD and PbO_2 electrode when niobium-supported BDD thin film (Nb/BDD) anode could be applied in a wide range of pH, reducing chemicals for pH adjustment [48].

In various systems, BDD allowed for higher removal rates of PPCPs than other anodes as higher quantities of *OH produced. Sirés et al. [80] indicated that the performance was demonstrated to be much more productive using a large surface area BDD anode than a Pt one, explained by a large number of active hydroxyl radicals BDD (*OH) and minimizing their parasitic reactions. Compared to the Pt and glassy carbon anodes, the BDD anode showed better efficiency for isothiazolin-3-one degradation [81]. BDD physisorbed *OH was observed to cause the combustion of ketoprofen into CO_2 and H_2O. The poor mineralization was attributed to the formation of chlorinated organic compounds that are refractory at both BDD and Pt anodes [82]. Omeprazole was primarily oxidized by *OH formed from water oxidation at the surface of the Pt or BDD [54]. It also can be seen that the BDD anode was superior to the Pt and PbO_2 electrodes for DEET abatement. At the same j value and temperature, the DEET abatement degradation in the order BDD, PbO_2, and Pt [52]. It also can be seen the higher oxidation power of BDD became evident in removing estrone than β-PbO_2 anode [83].

BDD electrode in a single compartment filter-press flow cell represented the conversion of cephalexin and its hydroxylated intermediates to CO_2 depended solely on their diffusion to the BDD surface. Due to the different types and quantities of electrogenerated oxidants, the oxidation rate of cephalexin using distinct salts as supporting electrolytes showed distinct rates; however, none of them were able to mineralize cephalexin and its intermediates, which only occurred through a diffusion mechanism on the surface of the BDD [84]. Due to the high concentration of *OH generated on the BDD surface, with the release of NH_4^+ and NO_3^- ions, nearly 50% of mineralization of paracetamol and diclofenac is always achieved [50].

Table 5. Selected results reported for PPCPs removal by EOP with dimensionally stable anode (DSA) anodes.

PPCPs	Initial C	Electrolyte	j/mA cm^{-2}	Reactors/Operational Parameters	Electrodes Anode	Electrodes Cathode	pH	Reaction Time (min)	Removal (%)	Ref.
Ceftazidime	5 mg/L	1 g/L Na$_2$SO$_4$	1.25	V reactor and electrolytic wastewater was 150 mL and 120 mL, respectively	Ti/TiO$_2$/SnO$_2$-Sb-Cu; (50 mm × 30 mm × 2 mm)	Pt wire; gap 4 cm	6	-	97.65	[75]
Iohexol	0.525 mg/L	0.1 M Na$_2$SO$_4$	38.1–45	Batch experiments, V 350 mL, pH 7.2, iohexol concentration 0.525 mg/L; j = 15, 30, and 45 mA/cm^2; pH (4.0, 7.0 ± 02, and 9.0)	Ti/RuO$_2$; 25 cm^2	SS; 0.5-mm gap	7.1	19.8–30	>90	[77]
Carboplatin	0.5 mg/L	0.1 M Na$_2$SO$_4$	30	One-compartment cell 350 mL; pH range 4–9; j = 15, 30 and 45 mA/cm^{-2};	Ti/RuO$_2$; 25 cm^2	SS plate; 25 cm^2 gap 0.5 cm	7	5	100.00	[59]
Methotrexate	0.5 mg/L	200 mg/L Na$_2$SO$_4$	30	One-compartment cell, V 350 mL, Na$_2$SO$_4$ (100, 200, 300 mg/L), pH range of 4–9; j = 15, 30 and 45 mA cm^{-2}	Ti/IrO$_2$-RuO$_2$; 25 cm^2	SS plate; 0.5 cm gap	7	5	95.00	[85]
Estriol	1000 µg/L	0.1M Na$_2$SO$_4$	20	Batch 3D electrolysis, an undivided rectangular reactor, V 300 mL, filled with approximately 50 g granular graphite particles and 70 g glass beads	Ti/IrO$_2$-RuO$_2$; 5 × 10 cm	Ti; 5 × 10 cm; gap could be adjusted	3–7	50	80.00	[69]
Sulfamethoxazole	200 mg/L	0.1 mol/L NaCl	≥20	Single compartment filter press-type flow cell reactor, flow rate: 425 mL/min	Ti/Ru$_{0.3}$Ti$_{0.7}$O; 14 cm^2	Ti plate; The same geometric area	3	30	>98	[86]
Ceftriaxone sodium	10 mg/L	0.1 mol/L Na$_2$SO$_4$	(The external potential of +2.0 V)	A cylindrical glass reactor made, fused and sealed at one end	TiO$_2$(40)/Nano-G	Titanium mesh; gap 2 cm	-	120	97.70	[87]
Clofibric acid	50 mg/L	50mM Na$_2$SO$_4$	33.6	250 mL undivided glass beaker containing 200 mL solution, T constant at 20 °C, constant current	Plate mixed metal oxide (DSA, Ti/RuO$_2$–IrO$_2$); 5.0 cm × 11.9 cm	SS; Same dimension; gap 4.0 cm	4	180	64.70 (TOC)	[88]

4.2.4. Other Electrodes

The Pt electrode showed better performance in sulfamethoxazole and diclofenac degradation wit electrolyte supports under the same conditions as the carbon electrode [89]. Compared to RuO_2/Ti, IrO_2/Ti, and $RuIrO_2$/Ti electrodes, Pt/Ti demonstrated that the removal efficiency of berberine was considerably higher [90].

Carbon nanotubes are recognized in wastewater as an advanced anode material for recalcitrant antibiotics for electrocatalysis oxidation. Cyclic voltammetry analysis of La_2O_3-CuO_2/carbon nanotube (CNT) showed a stronger catalytic activity of the modified electrode and stable working life with an efficiency of 90% to 1 mg/L ceftazidime within 30 min, which is much higher than that of pristine CNTs and DSA [91]. The addition of TiO_2 could promote the electron transfer and reusability of the CeO_2-ZrO_2/TiO_2/CNT electrode [92]. In three electrodes promoted by multiwall carbon nanotubes (MWCNTs) (MWCNT, MWCNT-COOH, and MWCNT-NH_2), concerning the electrode surface chemistry, MWCNT-NH_2, with the highest isoelectric point (4.70), is the most promising material due to improved reactant interactions [93].

4.3. Influence of Operational Parameters

4.3.1. Initial PPCPs Concentration

The initial drug concentration significantly influenced the rate of electrochemical decomposition and the process efficiency for both drugs, ifosfamide, and cyclophosphamide [94]. The higher degradation rate of ibuprofen achieved at relative lower initial concentrations at the initial ibuprofen concentrations ranges from 1.0 to 20.0 mg/L [66]. The concentration of parabens in the aqueous matrix was the element that, regardless of the aqueous matrix under investigation, exerts a more extraordinary effect on the target variable. An increase in the initial parabens concentration resulted in a decrease in the efficiency of removal [95] and the mineralization rate decreases when salicylic acid concentration rose from 200 mg/L. During bulk electrolyzes at a low j value and high salicylic acid concentration, salicylic acid was oxidized to aromatic compounds due to a low local concentration on the anode surface of electrogenerated *OH relative to salicylic acid. As bulk electrolyzes at a high j value and low salicylic acid concentration, the product was directly combusted to CO_2 due to a high local concentration on the anode surface of electrogenerated *OH relative to salicylic acid [96].

Interestingly, it could be seen that the efficiency improved with the increased concentration of paracetamol and diclofenac due to the gradual increase in the concentration of *OH to oxidize contaminants before participating in non-oxidizing reactions [50]. The removal of caffeine had two stages, depending on its concentration. At low concentrations, the efficiency significantly increased with j value, suggesting a crucial role of mediated oxidation processes [97].

4.3.2. Supporting Electrolytes

In the presence of NaCl as the supporting electrolyte, the degradation rate of PPCPs was favored. Experiments on RuO_2/Ti, IrO_2/Ti, $RuIrO_2$/Ti, and Pt Ti electrodes showed a constant reaction rate in NaCl solution three to five times higher than in Na_2SO_4 and the oxidation rate of berberines increased due to active chlorine formation [90]. Ambuludi et al. [98] indicated that the pseudo-first-order rate constant increased when NaCl replaced Na_2SO_4 as the electrolyte support and it was almost unaffected by the concentration of ibuprofen. Otherwise, the poor mineralization of ketoprofen was due to the formation of chlorinated organic compounds, which are refractory, at both BDD and Pt anodes in the presence of NaCl as supporting electrolyte while total mineralization using Na_2SO_4 as an electrolyte was achieved [82]. Indermuhle et al. [97] found using NaCl, compared to Na_2SO_4, caffeine could reach a faster degradation but more reaction intermediates are formed and the mechanism is consistent with other proposed (Figure 6).

Table 6. Selected results reported for PPCPs removal by EOP with BDD anodes.

PPCPs	Initial C	Electrolyte	j/mA cm^{-2}	Reactors/Operational Parameters	Electrodes Anode	Electrodes Cathode	pH	Reaction Time (min)	Removal (%)	Ref.
Atenolol	0.19 mmol/L	14 mmol/L Na$_2$SO$_4$	30	Double-jacket glass, one-compartment flow filter-press reactor, V 0.002 m^3, pH: 3 and 10, flow rate 3.33×10^{-5} m^3 s^{-1}, j (5, 10, 20 and 30 mAcm^{-2}), T = 25 °C	Nb/BDD500; 0.01 m^2	AISI 304L; Gap 0.02 m	10	120	100.00	[48]
Rifampicin	200 mg/L	0.5 mol/L Na$_2$SO$_4$	90	250-mL undivided open cell, equipped with magnetic stirring at 30 °C	BDD; 3.0 × 2.5 cm	Ti/Ru$_{0.3}$Ti$_{0.7}$O$_2$; 4.0 × 4.0 cm	3	180	95.00	[99]
Norfloxacin	100 mg/L	0.1 mol/L Na$_2$SO$_4$	10	One-compartment filter-press flow reactor, pH (3, 7, 10, and without specific control), j (10, 20, and 30 mA cm^{-2}), T (10, 25, and 40 °C)	BDD; Thickness of 2.9 μm	SS; area of 3.54 cm × 6.71 cm	not pH dependent	300	100.00	[100]
Estrone	500 μg/L	0.1 mol/L Na$_2$SO$_4$	10	A filter-press electrochemical reactor, 0.5L solution, flow rate (2.0, 3.0, 4.0, 5.0, 6.0, and 7.0 L/min), j (5, 10, and 25 mA cm^{-2}), pH (3.0, 7.0, and 10.0)	BDD; each face was 2.5 cm × 3.0 cm 15 cm^2	SS; (3.0 cm × 4.0 cm)	<=7	30	98.00	[83]
Paracetamol Diclofenac	50 mg/L, 100 mg/L	0.05 M Na$_2$SO$_4$	1.56–6.25	4L undivided filter flow press reactor, j (1.56 to 6.25 mAcm^{-2}), flow rate kept constant at 2 L/min	BDD; 64 cm^2	SS; Gap 2 cm	3	60	50.00 (TOC)	[50]
Methyl paraben	100 mg/L	0.05 mol/L K$_2$SO$_4$	10.8	One-compartment pyrex cell (400 mL) operated at 25 ± 1 °C in batch mode, j (1.35 to 21.6 mA cm^{-2})	BDD; 9.68 cm^{-2}	Titanium foil; The same area	5.7	300	100.00	[101]
Sulfonamides	50 mg/L	6.1 g/L Na$_2$SO$_4$	15	Undivided electrolytic cell, V 100 mL, pH (from 2.0 to 7.4), T (from 25 to 60 °C), and j (from 0.05 to 15 mA cm^{-2})	Si/BDD; 10 cm^2	SS; Gap 1cm	6.4	180	92.00	[102]

Table 6. Cont.

PPCPs	Initial C	Electrolyte	j/mA cm^{-2}	Reactors/Operational Parameters	Electrodes		pH	Reaction Time (min)	Removal (%)	Ref.
					Anode	Cathode				
Tetracycline	100 mg/L	5 g/L Na$_2$SO$_4$ or NaCl	25 to 300 A m^{-2}	Up-flow electrochemical cell, 20 cm^3, batch mode with recirculation; pH (2 to 12), j (25 to 300 A m^{-2})	BDD; 20 cm^2	SS; Gap 1 cm	5.6	30 min	100.00	[103]
Sulfachloropyridazine	0.2 mM	0.05 M Na$_2$SO$_4$	350 mA	An open, cylindrical and undivided glass cell 250 mL with magnetic stirring	BDD; 25 cm^2	Carbon-felt; 77 cm^2 (14.0 cm × 5.5 cm)	4.5	8h	95.00	[104]
Omeprazole	169 mg/L	0.05 Na$_2$SO$_4$	100	Undivided and cylindrical glass cell of 150 mL, with a double jacket, j = 33.3–150 mA cm^{-2}, T = 35 °C, stirred with 800 rpm	BDD; 3 cm^2	Carbon-PTFE air-diffusion; Gap 1 cm	7	360	78.00 (TOC)	[54]
Ibuprofen	0.2 mM	0.05 M Na$_2$SO$_4$	50–500 mA	Cylindrical, open, one-compartment cell 200 mL, at T (20 ± 2 °C)	BDD; 25 cm^2	Carbon-felt; 14 cm×5 cm each side, 0.5 cm width	3	480	>96 (TOC)	[98]

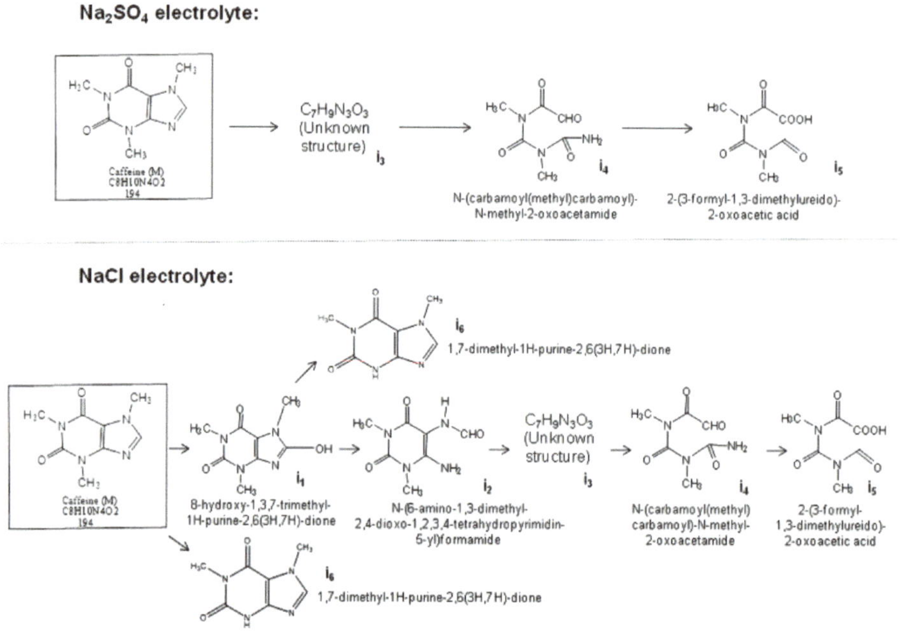

Figure 6. The mechanism model proposed for caffeine degradation by electrochemical oxidation with conductive-diamond electrodes using Na_2SO_4 or NaCl as the electrolyte. Reprinted from Indermuhle et al. [97], copyright © (2013) with permission from Elsevier.

In the presence of Na_2SO_4, the increasing concentration of Na_2SO_4 provided a higher rate of degradation of the anti-cancer drug carboplatin but further increased the concentration of Na_2SO_4, which did not offer a higher rate of degradation due to SO_4^{2-} excess [59]. Moreover, 0.1 M electrolyte-supporting Na_2SO_4 was found to be more active for sulfamethoxazole and diclofenac mineralization, with an efficiency of 15%–30% higher than 0.1 M electrolyte-supporting phosphate buffer on Pt and carbon electrodes [89].

Various inorganic ions have significant effects on removing certain PPCPs that were compared with a higher removal rate in the presence of chloride species than other ions. Acetaminophen, diclofenac, and sulfamethoxazole degradation showed high removal efficiencies, and faster reaction rates may correlate with the presence of chloride species, which may be due to the involvement of hypochlorite ions. Although all of the drugs were degraded by indirect electrochemical oxidation, cyclic voltammograms suggested that chloride species may have coexisted with *OH and have been converted into by-products of degradation [49], whereas ions Cl^- and PO_4^{3-} significantly increased the decomposition rate of ifosfamide [94].

4.3.3. Current Density, pH, Temperature, and Stirring Rate

Current density (j), pH, and temperature also among parameters that have been optimized and investigated in the EOP. Which factor most crucial for efficiency removal depends on the kinds of PPCPs, the material of electrodes and the nature of electrolytes applied. For naproxen removal, the current influence was the greatest among these variables, and the second was the salt concentration, the third flow rate and the fourth pH [105]. Domínguez et al. [106] also proved that the influence of the current was the greatest, then the concentration of salt and the flow rate, respectively, on carbamazepine degradation.

The j value shows a vital role in the removal efficiency with increasing removal efficiency when j increased in most cases of PPCPs [66,97,104] and other factors are dependent or not significant under certain operating conditions. Isothiazolin-3-ones degradation rate was faster as the j value applied increased but nearly independent of electrolyte pH [81]. Moreover, the complete removal of norfloxacin is dependent on pH. However, the removal increased with the temperature at 10 °C, 25 °C, and 40 °C may result from a gradual increase in the diffusion coefficient and the oxidation of byproducts under temperature conditions [100]. Interestingly, DEET degradation increased with increasing current density but was moderately affected by temperature (25–75 °C) [52]. Similarly, the salicylic acid mineralization rate increased at 25 °C with an increase of applied current, the pH impact was not significant [96]. This also can be seen in the case of ketoprofen [82], ifosfamide, and cyclophosphamide [94]. Interestingly, the carboplatin degradation rates increased significantly in the initial phases of electrolysis as j value increased on the Ti/RuO$_2$ electrode. However, a further increase in j did not affect the rate of degradation [59]. Sun et al. [107] found that pH decreased, the efficiency of chloramphenicol degradation increased, and maximum degradation was achieved at pH 2, Figure 7.

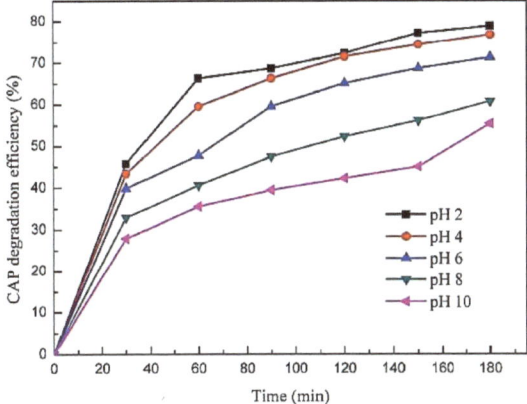

Figure 7. Effect of initial pH of wastewater on the chloramphenicol degradation efficiency of particle electrodes. Reprinted from Sun et al. [107], copyright © (2017) with permission from Elsevier.

Stirring increased the rate of mass transfer and PPCPs formed a contentious relationship on the electrode surface to increase the efficiency of removal. When the stirring speed was too slow, the mass transfer resistance would be the limitation. With the free radical produced from the electrode surface, PPCPs were unable to react quickly. It was also not possible to transfer the hydroxyls produced to the solution in time. On the other hand, the high stirring speed turned leads to short time for PPCPs touching the electrode surface, PPCPs could not be wholly oxidized and soon left the electrode surface. O$_2$ and H$_2$ bubbles produced from H$_2$O electrolysis would be more competitive to access molecule surface with extreme disturbance, resulting in reducing removal efficiency. The kinetic study of naproxen degradation at fix potential indicates that the rate of degradation increases with the stirring speed at 250 and 500 rpm [93]. For diffusion reactions, the stirring rate is an essential factor. The stirring rate showed a definite increase in the removal of ceftazidime and then decreased as the stirring speed between 150 and 200 rad min^{-1} [76].

4.3.4. Electrode Spacing and Fluid Velocity

The changes in the spacing of the electrodes would affect not only the mass transfer limitations but also the electron transport and electric resistance [108]. The effect of electrode spacing, however, depends on the direct or indirect oxidation. In the latter case, the electrode spacing should be matched with the diffusion length of *OH species. Duan, et al. [76] found that the oxidation of ceftazidime decreased as the spacing of the Ti/SnO$_2$-Cu electrode changed at 1, 2, and 3 cm under the current of 20 mA. As the spacing increased, the electrochemical resistance also increased while the charge in the electrolyte decreased. Xie et al. [65] (Figure 8) tested ofloxacin removal with the changes in electrode spacing. The reaction rate increased with the first-order pseudo constant changed as the distance decreased from 3 cm to 0.5 cm and the mass transfer coefficient increased.

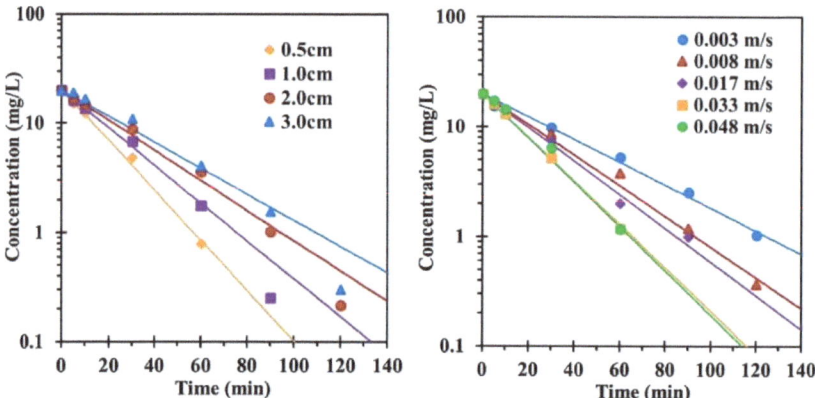

Figure 8. Effect of electrode spacing and fluid velocity on ofloxacin degradation. Anode surface area 10 cm^2, electrolyte concentration = 0.05 M Na$_2$SO$_4$ solution, j = 30 mA cm^{-2}, initial ofloxacin concentration 20 mg/L, voltage 6.2–6.3 V, initial pH 6.25, temperature 25 °C. Reprinted from Xie et al. [65], copyright © (2017) with permission from Elsevier.

It can be seen that the electrocatalytic oxidation process relied primarily on the high potential for direct oxidation on the electrode surface and the generation of free radicals for indirect oxidation of PPCPs. Consequently, the spacing increases, which leads to a loss in *OH production and oxidation power on anode surfaces. Diffusion efficiency also affects removal efficiency and so at a larger electrode spacing, the electrolysis process needs more time because of longer diffusion distance. Both electrode spacing and fluid velocity are critical since increasing velocities that lead to an increase in the rate of mass transfer while decreasing electrode spacing increases the surface area available for mass transfer [65].

4.4. Applications for Real Water and Wastewater Containing PPCPs

EOP is a promising technique with different degradation rates for the removal of PPCPs from water and WWTP effluents under optimal conditions concerning the ecological system [98,107,109,110].

Because of the presence of chloride ions in the effluent, oxidation in secondary treated wastewater was faster than in pure water [111]. Carbamazepine electrodegradation is feasible for WWT in several aqueous matrices [112], after 50 min of electrolysis time, caffeine was removed entirely in DIW and was almost removed in the wastewater sample may be related to the organic matter in wastewater. Having regard to these results, EOP is an effective method for further removal of caffeine from effluent from aerobic or anaerobic reactors that treat municipal wastewater, even though a high concentration of caffeine was used compared to low concentration in natural water. Compared to conventional methods for removing caffeine from urban wastewater, this approach appears to be more feasible for the following reasons: ease of operating, rapid removal of caffeine, and the effective efficiency of treatment [110]. The caffeine elimination obtained in real wastewater was found to be higher than in synthetic wastewater due to the contribution of electrogenerated oxidant species, such as hypochlorite [113], when sulfonamides and DEET removal were most efficient in the presence of municipal wastewater treatment plant (MWWTP) effluents [52,102].

4.5. Combined Systems

While EOP has been widely demonstrated for their ability to remove trace and persistent PPCPs in water and wastewater, complex water matrices could be found that inhibit their efficient operation. As a result, they may potentially reduce or fully retard the efficiency, requiring longer hydraulic retention time or higher volume capacity for compensation. System hybridization or combination of EOP with other water technologies is possible to overcome the operational problems associated with the complex water matrices.

Zaghdoudi et al. [114] investigated the possibility of coupling an electroreduction pretreatment before a biological process for dimetridazole removal. Direct electrolysis was initially conducted at the low potential to reduce amino derivatives formation and then azo dimer formation with a total degradation of dimetridazole achieved and the ratio of biochemical oxygen demand (BOD_5)/chemical oxygen demand (COD) increased. As mineralization yields of all electrolyzed solutions increased significantly, the enhancement of biodegradability was demonstrated during biological treatment. Nevertheless, the real mineralization yields should most likely be significantly higher if the contribution of titanocene, which is possibly biorecalcitrant, is not taken into account in the amount of TOC. Belkheiri et al. [115] examined the biodegradability improvement of tetracycline-containing solutions after an electrochemical pretreatment, as a large amount of the applied drugs are not metabolized and, therefore, can be found in wastewater. BOD_5 measurements verified biodegradability increased with the oxidation potential as the ratio of BOD_5/COD increased. Despite its chemical transformation, none of the reduced tetracycline solutions are biodegradable. Yahiaoui et al. [116] found after 5 h of electrochemical pre-treatment of tetracycline, the BOD_5/COD ratio increased considerably and confirmed during biological treatment, with 76% of dissolved organic carbon (DOC) removed.

Pharmaceutical degradation in conventional WWTPs is a problem because industrial sewage and hospital effluents contain low-concentration pharmaceuticals. Rodríguez-Nava et al. [117] found high efficiencies in removal without affecting activated sludge performance of integrating EOP with a biological system for simultaneous removal from wastewater of recalcitrant drugs (bezafibrate, gemfibrozil, indomethacin, and sulfamethoxazole). Drugs contained in wastewater without electrochemical pretreatment was persistent in the biological process and encouraged bulking formation. García-Gómez et al. [118] proved membrane bioreactor (MBR) high capacity to remove COD and low capacity for degradation (20%) of carbamazepine after 120 days, which presumably suggests that given the weak degradation and carbamazepine was not toxic to microorganisms. The EOP, on the other hand, was able to degrade carbamazepine completely.

In an exciting study for investigating pre- and post-treatment in one system to remove synthetic hospital wastewater fortified with four drug pollutants including carbamazepine, ibuprofen, estradiol, and venlafaxine by the combination of MBR and EOP, MBR alone treatment of wastewater showed a high percentage of ibuprofen and estradiol removal (about 90%), while carbamazepine and venlafaxine performed a low elimination (at around 10%). EOP as post-treatment, this allowed high removal (about 97%) of the four pharmaceutical pollutants and far more successful compared to EOP as pre-treatment [119]. The integration of electrochemical processes into MBR systems can utilize the mechanism of biodegradation, sorption, hydrolysis, and filtration on conventional MBR and electrocoagulation, electroosmosis, and electrophoresis on electrochemical processes that improve both the performance and the control of membrane fouling for eliminating recalcitrant micropollutants [120,121].

5. Conclusions

EOP is a promising technique with different degradation rates for the removal of PPCPs from water and wastewater, from synthetic or real, concerning the ecological system. There are numerous studies that have recently focused on the finding of electrode materials and optimal conditions, including initial PPCPs concentration, supporting electrolytes, j value, pH, temperature, stirring rate, and electrode spacing that are effective for removing a certain or groups of PPCPs with considering reduce operating cost. In terms of operational parameters, it was shown that the current influence was the greatest among these variables in some mentioned studies. Although the electrochemical process has recorded several influential factors, only some of them show a significant impact on real systems.

Studies showed that the EOP system depends heavily on the type of anode. BDD anode shows high performance on various kinds of PPCPs. The BDD anodes have been reported to produce higher organic oxidation rates and higher current efficiencies than other metal oxides commonly used. The development of BDD anodes and the enormous advantages of this electrode compared to others make this material was investigated on most of the works published in the literature. The performance of 3D electrolysis is much better, more cost-effective, and saves more energy consumption than traditional 2D electrolysis. The results validate 3D electrolysis in pretreatment or advanced treatment applications as a promising alternative method to remove PPCPs from secondary effluents.

Real field samples may contain other species of radical electrolytes that may participate in the electrochemical process and therefore act as interferences within the EOP system. It is therefore recommended that the electrochemical degradation process be the last step in the domestic water treatment since the technique also largely depends on the electrolytes in the water.

Toxicity evaluation is an essential environmental pollution control factor since the degradation by-products from the initial structure can be more toxic. It can be seen that in some kinds of PPCPs, intermediates are more toxic than the molecule of the parent, while others are less harmful. By evaluating toxicity, it helps significantly in optimizing treatment conditions to achieve the elimination of adverse effects of by-products.

EOP has widely demonstrated their ability to remove trace and persistent PPCPs in water and wastewater. Further, complex water matrices could be found that inhibit their efficient operation. System hybridization or combination of EOP with other water technologies is possible to overcome the operational problems associated with the complex water matrices.

Author Contributions: Conceptualization, K.C.D., Y.-P.T. and K.-F.C.; resources, Y.-P.T. and K.-F.C; writing—original draft preparation, K.C.D.; writing—review and editing, K.C.D., and C.-C.Y.; supervision, Y.-P.T. and K.-F.C.; project administration, C.-C.Y.; funding acquisition, Y.-P.T. All authors have read and agreed to the published version of the manuscript.

Funding: This research was funded by the Ministry of Science and Technology, Taiwan, grant number MOST 106-2221-E-260-003-MY3.

Conflicts of Interest: The authors declare no conflicts of interest.

References

1. Shen, R.; Andrews, S.A. Demonstration of 20 pharmaceuticals and personal care products (PPCPs) as nitrosamine precursors during chloramine disinfection. *Water Res.* **2011**, *45*, 944–952. [CrossRef] [PubMed]
2. Boxall, A.B.; Rudd, M.A.; Brooks, B.W.; Caldwell, D.J.; Choi, K.; Hickmann, S.; Innes, E.; Ostapyk, K.; Staveley, J.P.; Verslycke, T. Pharmaceuticals and personal care products in the environment: What are the big questions? *Environ. Health Perspect.* **2012**, *120*, 1221–1229. [CrossRef]
3. Deblonde, T.; Cossu-Leguille, C.; Hartemann, P. Emerging pollutants in wastewater: A review of the literature. *Int. J. Hyg. Environ. Health* **2011**, *214*, 442–448. [CrossRef]
4. Rajapaksha, A.U.; Vithanage, M.; Lim, J.E.; Ahmed, M.B.M.; Zhang, M.; Lee, S.S.; Ok, Y.S. Invasive plant-derived biochar inhibits sulfamethazine uptake by lettuce in soil. *Chemosphere* **2014**, *111*, 500–504. [CrossRef] [PubMed]
5. Helbling, D.E.; Hollender, J.; Kohler, H.-P.E.; Singer, H.; Fenner, K. High-throughput identification of microbial transformation products of organic micropollutants. *Environ. Sci. Technol. Water Treat.* **2010**, *44*, 6621–6627. [CrossRef] [PubMed]
6. Xia, K.; Bhandari, A.; Das, K.; Pillar, G. Occurrence and fate of pharmaceuticals and personal care products (PPCPs) in biosolids. *J. Environ. Qual.* **2005**, *34*, 91–104. [CrossRef]
7. Blair, B.D.; Crago, J.P.; Hedman, C.J.; Klaper, R.D. Pharmaceuticals and personal care products found in the Great Lakes above concentrations of environmental concern. *Chemosphere* **2013**, *93*, 2116–2123. [CrossRef]

8. Chiron, S.; Minero, C.; Vione, D. Photodegradation processes of the antiepileptic drug carbamazepine, relevant to estuarine waters. *Environ. Sci.* **2006**, *40*, 5977–5983. [CrossRef]
9. Tolls, J. Sorption of veterinary pharmaceuticals in soils: A review. *Environ. Sci. Technol. Water Treat.* **2001**, *35*, 3397–3406. [CrossRef]
10. Panizza, M.; Cerisola, G. Direct and mediated anodic oxidation of organic pollutants. *Chem. Rev.* **2009**, *109*, 6541–6569. [CrossRef]
11. Hollender, J.; Zimmermann, S.G.; Koepke, S.; Krauss, M.; McArdell, C.S.; Ort, C.; Singer, H.; von Gunten, U.; Siegrist, H. Elimination of organic micropollutants in a municipal wastewater treatment plant upgraded with a full-scale post-ozonation followed by sand filtration. *Environ. Sci. Technol. Water Treat.* **2009**, *43*, 7862–7869. [CrossRef]
12. Martinez-Huitle, C.A.; Ferro, S. Electrochemical oxidation of organic pollutants for the wastewater treatment: Direct and indirect processes. *Chem. Soc. Rev.* **2006**, *35*, 1324–1340. [CrossRef]
13. Kaestner, M.; Nowak, K.M.; Miltner, A.; Trapp, S.; Schaeffer, A. Classification and modelling of nonextractable residue (NER) formation of xenobiotics in soil–a synthesis. *Crit. Rev. Environ. Sci. Technol. Water Treat.* **2014**, *44*, 2107–2171. [CrossRef]
14. Kallenborn, R. *Perfluorinated Alkylated Substances (PFAS) in the Nordic Environment*; Nordic Council of Ministers: Copenhagen, Denmark, 2004.
15. Daughton, C.G.; Ternes, T.A. Pharmaceuticals and personal care products in the environment: Agents of subtle change? *Environ. Health Perspect.* **1999**, *107*, 907–938. [CrossRef] [PubMed]
16. Golet, E.M.; Alder, A.C.; Hartmann, A.; Ternes, T.A.; Giger, W. Trace determination of fluoroquinolone antibacterial agents in urban wastewater by solid-phase extraction and liquid chromatography with fluorescence detection. *Anal. Chem.* **2001**, *73*, 3632–3638. [CrossRef]
17. Lishman, L.; Smyth, S.A.; Sarafin, K.; Kleywegt, S.; Toito, J.; Peart, T.; Lee, B.; Servos, M.; Beland, M.; Seto, P. Occurrence and reductions of pharmaceuticals and personal care products and estrogens by municipal wastewater treatment plants in Ontario, Canada. *Sci. Total Environ.* **2006**, *367*, 544–558. [CrossRef]
18. Roberts, J.; Kumar, A.; Du, J.; Hepplewhite, C.; Ellis, D.J.; Christy, A.G.; Beavis, S.G. Pharmaceuticals and personal care products (PPCPs) in Australia's largest inland sewage treatment plant, and its contribution to a major Australian river during high and low flow. *Sci. Total Environ.* **2016**, *541*, 1625–1637. [CrossRef] [PubMed]
19. Yu, Y.; Wu, L.; Chang, A.C. Seasonal variation of endocrine disrupting compounds, pharmaceuticals and personal care products in wastewater treatment plants. *Sci. Total Environ.* **2013**, *442*, 310–316. [CrossRef] [PubMed]
20. Martín, J.; Camacho-Muñoz, D.; Santos, J.L.; Aparicio, I.; Alonso, E. Occurrence and ecotoxicological risk assessment of 14 cytostatic drugs in wastewater. *Water Air Soil Pollut.* **2014**, *225*, 1896. [CrossRef]
21. Ternes, T.A.; Bonerz, M.; Herrmann, N.; Teiser, B.; Andersen, H.R. Irrigation of treated wastewater in Braunschweig, Germany: An option to remove pharmaceuticals and musk fragrances. *Chemosphere* **2007**, *66*, 894–904. [CrossRef]
22. Qiu, G.; Song, Y.-H.; Zeng, P.; Duan, L.; Xiao, S. Characterization of bacterial communities in hybrid upflow anaerobic sludge blanket (UASB)–membrane bioreactor (MBR) process for berberine antibiotic wastewater treatment. *Bioresour. Technol.* **2013**, *142*, 52–62. [CrossRef]
23. Rosal, R.; Rodríguez, A.; Perdigón-Melón, J.A.; Petre, A.; García-Calvo, E.; Gómez, M.J.; Agüera, A.; Fernández-Alba, A.R. Occurrence of emerging pollutants in urban wastewater and their removal through biological treatment followed by ozonation. *Water Res.* **2010**, *44*, 578–588. [CrossRef]
24. Leclercq, M.; Mathieu, O.; Gomez, E.; Casellas, C.; Fenet, H.; Hillaire-Buys, D. Presence and fate of carbamazepine, oxcarbazepine, and seven of their metabolites at wastewater treatment plants. *Arch. Environ. Contam. Toxicol.* **2009**, *56*, 408. [CrossRef] [PubMed]
25. Lenz, K.; Hann, S.; Koellensperger, G.; Stefanka, Z.; Stingeder, G.; Weissenbacher, N.; Mahnik, S.N.; Fuerhacker, M. Presence of cancerostatic platinum compounds in hospital wastewater and possible elimination by adsorption to activated sludge. *Sci. Total Environ.* **2005**, *345*, 141–152. [CrossRef]
26. Liu, Y.; Wang, Z.; Yan, K.; Wang, Z.; Torres, O.L.; Guo, R.; Chen, J. A new disposal method for systematically processing of ceftazidime: The intimate coupling UV/algae-algae treatment. *Chem. Eng. J.* **2017**, *314*, 152–159. [CrossRef]

27. Junker, T.; Alexy, R.; Knacker, T.; Kümmerer, K. Biodegradability of 14C-labeled antibiotics in a modified laboratory scale sewage treatment plant at environmentally relevant concentrations. *Environ. Sci. Technol.* **2006**, *40*, 318–324. [CrossRef] [PubMed]
28. Watkinson, A.J.; Murby, E.J.; Costanzo, S.D. Removal of antibiotics in conventional and advanced wastewater treatment: Implications for environmental discharge and wastewater recycling. *Water Res.* **2007**, *41*, 4164–4176. [CrossRef]
29. Leung, H.W.; Minh, T.B.; Murphy, M.B.; Lam, J.C.; So, M.K.; Martin, M.; Lam, P.K.; Richardson, B.J. Distribution, fate and risk assessment of antibiotics in sewage treatment plants in Hong Kong, South China. *Environ. Int.* **2012**, *42*, 1–9. [CrossRef]
30. Xu, W.; Zhang, G.; Li, X.; Zou, S.; Li, P.; Hu, Z.; Li, J. Occurrence and elimination of antibiotics at four sewage treatment plants in the Pearl River Delta (PRD), South China. *Water Res.* **2007**, *41*, 4526–4534. [CrossRef]
31. Blair, B.; Nikolaus, A.; Hedman, C.; Klaper, R.; Grundl, T. Evaluating the degradation, sorption, and negative mass balances of pharmaceuticals and personal care products during wastewater treatment. *Chemosphere* **2015**, *134*, 395–401. [CrossRef]
32. Salgado, R.; Oehmen, A.; Carvalho, G.; Noronha, J.P.; Reis, M.A. Biodegradation of clofibric acid and identification of its metabolites. *J. Hazard. Mater.* **2012**, *241*, 182–189. [CrossRef]
33. Sun, Q.; Lv, M.; Hu, A.; Yang, X.; Yu, C.P. Seasonal variation in the occurrence and removal of pharmaceuticals and personal care products in a wastewater treatment plant in Xiamen, China. *J. Hazard. Mater.* **2014**, *277*, 69–75. [CrossRef] [PubMed]
34. He, K.; Soares, A.D.; Adejumo, H.; McDiarmid, M.; Squibb, K.; Blaney, L. Detection of a wide variety of human and veterinary fluoroquinolone antibiotics in municipal wastewater and wastewater-impacted surface water. *J. Pharm. Biomed. Anal.* **2015**, *106*, 136–143. [CrossRef]
35. Voets, J.P.; Pipyn, P.; Van Lancker, P.; Verstraete, W. Degradation of microbicides under different environmental conditions. *J. Appl. Bacteriol.* **1976**, *40*, 67–72. [CrossRef]
36. Prasse, C.; Schlusener, M.P.; Schulz, R.; Ternes, T.A. Antiviral drugs in wastewater and surface waters: A new pharmaceutical class of environmental relevance? *Environ. Sci. Technol.* **2010**, *44*, 1728–1735. [CrossRef] [PubMed]
37. Simonich, S.L.; Federle, T.W.; Eckhoff, W.S.; Rottiers, A.; Webb, S.; Sabaliunas, D.; De Wolf, W. Removal of fragrance materials during US and European wastewater treatment. *Environ. Sci. Technol.* **2002**, *36*, 2839–2847. [CrossRef] [PubMed]
38. Nakada, N.; Tanishima, T.; Shinohara, H.; Kiri, K.; Takada, H. Pharmaceutical chemicals and endocrine disrupters in municipal wastewater in Tokyo and their removal during activated sludge treatment. *Water Res.* **2006**, *40*, 3297–3303. [CrossRef]
39. Molins-Delgado, D.; Díaz-Cruz, M.S.; Barceló, D. Ecological risk assessment associated to the removal of endocrine-disrupting parabens and benzophenone-4 in wastewater treatment. *J. Hazard. Mater.* **2016**, *310*, 143–151. [CrossRef]
40. Zhang, X.; Zhao, H.; Du, J.; Qu, Y.; Shen, C.; Tan, F.; Chen, J.; Quan, X. Occurrence, removal, and risk assessment of antibiotics in 12 wastewater treatment plants from Dalian, China. *Environ. Sci. Pollut. Res. Int.* **2017**, *24*, 16478–16487. [CrossRef]
41. Kosma, C.I.; Lambropoulou, D.A.; Albanis, T.A. Occurrence and removal of PPCPs in municipal and hospital wastewaters in Greece. *J. Hazard. Mater.* **2010**, *179*, 804–817. [CrossRef]
42. Verlicchi, P.; Al Aukidy, M.; Zambello, E. Occurrence of pharmaceutical compounds in urban wastewater: Removal, mass load and environmental risk after a secondary treatment—A review. *Sci. Total Environ.* **2012**, *429*, 123–155. [CrossRef]
43. Huddleston, J.G.; Willauer, H.D.; Swatloski, R.P.; Visser, A.E.; Rogers, R.D. Room temperature ionic liquids as novel media for 'clean'liquid–liquid extraction. *Chem. Commun.* **1998**, 1765–1766. [CrossRef]
44. Mohammadhosseini, M.; Tehrani, M.S.; Ganjali, M.R. Preconcentration, determination and speciation of chromium (III) using solid phase extraction and flame atomic absorption spectrometry. *J. Chin. Chem. Soc.* **2006**, *53*, 549–557. [CrossRef]
45. Rajeshwar, K.; Ibanez, J.G. *Environmental Electrochemistry: Fundamentals and Applications in Pollution Sensors and Abatement*; Elsevier: Amsterdam, The Netherlands, 1997.
46. Wendt, H.; Kreysa, G. *Electrochemical Engineering: Science and Technology in Chemical and Other Industries*; Springer: Berlin/Heidelberg, Germany, 1999.

47. Periyasamy, S.; Muthuchamy, M. Electrochemical oxidation of paracetamol in water by graphite anode: Effect of pH, electrolyte concentration and current density. *J. Environ. Chem. Eng.* **2018**, *6*, 7358–7367. [CrossRef]
48. Da Silva, S.W.; do Prado, J.M.; Heberle, A.N.A.; Schneider, D.E.; Rodrigues, M.A.S.; Bernardes, A.M. Electrochemical advanced oxidation of Atenolol at Nb/BDD thin film anode. *J. Electroanal. Chem.* **2019**, *844*, 27–33. [CrossRef]
49. Liu, Y.-J.; Hu, C.-Y.; Lo, S.-L. Direct and indirect electrochemical oxidation of amine-containing pharmaceuticals using graphite electrodes. *J. Hazard. Mater.* **2019**, *366*, 592–605. [CrossRef]
50. García-Montoya, M.F.; Gutiérrez-Granados, S.; Alatorre-Ordaz, A.; Galindo, R.; Ornelas, R.; Peralta-Hernandez, J.M.; Chemistry, E. Application of electrochemical/BDD process for the treatment wastewater effluents containing pharmaceutical compounds. *J. Ind. Eng. Chem.* **2015**, *31*, 238–243. [CrossRef]
51. Mora-Gomez, J.; Ortega, E.; Mestre, S.; Pérez-Herranz, V.; García-Gabaldón, M. Electrochemical degradation of norfloxacin using BDD and new Sb-doped SnO_2 ceramic anodes in an electrochemical reactor in the presence and absence of a cation-exchange membrane. *Sep. Purif. Technol.* **2019**, *208*, 68–75. [CrossRef]
52. Chen, T.-S.; Chen, P.-H.; Huang, K.-L. Electrochemical degradation of N, N-diethyl-m-toluamide on a boron-doped diamond electrode. *J. Taiwan Inst. Chem. Eng.* **2014**, *45*, 2615–2621. [CrossRef]
53. Brillas, E.; Sires, I.; Arias, C.; Cabot, P.L.; Centellas, F.; Rodriguez, R.M.; Garrido, J.A. Mineralization of paracetamol in aqueous medium by anodic oxidation with a boron-doped diamond electrode. *Chemosphere* **2005**, *58*, 399–406. [CrossRef]
54. Cavalcanti, E.B.; Garcia-Segura, S.; Centellas, F.; Brillas, E. Electrochemical incineration of omeprazole in neutral aqueous medium using a platinum or boron-doped diamond anode: Degradation kinetics and oxidation products. *Water Res.* **2013**, *47*, 1803–1815. [CrossRef]
55. Dai, Q.; Xia, Y.; Sun, C.; Weng, M.; Chen, J.; Wang, J.; Chen, J. Electrochemical degradation of levodopa with modified PbO_2 electrode: Parameter optimization and degradation mechanism. *Chem. Eng. J.* **2014**, *245*, 359–366. [CrossRef]
56. Radjenovic, J.; Bagastyo, A.; Rozendal, R.A.; Mu, Y.; Keller, J.; Rabaey, K. Electrochemical oxidation of trace organic contaminants in reverse osmosis concentrate using RuO_2/IrO_2-coated titanium anodes. *Water Res.* **2011**, *45*, 1579–1586. [CrossRef]
57. Sopaj, F.; Rodrigo, M.A.; Oturan, N.; Podvorica, F.I.; Pinson, J.; Oturan, M.A. Influence of the anode materials on the electrochemical oxidation efficiency. Application to oxidative degradation of the pharmaceutical amoxicillin. *Chem. Eng. J.* **2015**, *262*, 286–294. [CrossRef]
58. Oaks, J.L.; Gilbert, M.; Virani, M.Z.; Watson, R.T.; Meteyer, C.U.; Rideout, B.A.; Shivaprasad, H.; Ahmed, S.; Chaudhry, M.J.I.; Arshad, M. Diclofenac residues as the cause of vulture population decline in Pakistan. *Nature* **2004**, *427*, 630. [CrossRef]
59. Barışçı, S.; Turkay, O.; Ulusoy, E.; Soydemir, G.; Seker, M.G.; Dimoglo, A. Electrochemical treatment of anti-cancer drug carboplatin on mixed-metal oxides and boron doped diamond electrodes: Density functional theory modelling and toxicity evaluation. *J. Hazard. Mater.* **2018**, *344*, 316–321. [CrossRef]
60. El-Ashtoukhy, E.-S.; Amin, N.; Abdelwahab, O. Treatment of paper mill effluents in a batch-stirred electrochemical tank reactor. *Chem. Eng. J.* **2009**, *146*, 205–210. [CrossRef]
61. Wang, Q.; Jin, T.; Hu, Z.; Zhou, L.; Zhou, M. TiO_2-NTs/SnO_2-Sb anode for efficient electrocatalytic degradation of organic pollutants: Effect of TiO_2-NTs architecture. *Sep. Purif. Technol.* **2013**, *102*, 180–186. [CrossRef]
62. Wu, W.; Huang, Z.-H.; Lim, T.-T. Recent development of mixed metal oxide anodes for electrochemical oxidation of organic pollutants in water. *Appl. Catal. A Gen.* **2014**, *480*, 58–78. [CrossRef]
63. Dai, Q.; Zhou, J.; Meng, X.; Feng, D.; Wu, C.; Chen, J. Electrochemical oxidation of cinnamic acid with Mo modified PbO_2 electrode: Electrode characterization, kinetics and degradation pathway. *Chem. Eng. J.* **2016**, *289*, 239–246. [CrossRef]
64. Zhao, W.; Xing, J.; Chen, D.; Jin, D.; Shen, J. Electrochemical degradation of Musk ketone in aqueous solutions using a novel porous Ti/SnO_2-Sb_2O_3/PbO_2 electrodes. *J. Electroanal. Chem.* **2016**, *775*, 179–188. [CrossRef]
65. Xie, R.; Meng, X.; Sun, P.; Niu, J.; Jiang, W.; Bottomley, L.; Li, D.; Chen, Y.; Crittenden, J. Electrochemical oxidation of ofloxacin using a TiO_2-based SnO_2-Sb/polytetrafluoroethylene resin-PbO_2 electrode: Reaction kinetics and mass transfer impact. *Appl. Catal. B Environ.* **2017**, *203*, 515–525. [CrossRef]
66. Wang, C.; Yu, Y.; Yin, L.; Niu, J.; Hou, L.-A. Insights of ibuprofen electro-oxidation on metal-oxide-coated Ti anodes: Kinetics, energy consumption and reaction mechanisms. *Chemosphere* **2016**, *163*, 584–591. [CrossRef] [PubMed]

67. Brillas, E.; Martínez-Huitle, C.A. Decontamination of wastewaters containing synthetic organic dyes by electrochemical methods. An updated review. *Appl. Catal. B Environ.* **2015**, *166*, 603–643. [CrossRef]
68. Alighardashi, A.; Aghta, R.S.; Ebrahimzadeh, H. Improvement of Carbamazepine Degradation by a Three-Dimensional Electrochemical (3-EC) Process. *Int. J. Environ. Res. Public Health* **2018**, *12*, 451–458. [CrossRef]
69. Shen, B.; Wen, X.-H.; Huang, X. Enhanced removal performance of estriol by a three-dimensional electrode reactor. *Chem. Eng. J.* **2017**, *327*, 597–607. [CrossRef]
70. Wang, Y.; Zhou, C.; Chen, J.; Fu, Z.; Niu, J. Bicarbonate enhancing electrochemical degradation of antiviral drug lamivudine in aqueous solution. *J. Electroanal. Chem.* **2019**, *848*, 113314. [CrossRef]
71. Wachter, N.; Aquino, J.M.; Denadai, M.; Barreiro, J.C.; Silva, A.J.; Cass, Q.B.; Rocha-Filho, R.C.; Bocchi, N. Optimization of the electrochemical degradation process of the antibiotic ciprofloxacin using a double-sided β-PbO$_2$ anode in a flow reactor: Kinetics, identification of oxidation intermediates and toxicity evaluation. *Environ. Sci. Pollut. Res.* **2019**, *26*, 4438–4449. [CrossRef]
72. Wang, C.; Yin, L.; Xu, Z.; Niu, J.; Hou, L.-A. Electrochemical degradation of enrofloxacin by lead dioxide anode: Kinetics, mechanism and toxicity evaluation. *Chem. Eng. J.* **2017**, *326*, 911–920. [CrossRef]
73. Wei, L.; Guo, S.; Yan, G.; Chen, C.; Jiang, X. Electrochemical pretreatment of heavy oil refinery wastewater using a three-dimensional electrode reactor. *Electrochim. Acta* **2010**, *55*, 8615–8620. [CrossRef]
74. Fortuny, A.; Font, J.; Fabregat, A. Wet air oxidation of phenol using active carbon as catalyst. *Appl. Catal. B Environ.* **1998**, *19*, 165–173. [CrossRef]
75. Li, X.; Duan, P.; Lei, J.; Sun, Z.; Hu, X. Fabrication of Ti/TiO$_2$/SnO$_2$-Sb-Cu electrode for enhancing electrochemical degradation of ceftazidime in aqueous solution. *J. Electroanal. Chem.* **2019**, *847*, 113231. [CrossRef]
76. Duan, P.; Hu, X.; Ji, Z.; Yang, X.; Sun, Z. Enhanced oxidation potential of Ti/SnO$_2$-Cu electrode for electrochemical degradation of low-concentration ceftazidime in aqueous solution: Performance and degradation pathway. *Chemosphere* **2018**, *212*, 594–603. [CrossRef] [PubMed]
77. Turkay, O.; Barisci, S.; Ulusoy, E.; Dimoglo, A. Electrochemical Reduction of X-ray Contrast Iohexol at Mixed Metal Oxide Electrodes: Process Optimization and By-product Identification. *Water Air Soil Pollut.* **2018**, *229*, 170. [CrossRef]
78. Chen, X.; Chen, G. Fabrication and application of Ti/BDD for wastewater treatment. *Synth. Diam. Film. Prep. Electrochem. Charact. Appl.* **2011**, 353–371. [CrossRef]
79. He, Y.; Huang, W.; Chen, R.; Zhang, W.; Lin, H.; Li, H. Anodic oxidation of aspirin on PbO$_2$, BDD and porous Ti/BDD electrodes: Mechanism, kinetics and utilization rate. *Sep. Purif. Technol.* **2015**, *156*, 124–131. [CrossRef]
80. Sirés, I.; Oturan, N.; Oturan, M.A. Electrochemical degradation of β-blockers. Studies on single and multicomponent synthetic aqueous solutions. *Water Res.* **2010**, *44*, 3109–3120. [CrossRef] [PubMed]
81. Kandavelu, V.; Yoshihara, S.; Kumaravel, M.; Murugananthan, M. Anodic oxidation of isothiazolin-3-ones in aqueous medium by using boron-doped diamond electrode. *Diam. Relat. Mater.* **2016**, *69*, 152–159. [CrossRef]
82. Murugananthan, M.; Latha, S.; Raju, G.B.; Yoshihara, S. Anodic oxidation of ketoprofen—An anti-inflammatory drug using boron doped diamond and platinum electrodes. *J. Hazard. Mater.* **2010**, *180*, 753–758. [CrossRef]
83. Brocenschi, R.F.; Rocha-Filho, R.C.; Bocchi, N.; Biaggio, S.R. Electrochemical degradation of estrone using a boron-doped diamond anode in a filter-press reactor. *Electrochim. Acta* **2016**, *197*, 186–193. [CrossRef]
84. Coledam, D.A.; Pupo, M.M.; Silva, B.F.; Silva, A.J.; Eguiluz, K.I.; Salazar-Banda, G.R.; Aquino, J.M. Electrochemical mineralization of cephalexin using a conductive diamond anode: A mechanistic and toxicity investigation. *Chemosphere* **2017**, *168*, 638–647. [CrossRef] [PubMed]
85. Barışçı, S.; Turkay, O.; Ulusoy, E.; Şeker, M.G.; Yüksel, E.; Dimoglo, A. Electro-oxidation of cytostatic drugs: Experimental and theoretical identification of by-products and evaluation of ecotoxicological effects. *Chem. Eng. J.* **2018**, *334*, 1820–1827. [CrossRef]
86. Hussain, S.; Gul, S.; Steter, J.R.; Miwa, D.W.; Motheo, A.J. Route of electrochemical oxidation of the antibiotic sulfamethoxazole on a mixed oxide anode. *Environ. Sci. Pollut. Res.* **2015**, *22*, 15004–15015. [CrossRef] [PubMed]
87. Guo, X.; Li, D.; Wan, J.; Yu, X. Preparation and electrochemical property of TiO$_2$/Nano-graphite composite anode for electro-catalytic degradation of ceftriaxone sodium. *Electrochim. Acta* **2015**, *180*, 957–964. [CrossRef]

88. Lin, H.; Wu, J.; Zhang, H. Degradation of clofibric acid in aqueous solution by an EC/Fe^{3+}/PMS process. *Chem. Eng. J.* **2014**, *244*, 514–521. [CrossRef]
89. Sifuna, F.W.; Orata, F.; Okello, V.; Jemutai-Kimosop, S. Comparative studies in electrochemical degradation of sulfamethoxazole and diclofenac in water by using various electrodes and phosphate and sulfate supporting electrolytes. *J. Environ. Sci. Health Part A* **2016**, *51*, 954–961. [CrossRef]
90. Tu, X.; Xiao, S.; Song, Y.; Zhang, D.; Zeng, P. Treatment of simulated berberine wastewater by electrochemical process with Pt/Ti anode. *Environ. Earth Sci.* **2015**, *73*, 4957–4966. [CrossRef]
91. Duan, P.; Yang, X.; Huang, G.; Wei, J.; Sun, Z.; Hu, X. La$_2$O$_3$-CuO$_2$/CNTs electrode with excellent electrocatalytic oxidation ability for ceftazidime removal from aqueous solution. *Colloids Surf. A Physicochem. Eng. Asp.* **2019**, *569*, 119–128. [CrossRef]
92. Duan, P.; Gao, S.; Li, X.; Sun, Z.; Hu, X. Preparation of CeO$_2$-ZrO$_2$ and titanium dioxide coated carbon nanotube electrode for electrochemical degradation of ceftazidime from aqueous solution. *J. Electroanal. Chem.* **2019**, *841*, 10–20. [CrossRef]
93. Díaz, E.; Stożek, S.; Patiño, Y.; Ordóñez, S. Electrochemical degradation of naproxen from water by anodic oxidation with multiwall carbon nanotubes glassy carbon electrode. *Water Sci. Technol. Water Treat.* **2019**, *79*, 480–488. [CrossRef]
94. Fabiańska, A.; Ofiarska, A.; Fiszka-Borzyszkowska, A.; Stepnowski, P.; Siedlecka, E.M. Electrodegradation of ifosfamide and cyclophosphamide at BDD electrode: Decomposition pathway and its kinetics. *Chem. Eng. J.* **2015**, *276*, 274–282. [CrossRef]
95. Domínguez, J.R.; Muñoz-Peña, M.J.; González, T.; Palo, P.; Cuerda-Correa, E.M. Parabens abatement from surface waters by electrochemical advanced oxidation with boron doped diamond anodes. *Environ. Sci. Pollut. Res.* **2016**, *23*, 20315–20330. [CrossRef]
96. Rabaaoui, N.; Allagui, M.S. Anodic oxidation of salicylic acid on BDD electrode: Variable effects and mechanisms of degradation. *J. Hazard. Mater.* **2012**, *243*, 187–192. [CrossRef] [PubMed]
97. Indermuhle, C.; Martin de Vidales, M.J.; Saez, C.; Robles, J.; Canizares, P.; Garcia-Reyes, J.F.; Molina-Diaz, A.; Comninellis, C.; Rodrigo, M.A. Degradation of caffeine by conductive diamond electrochemical oxidation. *Chemosphere* **2013**, *93*, 1720–1725. [CrossRef]
98. Ambuludi, S.L.; Panizza, M.; Oturan, N.; Ozcan, A.; Oturan, M.A. Kinetic behavior of anti-inflammatory drug ibuprofen in aqueous medium during its degradation by electrochemical advanced oxidation. *Environ. Sci. Pollut. Res. Int.* **2013**, *20*, 2381–2389. [CrossRef] [PubMed]
99. Da Silva Duarte, J.L.; Solano, A.M.S.; Arguelho, M.L.; Tonholo, J.; Martínez-Huitle, C.A.; e Silva, C.L.d.P. Evaluation of treatment of effluents contaminated with rifampicin by Fenton, electrochemical and associated processes. *J. Water Process Eng.* **2018**, *22*, 250–257. [CrossRef]
100. Coledam, D.A.; Aquino, J.M.; Silva, B.F.; Silva, A.J.; Rocha-Filho, R.C. Electrochemical mineralization of norfloxacin using distinct boron-doped diamond anodes in a filter-press reactor, with investigations of toxicity and oxidation by-products. *Electrochim. Acta* **2016**, *213*, 856–864. [CrossRef]
101. Steter, J.R.; Rocha, R.S.; Dionísio, D.; Lanza, M.R.; Motheo, A.J. Electrochemical oxidation route of methyl paraben on a boron-doped diamond anode. *Electrochim. Acta* **2014**, *117*, 127–133. [CrossRef]
102. Fabiańska, A.; Białk-Bielińska, A.; Stepnowski, P.; Stolte, S.; Siedlecka, E.M. Electrochemical degradation of sulfonamides at BDD electrode: Kinetics, reaction pathway and eco-toxicity evaluation. *J. Hazard. Mater.* **2014**, *280*, 579–587. [CrossRef]
103. Brinzila, C.; Monteiro, N.; Pacheco, M.; Ciríaco, L.; Siminiceanu, I.; Lopes, A. Degradation of tetracycline at a boron-doped diamond anode: Influence of initial pH, applied current intensity and electrolyte. *Environ. Sci. Pollut. Res.* **2014**, *21*, 8457–8465. [CrossRef]
104. Haidar, M.; Dirany, A.; Sirés, I.; Oturan, N.; Oturan, M.A. Electrochemical degradation of the antibiotic sulfachloropyridazine by hydroxyl radicals generated at a BDD anode. *Chemosphere* **2013**, *91*, 1304–1309. [CrossRef]
105. González, T.; Domínguez, J.R.; Palo, P.; Sánchez-Martín, J. Conductive-diamond electrochemical advanced oxidation of naproxen in aqueous solution: Optimizing the process. *J. Chem. Technol. Biotechnol. Adv.* **2011**, *86*, 121–127. [CrossRef]
106. Domínguez, J.R.; González, T.; Palo, P.; Sánchez-Martín, J. Electrochemical advanced oxidation of carbamazepine on boron-doped diamond anodes. Influence of operating variables. *Ind. Eng. Chem. Res.* **2010**, *49*, 8353–8359. [CrossRef]

107. Sun, Y.; Li, P.; Zheng, H.; Zhao, C.; Xiao, X.; Xu, Y.; Sun, W.; Wu, H.; Ren, M. Electrochemical treatment of chloramphenicol using Ti-Sn/γ-Al$_2$O$_3$ particle electrodes with a three-dimensional reactor. *Chem. Eng. J.* **2017**, *308*, 1233–1242. [CrossRef]
108. Hu, X.; Yu, Y.; Sun, Z. Preparation and characterization of cerium-doped multiwalled carbon nanotubes electrode for the electrochemical degradation of low-concentration ceftazidime in aqueous solutions. *Electrochim. Acta* **2016**, *199*, 80–91. [CrossRef]
109. Yang, W.; Zhou, M.; Oturan, N.; Li, Y.; Su, P.; Oturan, M.A. Enhanced activation of hydrogen peroxide using nitrogen doped graphene for effective removal of herbicide 2, 4-D from water by iron-free electrochemical advanced oxidation. *Electrochim. Acta* **2019**, *297*, 582–592. [CrossRef]
110. Al-Qaim, F.F.; Mussa, Z.H.; Othman, M.R.; Abdullah, M.P. Removal of caffeine from aqueous solution by indirect electrochemical oxidation using a graphite-PVC composite electrode: A role of hypochlorite ion as an oxidising agent. *J. Hazard. Mater.* **2015**, *300*, 387–397. [CrossRef] [PubMed]
111. Frontistis, Z.; Antonopoulou, M.; Yazirdagi, M.; Kilinc, Z.; Konstantinou, I.; Katsaounis, A.; Mantzavinos, D. Boron-doped diamond electrooxidation of ethyl paraben: The effect of electrolyte on by-products distribution and mechanisms. *J. Environ. Manag.* **2017**, *195*, 148–156. [CrossRef] [PubMed]
112. Palo, P.; Domínguez, J.R.; González, T.; Sánchez-Martin, J.; Cuerda-Correa, E.M. Feasibility of electrochemical degradation of pharmaceutical pollutants in different aqueous matrices: Optimization through design of experiments. *J. Environ. Sci. Health Part A* **2014**, *49*, 843–850. [CrossRef]
113. De Vidales, M.J.M.; Millán, M.; Sáez, C.; Pérez, J.F.; Rodrigo, M.A.; Cañizares, P. Conductive diamond electrochemical oxidation of caffeine-intensified biologically treated urban wastewater. *Chemosphere* **2015**, *136*, 281–288. [CrossRef]
114. Zaghdoudi, M.; Fourcade, F.; Soutrel, I.; Floner, D.; Amrane, A.; Maghraoui-Meherzi, H.; Geneste, F. Direct and indirect electrochemical reduction prior to a biological treatment for dimetridazole removal. *J. Hazard. Mater.* **2017**, *335*, 10–17. [CrossRef]
115. Belkheiri, D.; Fourcade, F.; Geneste, F.; Floner, D.; Aït-Amar, H.; Amrane, A. Feasibility of an electrochemical pre-treatment prior to a biological treatment for tetracycline removal. *Sep. Purif. Technol.* **2011**, *83*, 151–156. [CrossRef]
116. Yahiaoui, I.; Aissani-Benissad, F.; Fourcade, F.; Amrane, A. Removal of tetracycline hydrochloride from water based on direct anodic oxidation (Pb/PbO$_2$ electrode) coupled to activated sludge culture. *Chem. Eng. J.* **2013**, *221*, 418–425. [CrossRef]
117. Rodríguez-Nava, O.; Ramírez-Saad, H.; Loera, O.; González, I. Evaluation of the simultaneous removal of recalcitrant drugs (bezafibrate, gemfibrozil, indomethacin and sulfamethoxazole) and biodegradable organic matter from synthetic wastewater by electro-oxidation coupled with a biological system. *Environ. Technol.* **2016**, *37*, 2964–2974. [CrossRef] [PubMed]
118. García-Gómez, C.; Drogui, P.; Seyhi, B.; Gortáres-Moroyoqui, P.; Buelna, G.; Estrada-Alvgarado, M.I.; Álvarez, L.H. Combined membrane bioreactor and electrochemical oxidation using Ti/PbO$_2$ anode for the removal of carbamazepine. *J. Taiwan Inst. Chem. Eng.* **2016**, *64*, 211–219. [CrossRef]
119. Ouarda, Y.; Tiwari, B.; Azais, A.; Vaudreuil, M.A.; Ndiaye, S.D.; Drogui, P.; Tyagi, R.D.; Sauve, S.; Desrosiers, M.; Buelna, G.; et al. Synthetic hospital wastewater treatment by coupling submerged membrane bioreactor and electrochemical advanced oxidation process: Kinetic study and toxicity assessment. *Chemosphere* **2018**, *193*, 160–169. [CrossRef]
120. Ensano, B.M.B.; Borea, L.; Naddeo, V.; Belgiorno, V.; de Luna, M.D.G.; Balakrishnan, M.; Ballesteros, F.C., Jr. Applicability of the electrocoagulation process in treating real municipal wastewater containing pharmaceutical active compounds. *J. Hazard. Mater.* **2019**, *361*, 367–373. [CrossRef]
121. Borea, L.; Ensano, B.M.B.; Hasan, S.W.; Balakrishnan, M.; Belgiorno, V.; de Luna, M.D.G.; Ballesteros, F.C.; Naddeo, V. Are pharmaceuticals removal and membrane fouling in electromembrane bioreactor affected by current density? *Sci. Total Environ.* **2019**, *692*, 732–740. [CrossRef]

© 2020 by the authors. Licensee MDPI, Basel, Switzerland. This article is an open access article distributed under the terms and conditions of the Creative Commons Attribution (CC BY) license (http://creativecommons.org/licenses/by/4.0/).

Article

A Comparison of the Mechanism of TOC and COD Degradation in Rhodamine B Wastewater by a Recycling-Flow Two- and Three-dimensional Electro-Reactor System

Jin Ni [1], Huimin Shi [1], Yuansheng Xu [1] and Qunhui Wang [1,2,*]

[1] Department of Environmental Engineering, School of Energy and Environmental Engineering, University of Science and Technology Beijing, 30 Xueyuan Road, Haidian District, Beijing 10083, China; jolinxiaopang@163.com (J.N.); shihuimin99@hotmail.com (H.S.); xys519828120@163.com (Y.X.)
[2] Beijing Key Laboratory on Resource-oriented Treatment of Industrial Pollutants, University of Science and Technology Beijing, 30 Xueyuan Road, Beijing 10083, China
* Correspondence: wangqh59@sina.com

Received: 5 June 2020; Accepted: 23 June 2020; Published: 28 June 2020

Abstract: Dye wastewater, as a kind of refractory wastewater (with a ratio of biochemical oxygen demand (BOD) and chemical oxygen demand (COD) of less than 0.3), still needs advanced treatments in order to reach the discharge standard. In this work, the recycling-flow three-dimensional (3D) electro-reactor system was designed for degrading synthetic rhodamine B (RhB) wastewater as dye wastewater (100 mg/L). After 180 min of degradation, the removal of total organic carbon (TOC) and chemical oxygen demand (COD) of RhB wastewater were both approximately double the corresponding values in the recycling-flow two-dimensional (2D) electro-reactor system. Columnar granular activated carbon (CGAC), as micro-electrodes packed between anodic and cathodic electrodes in the recycling-flow 3D electro-reactor system, generated an obviously characteristic peak of anodic catalytic oxidation, increased the mass transfer rate and electrochemically active surface area (EASA) by 40%, and rapidly produced 1.52 times more hydroxyl radicals (·OH) on the surface of CGAC electrodes, in comparison to the recycling-flow 2D electro-reactor system. Additionally, the recycling-flow 3D electro-reactor system can maintain higher current efficiency (CE) and lower energy consumption (Es).

Keywords: recycling-flow; three-dimensional electro-reactor system; two-dimensional electro-reactor system; rhodamine B; wastewater treatment

1. Introduction

Nowadays, the unsafe disposal of dye wastewater, which still contains lots of complex pollutants and toxic matter, such as aromatic, chloric, and azo compounds, is seriously threating environmental and ecological systems and human health [1–5]. The reason is that dye wastewater, as a kind of refractory wastewater (with a ratio of biochemical oxygen demand (BOD) and chemical oxygen demand (COD) of less than 0.3), is severely difficult to degrade in order to fall under discharge standards in the activated sludge process using traditional and biological wastewater treatment methods [6]. Therefore, there is a definite urgent need for methods which efficiently degrade dye wastewater after biological treatment [7–10].

In comparison to the two-dimensional (2D) electro-reactor, consisting of an anode electrode, cathode electrode, and electrolyzer, the three-dimensional (3D) electro-reactor contains a certain number of small granular substances, such as activated carbon particles and diatomite particles, charged by an electric field to form micro-electrodes and then acquires an electrochemically oxidative ability with a third electrode, which is placed between the anode and the cathode electrodes [11–14]. Simultaneously, the degradation of dye pollutants from wastewater can occur on the surface of these small granular electrodes and anode and cathode electrodes in the 3D electro-reactor [15,16]. Hence, the 3D electro-reactor theoretically shows more brilliant promise in the advanced treatment of dye wastewater as effluent than the 2D electro-reactor.

Currently, the number of researchers [17–19] who pay attention to the study of 3D electrode technology is growing dramatically, but they are always focusing on the novel methods of granular electrode modification in order to improve the oxidative degradation ability of the 3D electro-reactor instead of the design and amendment of the 3D electro-reactor system. However, the processes of granular electrode modification normally require quite serious and extreme conditions, such as high temperature and pressure, and special materials, such as noble gases and metal [20,21].

As is well known, the fixed bed 3D electro-reactor system and the fluid bed 3D electro-reactor system are usually used to treat dye wastewater. The former system has the main disadvantage of low treatment efficiency due to extremely a high hydraulic retention time (HRT) and, meanwhile, the main disadvantage of the latter system is effluent's high total organic carbon (TOC) and COD above the discharged standard due to a low HRT [22–24]. Therefore, a recycling-flow 3D electro-reactor system, taking advantages of the strong oxidative degradation ability, high treatment efficiency, no secondary pollution, and being operated under normal temperature and atmosphere pressure, is designed for degrading dye wastewater in our work.

Choosing RhB wastewater as one kind of dye wastewater, this paper is going to analyze the mechanism of the TOC and COD degradation of RhB wastewater in the aspect of mass transfer, the electrochemically active surface area (EASA) of electrodes, the instant concentration of hydroxyl radicals (·OH), current efficiency (CE), and energy consumption (Es) in the recycling-flow 3D electro-reactor system, compared with the recycling-flow 2D electro-reactor system. Additionally, it carries on, finding out the mechanism of TOC and COD degradation with different voltages, with different electrolytes, and at different HRTs in the recycling-flow 3D electro-reactor system.

2. Materials and Methods

2.1. Materials

RhB as a dye has a molecular formula of $C_{28}H_{31}ClN_2O_3$ and a molecular weight of 479.01. Columnar granular activated carbon (CGAC) from coconut shells (average size: 1.50 mm) was purchased from Henan Lianhua Carbon Manufacturing Co. Two $Ti/RuO_2/TiO_2$ board electrodes (size: 60 mm × 100 mm × 2 mm) were provided by the Second Research Institute of the China Aerospace Science and Industry Group. Anhydrous sodium sulfate (Na_2SO_4), sodium chloride (NaCl), mercury(II) sulfate ($HgSO_4$), phosphoric acid (H_3PO_4), potassium dichromate ($K_2Cr_2O_7$), sulfuric acid (H_2SO_4), and potassium ferrocyanide ($K_4Fe(CN)_6$) were of analytical grade and used without any further purification. Silver sulfate (Ag_2SO_4) was not less than 99.7%. Ammonium iron(II) sulfate (($NH_4)_2Fe(SO_4)_2$) was not less than 99.5%. The ferroin indicator solution standard is Q/12NK4019-2011.

2.2. Experimental Setup and Procedure

A virtual diagram of the experimental setup is shown in Figure 1. The electro-reactor was a plexiglass rectangular tank (Organic Glass Factory, Beijing, China) with two $Ti/RuO_2/TiO_2$ board electrodes as the anode and cathode in the 2D electro-reactor. The anode and cathode were positioned vertically and parallel to each other with an inter-electrode gap of 30 mm. The CGAC electrodes were packed between the anode and cathode up to a height of 80 mm in the 3D electro-reactor; the liquid level equaled the height of the packed bed.

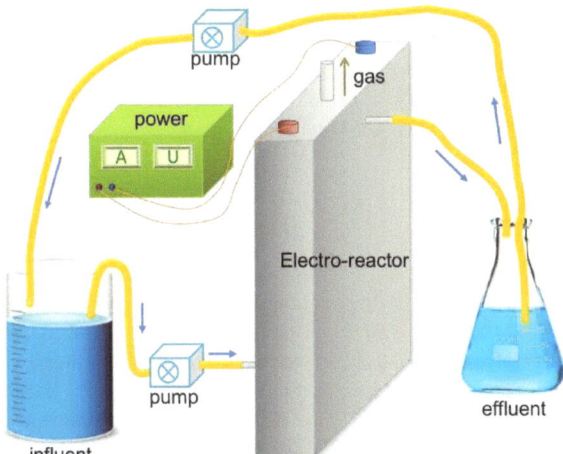

Figure 1. The virtual diagram of the recycling-flow 3D electro-reactor system.

In this recycling system, one peristaltic pump (Youji Keyi ZQ000S, Baoding, Hebei, China) was used to pump the influent RhB wastewater from a beaker into the bottom of the electro-reactor. Then the RhB wastewater flowed out from the upper outlet of the electro-reactor and was collected in an Erlenmeyer flask. Simultaneously, the other peristaltic pump was used to pump the effluent RhB wastewater from the Erlenmeyer flask back to the beaker.

Prior to commencing the electrochemical oxidation treatment, CGAC electrodes adsorbed RhB solution until becoming saturated in order to minimize the effect of adsorption on TOC and COD removal. All experiments used a digital DC power supply (DC 30 V/5 A; DH1716-6D). After completing the experiment, all treated samples were collected and filtered through 0.45 mm filters. The filtrate was then analyzed, as described in the next sub-sections.

2.3. Analytical Methods

TOC was measured by a Vario TOC analysis device. According to the instructions of the Vario TOC analysis device, when carbonaceous compounds are burned in an oxygen-rich environment, the carbon is completely converted into CO_2, and then the non-scattering infrared detector (NDIR) detects the amount of CO_2 and converts it into total carbon (TC) in the sample. After the sample is acidified by phosphoric acid (H_3PO_4) (1% v/v) and pH decreasing, the carbonate and bicarbonate in the sample are converted into CO_2, which is blown out and enters the NDIR, and then the detected amount of CO_2 is converted into total inorganic carbon (TIC). The value of TOC is TC minus TIC.

COD was measured by a microwave digestion method. According to the instructions of the Kedibo microwave digestion device, mercury(II) sulfate ($HgSO_4$) as a masking agent, 0.05 mol/L potassium dichromate ($K_2Cr_2O_7$) as a digestion solution, a mix of 10 g silver sulfate (Ag_2SO_4) and 1 L sulfuric acid (H_2SO_4) as a catalyst, ferroin solution as an indicator, and potassium ferrocyanide ($K_4Fe(CN)_6$) as a standard solution were used.

Electrochemical measurements were performed using a conventional three-electrode cell and a CHI 660E electrochemical workstation (CHI, Beijing, China). Ag/AgCl and $Ti/RuO_2/TiO_2$ board electrodes served as the reference and counter electrodes, respectively.

CE (%) is the rate of the efficient current and total current in a period and Es (KW·h/kg TOC) is the electricity consumption of 1 kg TOC degradation. They were calculated according to the following equations [25–27]:

$$CE = \frac{TOC_0 - TOC_t}{480I} FQ \qquad (1)$$

$$Es = \frac{UI}{60(TOC_0 - TOC_t)Q} \tag{2}$$

where TOC_0 (g/L) and TOC_t (g/L) correspond to the total organic carbon at $t = 0$ min and $t = t$ min, respectively. I is the average current (A), F is the Faraday constant (96,485 C/mol), Q is the flow rate of water (L/min), and U is the applied electric voltage (V).

3. Result and Discussion

3.1. Electrochemical Properties of TOC and COD Degradation in the Recycling-Flow Elecro-Reactor System

As shown in Figure 2, the TOC removal of RhB wastewater in the recycling-flow 3D electro-reactor system was always higher than that in the recycling-flow 2D electro-reactor system, and the highest TOC removal was 72.0%, which was 1.98 times that in the recycling-flow 2D electro-reactor system (36.3%). From the perspective of COD, the COD removal of RhB wastewater in the recycling-flow 3D electro-reactor system was much higher than that in the recycling-flow 2D electro-reactor system and the highest COD removal was up to 86.9%. By the 30th minute, COD removal in the recycling-flow 3D electro-reactor system had already reached 63.4%, meanwhile, the recycling-flow 2D electro-reactor system just reached 41.9% by the 180th minute.

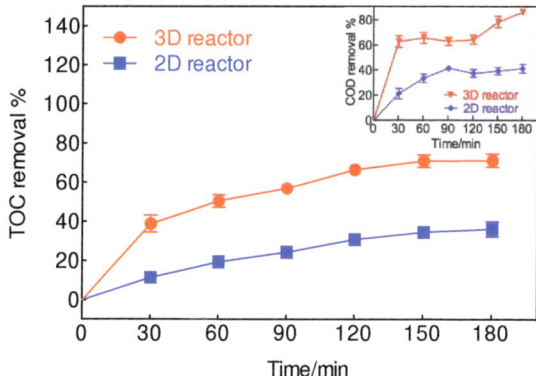

Figure 2. Total organic carbon (TOC) and chemical oxygen demand (COD) removal of rhodamine B (RhB) wastewater in the recycling-flow 3D and 2D electro-reactor systems (RhB wastewater initial concentration is 100 mg/L; volume is 500 mL; initial pH is 7; Na_2SO_4 as an electrolyte, initial concentration is 2 g/L; voltage is 5 V; hydraulic retention time (HRT) is 20 min).

It is normally considered that the mass transfer and EASA of the electrodes play major roles. Therefore, the higher TOC and COD removal in the recycling-flow 3D electro-reactor system is due to the presence of CGAC as conductive particles packed in the 3D electro-reactor and constitute a number of microelectrodes, which dramatically increase the area of the reaction electrode and benefit organic matter degradation easily and quickly in oxidative processes [28,29].

In order to obtain the mass transfer rate and EASA of the electrodes in the 2D and 3D electro-reactors, respectively, cyclic voltammograms (CVs) of the 2D and 3D electro-reactors were measured in a 0.05 mol/L $K_4Fe(CN)_6$ + 0.45 mol/L Na_2SO_4 solution at different scan rates from 0.015 V/s to 0.1 V/s by a CHI 660E electrochemical workstation (CHI, China). The 3D electro-reactor had a characteristic peak of anodic catalytic oxidation, as illustrated in Figure 3. Taking a sweep rate of 0.1 V/s (red line) as an example, when the potential was 0.7 V, it should be 0.2 A if the potential and the corresponding current were in a linear relationship (demonstrated in Figure 3b). However, the Figure 3a curve shows that the corresponding current is as high as 0.35 A at the potential of 0.7 V. This was because the corresponding current incurred a mutation at 0.7 V, which is called a characteristic peak of anodic catalytic oxidation.

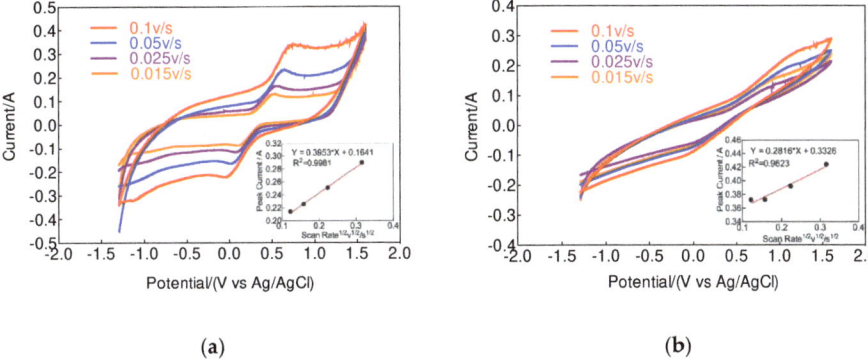

Figure 3. Cyclic voltammograms of the 3D and 2D electro-reactors in a 50 mmol/L K$_4$Fe(CN)$_6$ + 0.45 mol/L Na$_2$SO$_4$ solution at different scan rates. (**a**) 3D electro-reactor, (**b**) 2D electro-reactor. Insets show the plots of the peak current vs. the square root of the scan rate.

It was found that both the anodic peak current and the cathodic peak current increased as the scan rate increased, indicating a reversible electrochemical reaction of the [Fe(CN)$_6$]$^{4-}$/[Fe(CN)$_6$]$^{3-}$ redox couple. At the same time, Figure 3 (insets) shows a brilliant linear relationship between the oxidation peak current (I_p) and the square root of the scan rate ($v^{1/2}$), according to the following equation [21,30,31]:

$$I_p = (2.69 \times 10^5) n^{2/3} A D_R^{1/2} C_R v^{1/2} \tag{3}$$

where n is the number of transferred electrons, A is the EASA (cm^2), D_R is the diffusion coefficient of the reduced species (cm^2/s), C_R is the bulk reduced species concentration (mmol/L), and v is the scan rate (mV/s). The above equation can be simplified into the following equation:

$$I_p = k v^{1/2} \tag{4}$$

where k is a coefficient only relevant to A and D_R because n and C_R are constant in this study.

As shown in the insets of Figure 3a,b, the slopes of the linear relationship between I_p and $v^{1/2}$, named as the value of k, representing the mass transfer rate, were obtained according to the linear fitting of the plots. As expected, the k value of the 3D electro-reactor (0.3953) was larger than the corresponding value of the 2D electro-reactor (0.2816). This demonstrates that the 3D electro-reactor had greater mass transfer properties than the 2D electro-reactor [32,33].

In addition, the EASA can also be derived from the k value by assuming that the D_R value of [Fe(CN)$_6$]$^{4-}$ is constant in this study. The obtained EASA value of the 3D electro-reactor is also higher than that of the 2D electro-reactor. In particular, it was 1.40 times the corresponding value of the 2D electro-reactor. The higher EASA of the 3D electro-reactor means that the 3D electro-reactor will provide much more electrochemically active sites for RhB oxidation, and thus will be beneficial in improving the oxidation of organics on the electrode surface [34].

Based on the mass transfer rate and EASA result and discussion above, the 3D electro-reactor has been demonstrated to possess a much higher electro-catalytic activity for degrading organic matter (RhB) than the 2D electro-reactor.

It is well known that the hydroxyl radical (·OH), of which the oxidation potential (2.8 eV) is the second highest and regarded as a powerful oxidizing chemical in nature, plays an important role in the electrochemical oxidation of RhB wastewater [35]. Therefore, the production of ·OH was detected by using the electron paramagnetic resonance (EPR) technique by adding the ·OH scavenger 5,5-dimenthyl,1-pyrroline-N-oxide (DMPO) to further reveal the underlying mechanism of the TOC and COD degradation of RhB wastewater. As expected, typical EPR spectra of the DMPO–·OH adduct with a 1:2:2:1 quartet were acquired in the 3D and 2D electro-reactors when current and voltage were applied, as shown in Figure 4. It is worth noting that the 3D electro-reactor achieved the higher EPR intensity, which was nearly 2.52 times higher than the corresponding value of the 2D electro-reactor and demonstrated that the electro-generation of ·OH occurred on the CGAC electrodes and the usage of the CGAC electrodes could enhance ·OH generation effectively in electrochemical oxidation processes [36].

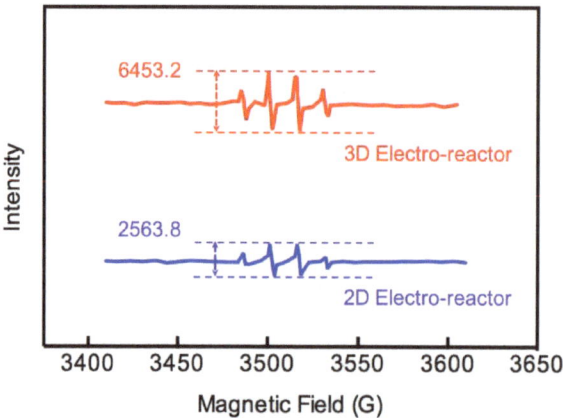

Figure 4. Dimenthyl,1-pyrroline-N-oxide (DMPO) spin trapping the electron paramagnetic resonance (EPR) spectra of hydroxyl radicals (OH) in the 3D and 2D electro-reactors.

Overall, the degradation of organic matter (RhB) by the 3D electro-reactor depends on more ·OH being adsorbed on the surface of CGAC electrodes, forming more electrochemically active sites for RhB oxidation. However, the 2D electro-reactor can just undergo less ·OH adsorption and the degradation strength is also reduced due to the absence of CGAC as conductive particles.

The CE and Es of the recycling-flow 3D and 2D electro-reactor systems are shown in Figure 5. Overall, the CE of the recycling-flow 3D and 2D electro-reactor systems both went up from 8.3% and 2.5% to 14.4% and 7.6%, respectively, and the Es of the recycling-flow 3D and 2D electro-reactor systems both decreased from 202.6 KW·h/kg TOC and 658.1 KW·h/kg TOC to 116.3 KW·h/kg TOC and 219.9 KW·h/kg TOC, respectively, during the electrochemical processes. The CE of the recycling-flow 3D electro-reactor system was always nearly twice as much as the corresponding value of the recycling-flow 2D electro-reactor system in the treatment period. Additionally, the lowest CE value of the recycling-flow 3D electro-reactor system, by the 30th minute, had reached 8.3%, which approximately equaled the highest CE value of the recycling-flow 2D electro-reactor system (7.6%).

From the perspective of Es, by the 30th minute, the Es of the recycling-flow 3D electro-reactor system was 202.6 KW·h/kg TOC, which was less than one third the corresponding value of the recycling-flow 2D electro-reactor system (658.1 KW·h/kg TOC). Although the Es of the recycling-flow 2D electro-reactor system reduced slightly, the minimum Es was still as high as 219.9 KW·h/kg TOC, which was almost double the corresponding value of the recycling-flow 3D electro-reactor system (116.3 KW·h/kg TOC).

This verified that the existence of CGAC electrodes in the recycling-flow 3D electro-reactor system increased the mass transfer rate and EASA, rapidly generated much more ·OH on the surface of CGAC electrodes, and improved CE and reduced Es.

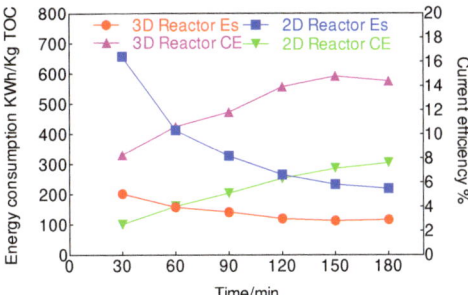

Figure 5. Energy consumption (Es) and current efficiency (CE) of the recycling-flow 3D and 2D electro-reactor systems (RhB wastewater initial concentration is 100 mg/L; volume is 500 mL; initial pH is 7, Na_2SO_4 as an electrolyte, initial concentration is 2 g/L; voltage is 5 V; HRT is 20 min; current density is 60 mA/cm^2).

3.2. Mechanism of TOC and COD Degradation with Different Electrolytes in the Recycling-Flow Electro-Reactor System

An electrolyte is frequently added to wastewater to enhance the conductivity of the solution and reduce impedance in an electrochemical reaction. It is actually necessary to research the effect of the electrolyte on the degradation of RhB wastewater, as shown in Figure 6.

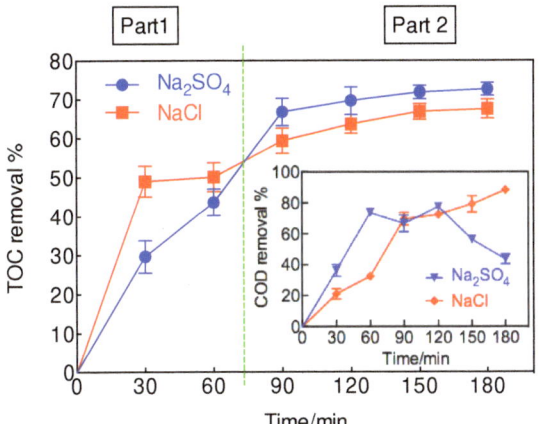

Figure 6. TOC and COD removal of RhB wastewater in the recycling-flow 3D electro-reactor system (NaCl and Na_2SO_4 as electrolytes, initial concentration is 2 g/L; RhB wastewater initial concentration is 100 mg/L; volume is 500 mL; initial pH is 7; voltage is 7 V; HRT is 20 min).

From the beginning to the 75th min as Part 1, the TOC removal of RhB wastewater (NaCl) was higher than that of RhB wastewater (Na_2SO_4). The reason was that Cl^- in the RhB wastewater (NaCl) dramatically generated lots of active chlorine (Cl), which reacted with ·OH in a synergistic process to increase TOC removal through electrochemical reaction processes. However, from the 75th min to the 180th min as Part 2, the TOC removal of RhB wastewater (Na_2SO_4) started to go beyond that of RhB wastewater (NaCl). The former and the latter peak rates reached 72.8% and 67.3%, respectively. As some active chlorine (Cl) was converted into chlorine (Cl_2) in the RhB wastewater (NaCl), Cl_2 partially escaped into the air. The electrons lost during the formation of Cl_2 could no longer be used due to the balance of electron gain and loss. Therefore, RhB wastewater (NaCl) had a final TOC removal lower than that of RhB wastewater (Na_2SO_4) [37].

Additionally, the COD removal of RhB wastewater (NaCl) increased gradually and then reached 89.3%. Meanwhile, the COD removal of RhB wastewater (Na$_2$SO$_4$) first increased and then dropped to 43.9%. The highest value could just reach 77.6%. RhB decomposed into small molecules which could not be oxidized by potassium dichromate (K$_2$Cr$_2$O$_7$) in the solution (Na$_2$SO$_4$) from the 60th min, whilst a side reaction occurred in the solution (NaCl) to generate ClO$^-$ which had strong oxidizing properties, and enhanced the ability to degrade COD [38].

Obviously, the CE of the recycling-flow 3D electro-reactor system with NaCl was higher than the corresponding value of the recycling-flow 3D electro-reactor system with Na$_2$SO$_4$ before the 75th minute, but the CE of the recycling-flow 3D electro-reactor system with Na$_2$SO$_4$ was higher than the corresponding value of the recycling-flow 3D electro-reactor system with NaCl up to the 180th min, as illustrated in Figure 7. Overall, the CE of the recycling-flow 3D electro-reactor systems with NaCl and Na$_2$SO$_4$ both went up smoothly and then got to the highest value (12.2% and 11.3%, respectively).

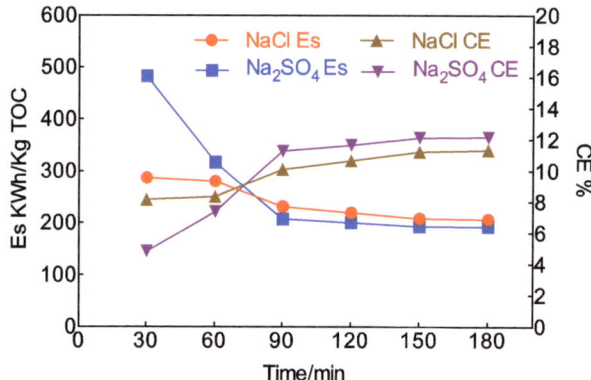

Figure 7. Es and CE of the recycling-flow 3D electro-reactor system with NaCl and Na$_2$SO$_4$ as electrolytes (NaCl and Na$_2$SO$_4$ initial concentration is 2 g/L; RhB wastewater initial concentration is 100 mg/L; volume is 500 mL; initial pH is 7; voltage is 7 V; HRT is 20 min; current density is 60 mA/cm^2).

Meanwhile, the Es of the recycling-flow 3D electro-reactor system with Na$_2$SO$_4$ was higher than the corresponding value of the recycling-flow 3D electro-reactor system with NaCl up to the 80th min. In particular, the Es of the recycling-flow 3D electro-reactor system with Na$_2$SO$_4$ was 483.5 KW·h/kg TOC, which was 68.3% higher than the corresponding value of the recycling-flow 3D electro-reactor system with NaCl (287.2 KW·h/kg TOC) by the 30th min. However, the Es of the recycling-flow 3D electro-reactor system with Na$_2$SO$_4$ started to be lower than the corresponding value of the recycling-flow 3D electro-reactor system with NaCl from the 80th min. In addition, the Es of the recycling-flow 3D electro-reactor systems with NaCl and Na$_2$SO$_4$ both declined to the lowest value (192.7 KW·h/kg TOC and 206.7 KW·h/kg TOC, respectively) in this period.

These again indicate that NaCl, as an electrolyte, had a high conductive efficiency in the former period due to active chlorine (Cl) generation and then changed to low conductive efficiency due to Cl$_2$ escaping into the air in the latter period from CE and Es [22].

3.3. Mechanism of TOC and COD Degradation in Different Voltages in the Recycling-Flow Electro-Reactor System

As seen in Figure 8, the COD removal curve shows a smooth increase and then reaches 30.1% with the voltage of 3 V. In addition, the COD removal of RhB wastewater rose dramatically with the voltages of 5 V and 7 V up to the 60th min and then kept nearly flat between the 60th min and 120th min. Finally, COD removal with 5 V carried on increasing to 86.9%, whilst the corresponding value with 7 V started decreasing to 43.9%. This could be explained by the equation below:

$$COD\ removal(\%) = \frac{COD_0 - COD_t}{COD_0} \times 100\% = 1 - \frac{COD_t}{COD_0} \times 100\% \tag{5}$$

where COD_0 is the initial COD of RhB wastewater and COD_t is the COD of RhB wastewater after treating t minutes.

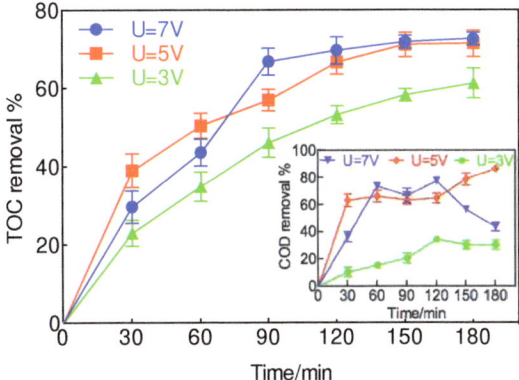

Figure 8. TOC and COD removal of RhB wastewater with different voltages in the recycling-flow 3D electro-reactor system (RhB wastewater initial concentration is 100 mg/L; volume is 500 mL; initial pH is 7; Na$_2$SO$_4$ as an electrolyte, initial concentration is 2 g/L; HRT is 20 min).

The COD_0 value was smaller than the actual value due to some macromolecular substances being unable to be oxidized by K$_2$Cr$_2$O$_7$ in the initial RhB wastewater, and then more macromolecules could be oxidized into smaller molecules with the voltage of 7 V set in this experiment, compared with the voltage of 5 V, which were easily oxidized by K$_2$Cr$_2$O$_7$, resulting in a COD_t value and $COD\ removal$ that are greater simultaneously [39].

From the perspective of TOC, as shown in Figure 8, the TOC removal of RhB wastewater grew the most slowly and the final TOC removal just arrived at 61.0% when the voltage was 3 V. TOC removal was always higher with the voltages of 7 V and 5 V, and the highest values were basically equivalent (72.2% and 72.0%, respectively), which were nearly 1.18 times the corresponding value with the voltage of 3V.

Interestingly, from the 0th min to the 70th min, TOC removal with the voltage of 5 V was higher than that with the voltage of 7 V. TOC degradation processes are illustrated in Figure 9. First of all, TOC was converted into TIC and then TIC was converted into CO$_2$ and H$_2$O. The concentration of ·OH in the RhB wastewater was higher, so that oxidation was stronger between the board electrodes with 7 V. TIC was quickly oxidized into CO$_2$ and H$_2$O and TOC was converted into TIC as main processes. With the voltage of 5 V, the oxidation was weaker [40–42]. Electron transfer processes were normally used to convert TOC into TIC whilst only some of TIC was converted into CO$_2$ and H$_2$O. Hence, TOC removal was higher with 5 V. After the 70th min, TOC removal with the voltage of 7 V was beyond the corresponding value with the voltage of 5 V. TIC from TOC was basically converted into CO$_2$ and H$_2$O with the voltage of 7 V, while TIC just started being rapidly converted into CO$_2$ and H$_2$O as main processes with the voltage of 5 V. Hence, TOC removal was higher with 7 V. The final TOC removal with 7 V was slightly higher than the corresponding value with 5 V since the potential of 7 V was higher than 5 V up to the 180th min [43,44].

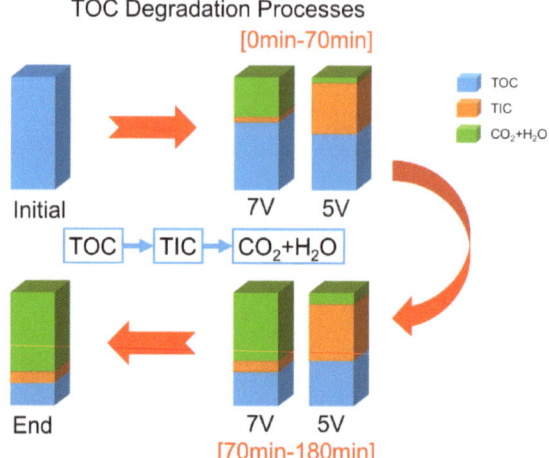

Figure 9. The diagram of TOC degradation processes in RhB wastewater with voltages of 7 V and 5 V.

CE with the voltage of 3 V rose the fastest from 8.0% by the 30th minute to 21.8% by the 180th minute, which increased by nearly two times, meanwhile, the CEs with 5 V and 7 V both showed gradually increasing trends to 14.1% and 12.2%, respectively, as indicated in Figure 10. Plus, the Es with different voltages (3 V, 5 V, and 7 V) gradually decreased by 63.1%, 42.6%, and 60.2%, respectively, from the 30th min up to the 180th min. In the whole electrolysis process, the higher the voltage was, the higher the Es was. At the 180th min, Es with the voltage of 3 V was 46.1 KW·h/kg TOC, which was 39.6% of the corresponding value with 5 V (116.3 KW·h/kg TOC) and 23.9% of the corresponding value with 7 V (192.7 KW·h/kg TOC).

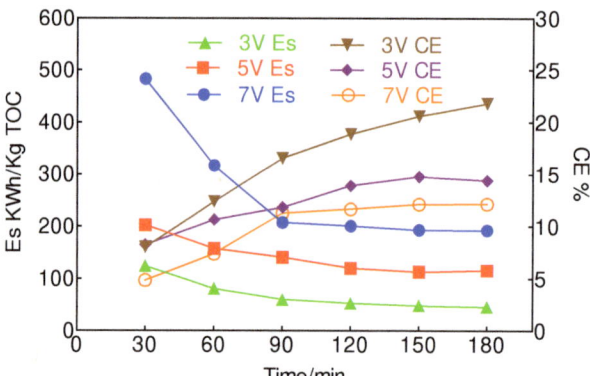

Figure 10. Es and CE of the recycling-flow 3D electro-reactor system with different voltages (RhB wastewater initial concentration is 100 mg/L; volume is 500 mL; initial pH is 7; Na_2SO_4 as an electrolyte, initial concentration is 2 g/L; HRT is 20 min; current density is 60 mA/cm^2).

3.4. Mechanism of TOC and COD Degradation at Different HRTs in the Recycling-Flow Electro-Reactor System

HRT refers to the residence time of wastewater in the reactor, which can be calculated according to the following equation [45]:

$$HRT = \frac{V}{Q} \tag{6}$$

where V is the reactor volume or pool capacity and Q is the influent flow rate.

In this experiment, the flow rate was controlled by operating the pumps in order to study the effect of HRT on the degradation of RhB wastewater.

COD removal curves were almost growing coincidently at different HRTs (20 min, 40 min, and 60 min), as illustrated in Figure 11. Finally, COD removal (86.9%) at HRT = 20 min was slightly higher than the corresponding values at HRT = 40 min (80.1%) and HRT = 60 min (83.4%). This indicated that HRT had little effect on the degradation of COD in the RhB wastewater.

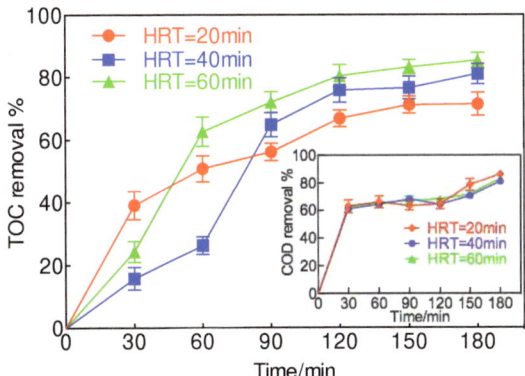

Figure 11. TOC and COD removal of RhB wastewater at different HRTs in the recycling-flow 3D electro-reactor system (RhB wastewater initial concentration is 100 mg/L; volume is 500 mL; initial pH is 7; Na_2SO_4 as an electrolyte, initial concentration is 2 g/L; voltage is 5 V).

TOC removal at different HRTs (20 min, 40 min, and 60 min) gradually increased and then reached the maximum value at the 180th min. The highest TOC removal (85.0%) at HRT = 60 min was 13.8% and 6.8% higher than the corresponding value (71.2% and 78.2%) at HRT = 20 min and HRT = 40 min, respectively. In the comparisons of TOC removal, it can be concluded that HRT = 60 min had a better removal effect.

However, HRT is equal to V/Q and the pool capacity of HRT = 60 min is three times that of HRT = 20 min, which means that the construction cost must be more, as the influent flow rate is constant. In addition, the initial concentration of RhB wastewater was 100 mg/L and the corresponding COD was 215 mg/L. The final COD removal (86.9%) at HRT = 20 min was that of the COD of the effluent, which was 28.2 mg/L after treatment, which reached the Grade A standard (< 50 mg/L) of the Pollutant Discharge Standard for Urban Sewage Treatment Plants [46]. In summary, HRT = 20 min, with less construction cost, is the optimal HRT.

The CE at HRT = 20 min (8.3–14.8%) was always higher than that at HRT = 40 min (CE: 2.4–12.4%) and HRT = 60 min (CE: 2.5–8.7%), as shown in Figure 12. The maximum CE (14.8%) at HRT = 20 min was 1.2 times and 1.7 times the corresponding values at HRT = 40 min (12.4%) and HRT = 60 min (8.7%), respectively. From the perspective of Es, Es at HRT = 20 min was always lower than that at HRT = 40 min and HRT = 60 min, and the minimum Es at HRT = 20 min, 40 min, and 60 min were 116.3 KW·h/kg TOC, 135.3 KW·h/kg TOC, and 192.4 KW·h/kg TOC, respectively.

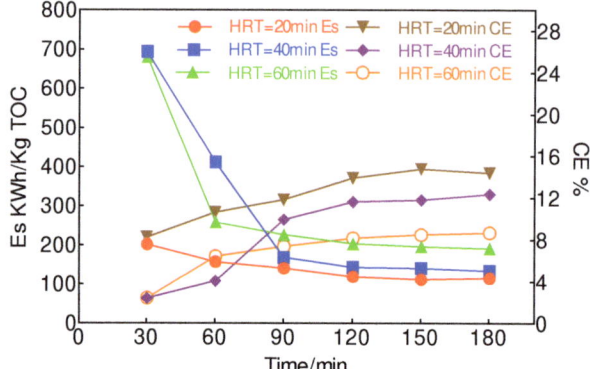

Figure 12. Es and CE of the recycling-flow 3D electro-reactor system at different HRTs (RhB wastewater initial concentration is 100 mg/L; volume is 500 mL; initial pH is 7; Na_2SO_4 as an electrolyte, initial concentration is 2 g/L; voltage is 5 V; current density is 60 mA/cm^2).

4. Conclusions

In conclusion, CGAC, as micro-electrodes between anodic and cathodic electrodes in the recycling-flow 3D electro-reactor system, generated an obviously characteristic peak of anodic catalytic oxidation, increased the mass transfer rate and EASA by 40%, and rapidly produced 1.52 times more ·OH on the surface of CGAC electrodes so that the TOC and COD removal of RhB wastewater were both approximately double the corresponding values in the recycling-flow 2D electro-reactor system after 3 h of treatment in the same experimental conditions, with higher a CE and lower Es.

Treating RhB wastewater in the recycling-flow 3D electro-reactor system, Na_2SO_4 as an electrolyte was more beneficial for TOC degradation, and got higher CE and lower Es, in long electrolyzing times (more than 75 min and less than 180 min). On the contrary, NaCl as an electrolyte could improve COD removal more, and get higher a CE and lower Es in short electrolyzing times (less than 75 min). Plus, TOC and COD removal were the best in the proper voltage (5 V), not the highest one or lowest one. Normally, the higher the voltage, the lower the CE and the more the Es. The HRT condition had little effect on COD removal but high HRT was good for TOC degradation.

Author Contributions: Data curation, J.N., H.S. and Y.X.; Writing—original draft, J.N.; Writing—review and editing, Supervision, Resources, Q.W. All authors have read and agreed to the published version of the manuscript.

Funding: This research received no external funding.

Acknowledgments: The authors appreciate the constructive suggestions from reviewers and editors that helped improve this paper and the support from National Environmental and Energy Base for International Science and Technology Cooperation.

Conflicts of Interest: The authors declare no conflict of interest.

References

1. Shannon, M.A.; Bohn, P.W.; Elimelech, M.; Georgiadis, J.G.; Marinas, B.J.; Mayes, A.M. Science and technology for water purification in the coming decades. *Nature* **2008**, *452*, 301–310. [CrossRef]
2. Holkar, C.R.; Jadhav, A.J.; Pinjari, D.V.; Mahamuni, N.M.; Pandit, A.B. A critical review on textile wastewater treatments: Possible approaches. *J. Environ. Manag.* **2016**, *182*, 351–366. [CrossRef]
3. Massoud, M.A.; Tarhini, A.; Nasr, J.A. Decentralized approaches to wastewater treatment and management: Applicability in developing countries. *J. Environ. Manag.* **2009**, *90*, 652–659. [CrossRef]
4. Naumczyka, J.H.; Kucharskab, M.A.; Ładyńskab, J.A.; Wojewódkaa, D. Electrochemical oxidation process in application to raw and biologically pre-treated tannery wastewater. *Desalin. Water Treat.* **2019**, *162*, 166–175. [CrossRef]

5. Cui, M.-H.; Gao, J.; Wang, A.-J.; Sangeetha, T. Azo dye wastewater treatment in a bioelectrochemical-aerobic integrated system: Effect of initial azo dye concentration and aerobic sludge concentration. *Desalin. Water Treat.* **2019**, *165*, 314–320. [CrossRef]
6. Zhao, R.; Zhao, H.; Dimassimo, R.; Xu, G. Pilot scale study of sequencing batch reactor (sbr) retrofit with integrated fixed film activated sludge (ifas): Nitrogen removal and design consideration. *Environ. Sci. Water Res. Technol.* **2018**, *4*, 569–581. [CrossRef]
7. Kim, S.; Kim, J.; Kim, S.; Lee, J.; Yoon, J. Electrochemical lithium recovery and organic pollutant removal from industrial wastewater of a battery recycling plant. *Environ. Sci. Water Res. Technol.* **2018**, *4*, 175–182. [CrossRef]
8. Soares, P.A.; Batalha, M.; Souza, S.M.A.G.U.; Boaventura, R.A.R.; Vilar, V.J.P. Enhancement of a solar photo-fenton reaction with ferric-organic ligands for the treatment of acrylic-textile dyeing wastewater. *J. Environ. Manag.* **2015**, *152*, 120–131. [CrossRef] [PubMed]
9. Jorfi, S.; Barzegar, G.; Ahmadi, M.; Soltani, R.D.C.; Takdastan, A.; Saeedi, R.; Abtahi, M. Enhanced coagulation-photocatalytic treatment of acid red 73 dye and real textile wastewater using uva/synthesized mgo nanoparticles. *J. Environ. Manag.* **2016**, *177*, 111–118. [CrossRef] [PubMed]
10. Le Luua, T.; Tiena, T.T.; Duongb, N.B.; Phuongb, N.T.T. Study of the treatment of tannery wastewater after biological pretreatment by using electrochemical oxidation on bdd/ti anode. *Desalin. Water Treat.* **2019**, *137*, 194–201. [CrossRef]
11. Li, X.; Zhu, W.; Wang, C.; Zhang, L.; Qian, Y.; Xue, F.; Wu, Y. The electrochemical oxidation of biologically treated citric acid wastewater in a continuous-flow three-dimensional electrode reactor (ctder). *Chem. Eng. J.* **2013**, *232*, 495–502. [CrossRef]
12. Zhang, C.; Jiang, Y.; Li, Y.; Hu, Z.; Zhou, L.; Zhou, M. Three-dimensional electrochemical process for wastewater treatment: A general review. *Chem. Eng. J.* **2013**, *228*, 455–467. [CrossRef]
13. Feng, Y.; Yang, L.; Liu, J.; Logan, B.E. Electrochemical technologies for wastewater treatment and resource reclamation. *Environ. Sci. Water Res. Technol.* **2016**, *2*, 800–831. [CrossRef]
14. Yousefi, Z.; Zafarzadeh, A.; Mohammadpour, R.A.; Zarei, E.; Mengelizadeh, N.; Ghezel, A. Electrochemical removal of acid red 18 dye from synthetic wastewater using a three-dimensional electrochemical reactor. *Desalin. Water Treat.* **2019**, *165*, 352–361. [CrossRef]
15. Liu, Y.; Yu, Z.; Hou, Y.; Peng, Z.; Wang, L.; Gong, Z.; Zhu, J.; Su, D. Highly efficient pd-fe/ni foam as heterogeneous fenton catalysts for the three-dimensional electrode system. *Catal. Commun.* **2016**, *86*, 63–66. [CrossRef]
16. Qiying, L.; Hongyu, S.; Xibo, L.; Junwu, X.; Fei, X.; Limin, L.; Jun, L.; Shuai, W. Ultrahigh capacitive performance of three-dimensional electrode nanomaterials based on α-mno$_2$ nanocrystallines induced by doping au through Å-scale channels. *Nano Energy* **2016**, *21*, 39–50.
17. Yu, X.; Hua, T.; Liu, X.; Yan, Z.; Xu, P.; Du, P. Nickel-based thin film on multiwalled carbon nanotubes as an efficient bifunctional electrocatalyst for water splitting. *Acs Appl. Mater. Interfaces* **2014**, *6*, 15395–15402. [CrossRef]
18. Yu, X.; Sun, Z.; Yan, Z.; Xiang, B.; Liu, X.; Du, P. Direct growth of porous crystalline nico2o4 nanowire arrays on a conductive electrode for high-performance electrocatalytic water oxidation. *J. Mater. Chem. A* **2014**, *2*, 20823–20831. [CrossRef]
19. Yu, X.; Xu, P.; Hua, T.; Han, A.; Liu, X.; Wu, H.; Du, P. Multi-walled carbon nanotubes supported porous nickel oxide as noble metal-free electrocatalysts for efficient water oxidation. *Int. J. Hydrog. Energy* **2014**, *39*, 10467–10475. [CrossRef]
20. Wu, W.; Huang, Z.H.; Lim, T.T. Enhanced electrochemical oxidation of phenol using hydrophobic tio2-nts/sno2-sb-ptfe electrode prepared by pulse electrodeposition. *RSC Adv.* **2015**, *5*, 32245–32255. [CrossRef]
21. Li, X.; Wu, Y.; Zhu, W.; Xue, F.; Qian, Y.; Wang, C. Enhanced electrochemical oxidation of synthetic dyeing wastewater using sno$_2$-sb-doped tio$_2$-coated granular activated carbon electrodes with high hydroxyl radical yields. *Electrochim. Acta* **2016**, *220*, 276–284. [CrossRef]
22. Liu, W.; Ai, Z.; Zhang, L. Design of a neutral three-dimensional electro-fenton system with foam nickel as particle electrodes for wastewater treatment. *J. Hazard. Mater.* **2012**, *243*, 257–264. [CrossRef] [PubMed]
23. Chen, J.-y.; Li, N.; Zhao, L. Three-dimensional electrode microbial fuel cell for hydrogen peroxide synthesis coupled to wastewater treatment. *J. Power Sources* **2014**, *254*, 316–322. [CrossRef]

24. Hao, R.; Li, S.; Li, J.; Meng, C. Denitrification of simulated municipal wastewater treatment plant effluent using a three-dimensional biofilm-electrode reactor: Operating performance and bacterial community. *Bioresour. Technol.* **2013**, *143*, 178–186. [CrossRef]
25. Neti, N.R.; Misra, R. Efficient degradation of reactive blue 4 in carbon bed electrochemical reactor. *Chem. Eng. J.* **2012**, *184*, 23–32. [CrossRef]
26. Pang, T.; Wang, Y.; Yang, H.; Wang, T.; Cai, W. Dynamic model of organic pollutant degradation in three dimensional packed bed electrode reactor. *Chemosphere* **2018**, *206*, 107–114. [CrossRef]
27. Liu, Z.; Wang, F.; Li, Y.; Xu, T.; Zhu, S. Continuous electrochemical oxidation of methyl orange waste water using a three-dimensional electrode reactor. *J. Environ. Sci.* **2011**, *23*, S70–S73. [CrossRef]
28. Zheng, T.; Wang, Q.; Shi, Z.; Fang, Y.; Shi, S.; Wang, J.; Wu, C. Advanced treatment of wet-spun acrylic fiber manufacturing wastewater using three-dimensional electrochemical oxidation. *J. Environ. Sci.* **2016**, *50*, 21–31. [CrossRef]
29. Zhao, H.Z.; Sun, Y.; Xu, L.N.; Ni, J.R. Removal of acid orange 7 in simulated wastewater using a three-dimensional electrode reactor: Removal mechanisms and dye degradation pathway. *Chemosphere* **2010**, *78*, 46–51. [CrossRef]
30. Wei, L.; Guo, S.; Yan, G.; Chen, C.; Jiang, X. Electrochemical pretreatment of heavy oil refinery wastewater using a three-dimensional electrode reactor. *Electrochim. Acta* **2010**, *55*, 8615–8620. [CrossRef]
31. Jung, K.-W.; Hwang, M.-J.; Park, D.-S.; Ahn, K.-H. Performance evaluation and optimization of a fluidized three-dimensional electrode reactor combining pre-exposed granular activated carbon as a moving particle electrode for greywater treatment. *Sep. Purif. Technol.* **2015**, *156*, 414–423. [CrossRef]
32. Chi, Z.; Wang, Z.; Liu, Y.; Yang, G. Preparation of organosolv lignin-stabilized nano zero-valent iron and its application as granular electrode in the tertiary treatment of pulp and paper wastewater. *Chem. Eng. J.* **2018**, *331*, 317–325. [CrossRef]
33. Can, W.; Yao-Kun, H.; Qing, Z.; Min, J. Treatment of secondary effluent using a three-dimensional electrode system: Cod removal, biotoxicity assessment, and disinfection effects. *Chem. Eng. J.* **2014**, *243*, 1–6. [CrossRef]
34. Li, X.-Y.; Xu, J.; Cheng, J.-P.; Feng, L.; Shi, Y.-F.; Ji, J. Tio2-sio2/gac particles for enhanced electrocatalytic removal of acid orange 7 (ao7) dyeing wastewater in a three-dimensional electrochemical reactor. *Sep. Purif. Technol.* **2017**, *187*, 303–310. [CrossRef]
35. Zhang, B.; Hou, Y.; Yu, Z.; Liu, Y.; Huang, J.; Qian, L.; Xiong, J. Three-dimensional electro-fenton degradation of rhodamine b with efficient fe-cu/kaolin particle electrodes: Electrodes optimization, kinetics, influencing factors and mechanism. *Sep. Purif. Technol.* **2019**, *210*, 60–68. [CrossRef]
36. Chen, H.; Feng, Y.; Suo, N.; Long, Y.; Li, X.; Shi, Y.; Yu, Y. Preparation of particle electrodes from manganese slag and its degradation performance for salicylic acid in the three-dimensional electrode reactor (tde). *Chemosphere* **2019**, *216*, 281–288. [CrossRef]
37. He, W.; Ma, Q.; Wang, J.; Yu, J.; Bao, W.; Ma, H.; Amrane, A. Preparation of novel kaolin-based particle electrodes for treating methyl orange wastewater. *Appl. Clay Sci.* **2014**, *99*, 178–186. [CrossRef]
38. Zhan, J.; Li, Z.; Yu, G.; Pan, X.; Wang, J.; Zhu, W.; Han, X.; Wang, Y. Enhanced treatment of pharmaceutical wastewater by combining three-dimensional electrochemical process with ozonation to in situ regenerate granular activated carbon particle electrodes. *Sep. Purif. Technol.* **2019**, *208*, 12–18. [CrossRef]
39. Nidheesh, P.V.; Gandhimathi, R. Trends in electro-fenton process for water and wastewater treatment: An overview. *Desalination* **2012**, *299*, 1–15. [CrossRef]
40. Yu, X.; Zhou, M.; Ren, G.; Ma, L. A novel dual gas diffusion electrodes system for efficient hydrogen peroxide generation used in electro-fenton. *Chem. Eng. J.* **2015**, *263*, 92–100. [CrossRef]
41. Wang, C.-T.; Chou, W.-L.; Chung, M.-H.; Kuo, Y.-M. Cod removal from real dyeing wastewater by electro-fenton technology using an activated carbon fiber cathode. *Desalination* **2010**, *253*, 129–134. [CrossRef]
42. Wang, C.T.; Hu, J.L.; Chou, W.L.; Kuo, Y.M. Removal of color from real dyeing wastewater by electro-fenton technology using a three-dimensional graphite cathode. *J. Hazard. Mater.* **2008**, *152*, 601–606. [CrossRef]
43. Lei, H.; Li, H.; Li, Z.; Li, Z.; Chen, K.; Zhang, X.; Wang, H. Electro-fenton degradation of cationic red x-grl using an activated carbon fiber cathode. *Process Saf. Environ. Prot.* **2010**, *88*, 431–438. [CrossRef]
44. Pérez, J.F.; Sabatino, S.; Galia, A.; Rodrigo, M.A.; Llanos, J.; Sáez, C.; Scialdone, O. Effect of air pressure on the electro-fenton process at carbon felt electrodes. *Electrochim. Acta* **2018**, *273*, 447–453. [CrossRef]

45. Huang, Z.; Ong, S.L.; Ng, H.Y. Submerged anaerobic membrane bioreactor for low-strength wastewater treatment: Effect of hrt and srt on treatment performance and membrane fouling. *Water Res.* **2011**, *45*, 705–713. [CrossRef] [PubMed]
46. Zhang, Q.H.; Yang, W.N.; Ngo, H.H.; Guo, W.S.; Jin, P.K.; Dzakpasu, M.; Yang, S.J.; Wang, Q.; Wang, X.C.; Ao, D. Current status of urban wastewater treatment plants in china. *Environ. Int.* **2016**, *92*, 11–22. [CrossRef]

© 2020 by the authors. Licensee MDPI, Basel, Switzerland. This article is an open access article distributed under the terms and conditions of the Creative Commons Attribution (CC BY) license (http://creativecommons.org/licenses/by/4.0/).

Article

Comparing the Effects of Types of Electrode on the Removal of Multiple Pharmaceuticals from Water by Electrochemical Methods

Yu-Jung Liu [1], Yung-Ling Huang [2], Shang-Lien Lo [1] and Ching-Yao Hu [2],*

[1] Graduate Institute of Environmental Engineering, National Taiwan University, Taipei 106, Taiwan; yujungliu77@gmail.com (Y.-J.L.); sllo@ntu.edu.tw (S.-L.L.)
[2] School of Public Health, Taipei Medical University, Taipei 110, Taiwan; m508101008@tmu.edu.tw
* Correspondence: cyhu@tmu.edu.tw

Received: 20 July 2020; Accepted: 18 August 2020; Published: 19 August 2020

Abstract: Considering the lack of information on simultaneously removing multiple pharmaceuticals from water or wastewater by electrochemical methods, this study aimed to investigate the removal of multiple pharmaceuticals by electro-coagulation and electro-oxidation based on two types of electrodes (aluminum and graphite). The synthetic wastewater contained a nonsteroidal anti-inflammatory drug (diclofenac), a sulfonamide antibiotic (sulfamethoxazole) and a β-blocker (atenolol). The pharmaceutical removal with electro-oxidation was much higher than those with the electro-coagulation process, which was obtained from a five-cell graphite electrode system, while the removal of pharmaceuticals with aluminum electrodes was about 20% (20 μM). In the electro-coagulation system, pharmaceutical removal was mainly influenced by the solubility or hydrophilicity of the compound. In the electro-oxidation system, the removal mechanism was influenced by the dissociation status of the compounds, which are attracted to the anode due to electrostatic forces and have a higher mass transformation rate with the electro-oxidation process. Therefore, atenolol, which was undissociated, cannot adequately be eliminated by electro-oxidation, unless the electrode's surface is large enough to increase the mass diffusion rate.

Keywords: β-blockers; electro-coagulation; electro-oxidation; nonsteroidal anti-inflammatory drugs (NSAIDs); sulfonamide antibiotics

---

1. Introduction

The presence of emerging pharmaceutical contaminants has drawn much attention in recent years; however, these substances are inefficiently removed by conventional unit operations utilized by most wastewater treatment plants (WWTPs) due to their intricate properties like high water solubility and poor biodegradability [1–6]. Among such pharmaceuticals, nonsteroidal anti-inflammatory drugs (NSAIDs), beta-blockers, and antibiotics are widely used groups, and thus these groups are most often found in wastewater, which may co-exist in water bodies and soil environment [1,6,7]. They may induce adverse effects on aquatic systems [6,8–11].

An electrochemical process was applied to broad applicability like textile, cellulose, paper factories, laundry, and various kinds of different characteristic wastewater [12–18]. This technology can be operated by separation and degradation, called electro-coagulation and electrochemical oxidation, which depends on the characteristics of the electrodes [18–20]. In a separation system, electro-coagulation involves applying electric current to sacrificial electrodes where coagulants and gas bubbles are generated in situ by the current, and destabilizes, suspends, emulsifies or dissolves pollutants in an aqueous medium, or floats pollutants to the surface by tiny bubbles of hydrogen and oxygen gases generated from water electrolysis [21]. Additionally, the stable complexation between pollutants and flocs might be due to the structure of the pollutants, which might be the assumption for high selectivity by the metal adsorption capacity of the ligand-based-like flocs [22–24]. In a degradation system, electro-oxidation occurs through two routes, (1) direct oxidation, where the pollutants eliminated at the anode surface; (2) indirect oxidation, where mediators (reactive oxygen species or active chlorine species) are anodically generated to carry out the reaction [20,25]. The electrochemical methods have the advantages of simple equipment, flexible operation, and being chemical free [26,27]. Although it requires wastewater under high conductivity and energy consumption, it still has high treatment efficiency with chloride without any secondary pollutants, and the reusability of chloride with relatively low costs, which is the concept of green chemistry [21,26–30].

Previous studies have focused on single pharmaceutical removal efficiencies by using electro-coagulation and electro-oxidation processes [19,20,31]. However, many pharmaceuticals are simultaneously present in the environment. To the best of our knowledge from reviewing the literature, less reports are available to date on removing multiple pharmaceuticals from water or wastewater using an electrochemical process. Accordingly, this study aimed to evaluate the effect of laboratory-scale electrochemical treatments of wastewater on removing multiple pharmaceuticals from synthetic water and spiked wastewater. We evaluated the effectiveness of aluminum and graphite electrodes in monopolar and bipolar arrangements under consideration for electrochemical utilization methods to eliminate a beta-blocker (atenolol; ATE), an NSAID (diclofenac; DIC), and a sulfonamide antibiotic (sulfamethoxazole; SMX).

2. Materials and Methods

2.1. Materials

Characteristics of the target compounds are shown in Table 1. All the chemicals were an analytical grade (≥98%). DIC ($C_{14}H_{11}Cl_2NO_2$) and SMX ($C_{10}H_{11}N_3O_3S$) were purchased from Sigma-Aldrich; ATE ($C_{14}H_{22}N_2O_3$) was from TCI; and sodium chloride (NaCl) (99.5%; Wako) was used as the supporting electrolyte to prevent passive film generated during the reaction, which may increase the resistance and diminish the release of the coagulants from the electrode [21,32,33]. Monopotassium phosphate (KH_2PO_4) (99%; Showa), methanol (high-performance liquid chromatography (HPLC) grade; Scharlau), and acetonitrile (HPLC grade; J.T. Baker) were used in the HPLC analyses. Stock solutions were prepared with deionized Milli-Q water obtained from a Merck-Millipore system and were stored in amber glass bottles at 4 °C. The standard solutions of target compounds at various concentrations were prepared by diluting the stock solutions prior to use. The anodes and cathodes used were 99% pure aluminum as the sacrificial electrodes, and graphite as the non-sacrificial electrodes.

Table 1. Characteristics of the target compounds.

Categories	Compound (CAS Number)	Structure	MW (g mol^{-1})	Solubility (mg/L)	pK$_a$	Log K$_{ow}$
β-blocker	Atenolol (ATE) (29122-68-7)		266.34	300	9.6	0.16
Nonsteroidal anti-inflammatory drug (NSAID)	Diclofenac (DIC) (15307-86-5)		296.13	2.37	4.15	4.51
Sulfonamide antibiotic	Sulfamethoxazole (SMX) (723-46-6)		253.28	370	5.7	0.89

2.2. Electrochemical (EC) Experiments and Analyses

EC experiments were carried out in a 1 L reactor, which is a double layered cylindrical glass container, with a diameter of 10 cm and a height of 20 cm, as schematically shown in Figure 1. The cell was operated with 2 and 6 electrodes held vertically, respectively; each electrode was 1 mm thick and had an effective area of 125 cm^2; the distance between the electrodes was 100 and 24 mm, respectively. All electrochemical experiments were carried out under potent stirring with a magnetic bar at 260 rpm and a constant current intensity (I = 0.5 A) provided by direct current (DC) (GPR-30H10D, Good Will Instrument). Desired concentrations of the pharmaceutical stock solutions (10 mM) were made by adding proper amounts of pharmaceuticals to distilled water. The volume (V) of the solution of each batch was 1 L with 100 µM of each pharmaceutical (ATE, DIC, and SMX) and 0.01 M NaCl as the supporting electrolyte. Solutions were maintained at 25 ± 1 °C in a water-bath. Samples were taken at assigned time intervals following the achievement of the EC process. The total reaction time was set to 20 or 40 min, which depended on what was required for the electrode cells to achieve a suitable removal efficiency. Pharmaceutical concentrations were investigated by high-performance liquid chromatography (HPLC; Hitachi, L-7200, Japan) with a diode array detector (DAD; Hitachi, L-7455, Japan), equipped with a C$_{18}$ column (RP-18 GP 150 × 4.6 mm, 5 µm, Mightysil). HPLC–DAD is a relatively cheap and simple operation technique for screening and analyzing purposes on gradient elution [34]. Mobile phase A was KH$_2$PO$_4$, B was acetonitrile, and the specific conditions are shown in Table 2. The flow rate of mobile phases A and B were both set to 1 mL min^{-1}. The injected volume of each sample was 20 µl. The deviations of all analyses were within 5%. Quality assurance/quality control (QA/QC) standards were prepared by diluting and combining the 0.01 M pharmaceutical stock solution accordingly, and to monitor HPLC–DAD performances.

Table 2. Conditions of HPLC–diode array detector (DAD).

Time (min)	Mobile Phase	
	A KH$_2$PO$_4$ (%)	B Acetonitrile (%)
0	80	20
3	80	20
4	60	40
10.5	60	40
11.5	40	60
18.5	40	60
19.5	80	20

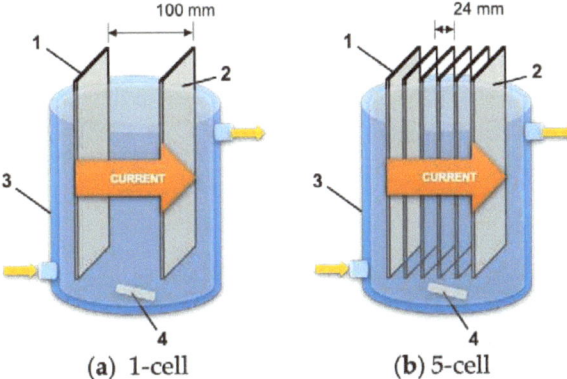

(a) 1-cell (b) 5-cell

Figure 1. Profile of the batch reactor: (a) 1 cell and (b) 5 cell. 1: anode (aluminum, graphite). 2: cathode (aluminum, graphite). 3: Double layered cylindrical glass container (V = 1 L) with a water bath. 4: magnetic stirring bar.

3. Results and Discussion

3.1. Removal of ATE, DIC, and SMX with Different Electrode Systems

Three electrode arrangements were investigated for single pharmaceutical removal in the batch electrochemical method with sacrificial and non-sacrificial anodes, under conditions of the present study, and two mechanisms were responsible for the removal of the pollutants: electro-coagulation, and electro-oxidation. We choose aluminum electrodes used as model sacrificial anode; graphite electrodes used as non-sacrificial anode. Figure 2a shows that removal rates of DIC, SMX, and ATE by the aluminum electrode system were 17.9%, 4.8%, and 2.3%, respectively. These results are similar to those of our previous study [19]; the substance with lipophilic characteristic could be removed either by precipitation of the flocs or by flotation with hydrogen gas during the electro-coagulation reaction, which depended on adsorption [35,36] and charge neutralization on the surface of hydrogen bubbles and aluminum hydroxide [37]. Additionally, it might be due to the soft atoms that are favorable for stable complexation mechanism, and this might be the main assumption for the high selectivity by the Al (II) adsorption capacity of the ligand-based flocs [22].

The main mechanism of pharmaceutical removal in the graphite electrode system should be electrochemical oxidation, which involves two procedures when an electric current is passed through a non-sacrificial electrode. The first one is the direct electrolysis of the pharmaceuticals on the surface of the anodes. The other one is indirect oxidation which involves a strong oxidant (•OH or HOCl) produced by the electrolysis of chloride or other compounds [20]. The oxidant is then transferred to the bulk solution to degrade the pharmaceuticals, and this process may cause the structure of the initial pollutant to be converted into byproducts [18,20]. DIC, SMX, and ATE, as shown in Figure 2b, were reduced by 100%, 99.8%, and 85%, respectively. The higher removal of DIC and SMX may have been due to the dissociation of the two compounds in the reaction conditions. As shown in Table 1, pKa values of both DIC and SMX are relatively smaller than that of ATE, which means the two compounds are dissociated in a neutral condition but ATE is not. The mass transfer rate of anions should be much higher than that of neutral compounds because of electrostatic forces. The removal efficiencies of ionic compounds such as DIC and SMX by electro-oxidation were higher than that of the neutral compound of ATE. The schemes of the mechanism of electro-coagulation and electro-oxidation are shown in Figure 3.

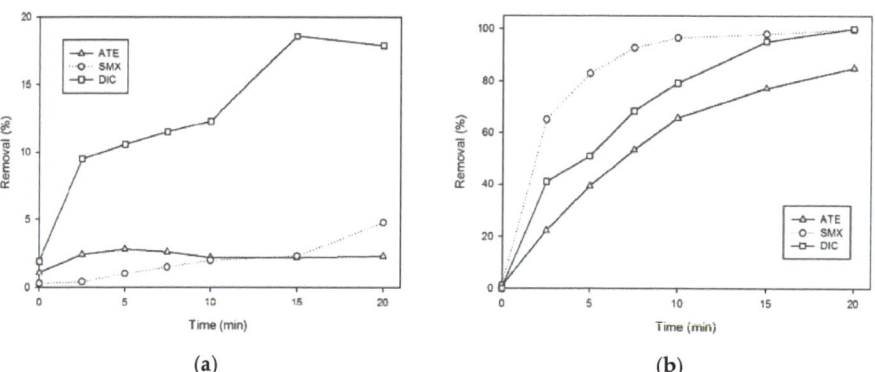

Figure 2. Removal efficiencies of single pharmaceuticals in a 1 L batch reactor with a 5-cell system of (a) aluminum or (b) graphite ([Phar.]$_0$ = 100 µM, [NaCl] = 0.01 M, I = 0.5 A).

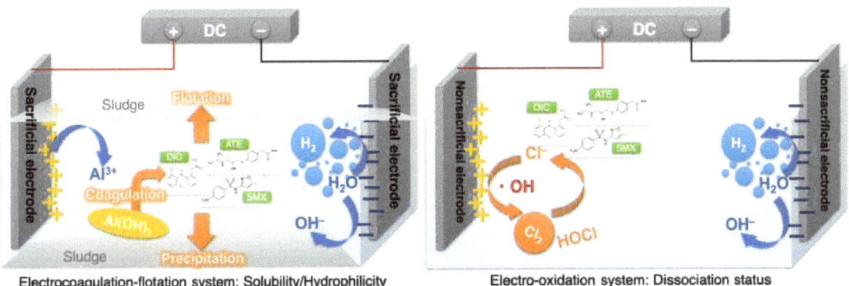

Figure 3. Schemes of the mechanism of electro-coagulation and electro-oxidation.

3.2. Removal Efficiencies of Multiple Pharmaceuticals

The above observations reveal that the removal efficiencies of single pharmaceuticals changed with different electrodes, especially in the graphite electrode system. To utilize this system in a practical way, as shown in Figure 4, the influences of the electrode cell size in the triple-pharmaceutical solutions should be considered. The removal of multiple pharmaceuticals in a 1-cell system was inefficient, especially for ATE. Compared to the one-cell electrode, five-cell electrodes were significantly more effective at removing multiple pharmaceuticals from the water, with removal rates of SMX, DIC, and ATE of 75%, 68%, and 55%, respectively. The reason could also be due to the dissociation of the selected pharmaceuticals. On the other hand, the synergistic effects of direct and mediated oxidation may be another possibility causing this circumstance. Hydroxyl radicals (•OH) produced by the oxidation of water are known to be a very active reagent and lead to the formation of higher-state oxides or superoxides (Equations (1) and (2)). Using sodium chloride (NaCl) as the electrolyte provided the formation of hypochlorous acid (HOCl) (Equation (3)). Accordingly, the oxidation of pharmaceuticals occurred more effectively since •OH and HOCl were generated in the five-cell electrode system:

$$M + H_2O \rightarrow M(\bullet OH) + H^+ + e^- \tag{1}$$

$$M(\bullet OH) \rightarrow MO + H^+ + e^- \tag{2}$$

$$Cl^- + H_2O \rightarrow HOCl + H^+ + 2e^- \tag{3}$$

To clarify the interactions between multiple pharmaceuticals in the electro-oxidation reaction, variations in removal rates of double pharmaceuticals with the reaction times in the graphite electrode system are shown in Figure 4. When ATE and SMX existed simultaneously (Figure 5a), only SMX could effectively be removed, at 69.9%. The same tendency is shown in Figure 5b of DIC and ATE concurrently. Elimination rates of DIC and ATE were 94.8% and 63.4%. Figure 5c shows that elimination rates of SMX and DIC were 75.2% and 73.7%.

As a result of the behavior described above, the structure and characteristics of the pharmaceuticals were the most important factors. Difficulty with cleavage affected the removal of multiple pharmaceuticals, which is related to competitive oxidation [38]. The S–N bond, isoxazole ring, and benzene ring of SMX and DIC were easily cleaved during oxidation [39]. Moreover, it was found that DIC and SMX were more effectively removed than ATE. After the reaction, the pH of the solution was about 7 ± 0.8, but the pKa of ATE was 9.6. This observation indicated that ATE was still in a molecular state and had not reached the anode; DIC and SMX were dissociated into ionic compounds and were attracted to the anode, and thus were rapidly degraded through the electrochemical oxidation system.

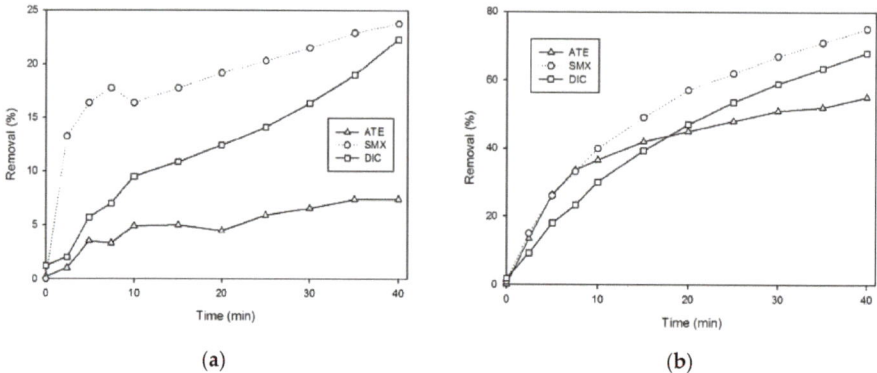

Figure 4. Removal efficiencies of triple pharmaceuticals in a 1 L batch reactor with a graphite electrode and a (**a**) 1-cell or (**b**) 5-cell system ([Phar.]$_0$ = 100 µM, [NaCl] = 0.01 M, I = 0.5 A).

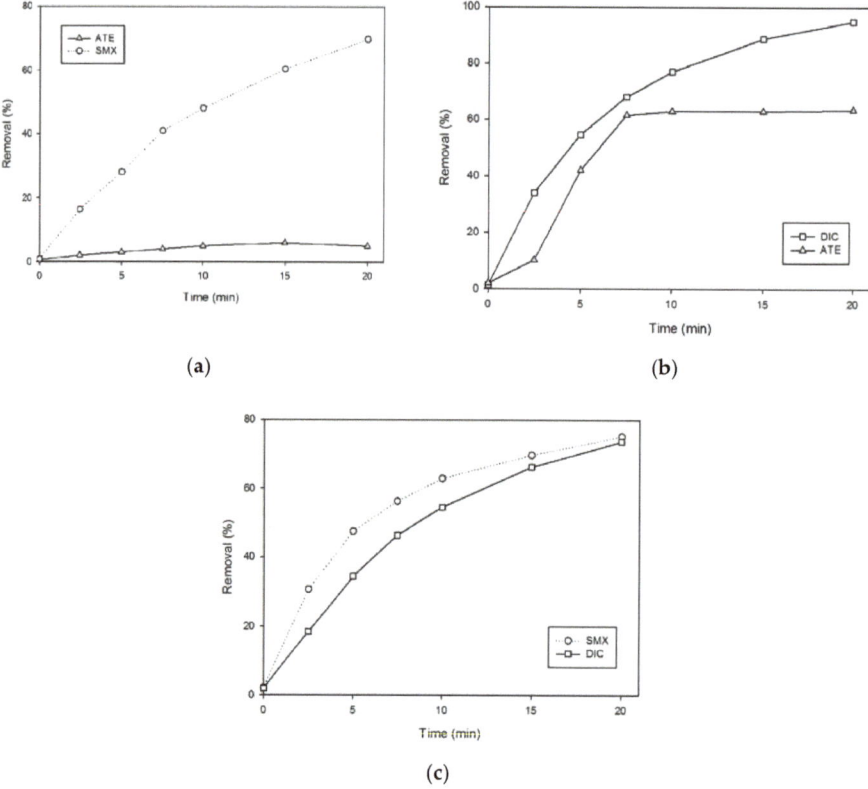

Figure 5. Removal efficiencies of dual pharmaceuticals in a 1 L batch reactor with 5-cell graphite electrodes ([Phar.]$_0$ = 100 µM, [NaCl] = 0.01 M, I = 0.5 A). (**a**) ATE + SMX; (**b**) DIC + ATE; (**c**) SMX + DIC.

3.3. Influences of Multiple Pharmaceuticals on Hospital Wastewater Matrixes

To simulate pharmaceutical interactions in real wastewater matrixes, we spiked our target compounds into an actual hospital effluent. Table 3 shows the characteristics of the hospital effluent. Figure 6 shows the performances of multiple pharmaceuticals in the electro-oxidation system with graphite electrodes. Martins et al. indicated that if a solution has large amounts of organic compounds or complex matrixes, pharmaceuticals may be adsorbed or attached onto other species or structures, which decreases the removal efficiency [40]. These results reflect the same circumstances of our experiment outcomes. Compared to laboratory outcomes, the declines in ATE, DIC, and SMX in an actual hospital effluent were only 11%, 47.8%, and 43%, respectively.

Table 3. Characteristics of hospital effluent.

Parameters	Hospital Effluent
Temperature (°C)	25.0
Dissolved oxygen (DO) (ppm)	3.35
pH	7.20
Turbidity (NTU)	10.9
Suspended solids (SS) (mg/L)	0.018
Chemical oxygen demand (COD) (mg/L)	41.5
Biochemical oxygen demand (BOD) (mg/L)	4.21

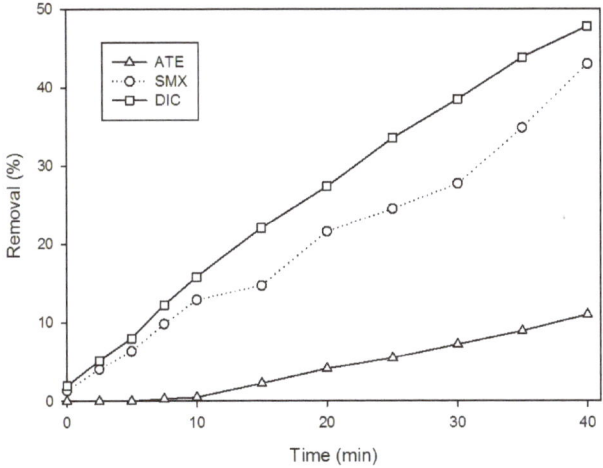

Figure 6. Removal efficiencies of spiked pharmaceuticals in actual hospital effluent matrices ([Phar.]$_s$ = 100 μM, [NaCl] = 0.01 M, I = 0.5 A).

4. Conclusions

To develop promising treatments for removing pharmaceuticals from sewage, the ability of a physicochemical process, electrochemical treatment, to remove selected pharmaceuticals was assessed. Aluminum was used for the active electrodes with the electrocoagulation–flotation process; graphite was used for the inert electrodes with electro-oxidation. The electrochemical process with the graphite electrode system removed the pharmaceuticals in water more effectively with multiple electrodes by the electro-oxidation process. Removal efficiencies were affected by the hydrophobic character of the pharmaceuticals under the electrocoagulation–flotation system, and the uneven electron structural characteristics of the pharmaceuticals under the electro-oxidation system. According to these results, the degradation of pharmaceuticals was improved and more efficient than the separation of medicines.

Author Contributions: Y.-L.H. carried out the experiments. Y.-J.L. processed the experiments, analyzed the data, and wrote the manuscript in consultation with S.-L.L. and C.-Y.H. S.-L.L. devised the conceptual ideas, discussed the results, and commented on the manuscript. C.-Y.H. conceived the original idea, helped supervise the project, and final approval of the version to be published. All authors have read and agreed to the published version of the manuscript.

Funding: This research received the funding from Ministry of Science and Technology, Taiwan, ROC (under contract no. MOST 107-2221-E-038-001).

Conflicts of Interest: The authors declare no conflict of interest.

Abbreviations

ATE	Atenolol
DAD	Diode Array Detector
DC	Direct Current
DIC	Diclofenac
EC	Electrochemical
HOCl	Hypochlorous Acid
HPLC	High-Performance Liquid Chromatography
I	Current
KH_2PO_4	Monopotassium Phosphate
NaCl	Sodium Chloride
NSAIDs	Nonsteroidal Anti-Inflammatory Drugs
•OH	Hydroxyl Radicals
[Phar]$_0$	Initial Concentration of Pharmaceutical
[Phar]$_s$	Initial Concentration of Spiked Pharmaceuticals
QA/QC	Quality Assurance/Quality Control
SMX	Sulfamethoxazole

References

1. Rivera-Utrilla, J.; Sánchez-Polo, M.; Ferro-García, M.Á.; Prados-Joya, G.; Ocampo-Pérez, R. Pharmaceuticals as emerging contaminants and their removal from water. A review. *Chemosphere* **2013**, *93*, 1268–1287. [CrossRef] [PubMed]
2. Crouse, B.A.; Ghoshdastidar, A.J.; Tong, A.Z. The presence of acidic and neutral drugs in treated sewage effluents and receiving waters in the Cornwallis and Annapolis River watersheds and the Mill CoveSewage Treatment Plant in Nova Scotia, Canada. *Environ. Res.* **2012**, *112*, 92–99. [CrossRef] [PubMed]
3. Li, Y.; Zhang, S.; Zhang, W.; Xiong, W.; Ye, Q.; Hou, X.; Wang, C.; Wang, P. Life cycle assessment of advanced wastewater treatment processes: Involving 126 pharmaceuticals and personal care products in life cycle inventory. *J. Environ. Manag.* **2019**, *238*, 442–450. [CrossRef] [PubMed]
4. Mestre, A.S.; Pires, R.A.; Aroso, I.; Fernandes, E.M.; Pinto, M.L.; Reis, R.L.; Andrade, M.A.; Pires, J.; Silva, S.P.; Carvalho, A.P. Activated carbons prepared from industrial pre-treated cork: Sustainable adsorbents for pharmaceutical compounds removal. *Chem. Eng. J.* **2014**, *253*, 408–417. [CrossRef]
5. Phoon, B.L.; Ong, C.C.; Mohamed Saheed, M.S.; Show, P.-L.; Chang, J.-S.; Ling, T.C.; Lam, S.S.; Juan, J.C. Conventional and emerging technologies for removal of antibiotics from wastewater. *J. Hazard. Mater.* **2020**, *400*, 122961. [CrossRef]
6. Khan, A.H.; Khan, N.A.; Ahmed, S.; Dhingra, A.; Singh, C.P.; Khan, S.U.; Mohammadi, A.A.; Changani, F.; Yousefi, M.; Alam, S.; et al. Application of advanced oxidation processes followed by different treatment technologies for hospital wastewater treatment. *J. Clean. Prod.* **2020**, *269*, 122411. [CrossRef]
7. Petrovic, M.; Hernando, M.D.; Diaz-Cruz, M.S.; Barcelo, D. Liquid chromatography-tandem mass spectrometry for the analysis of pharmaceutical residues in environmental samples: A review. *J. Chromatogr. A* **2005**, *1067*, 1–14. [CrossRef]
8. Verlicchi, P.; Al Aukidy, M.; Zambello, E. Occurrence of pharmaceutical compounds in urban wastewater: Removal, mass load and environmental risk after a secondary treatment—A review. *Sci. Total Environ.* **2012**, *429*, 123–155. [CrossRef]
9. Berninger, J.P.; Brooks, B.W. Leveraging mammalian pharmaceutical toxicology and pharmacology data to predict chronic fish responses to pharmaceuticals. *Toxicol. Lett.* **2010**, *193*, 69–78. [CrossRef]
10. Camacho-Muñoz, D.; Martín, J.; Santos, J.L.; Aparicio, I.; Alonso, E. Occurrence, temporal evolution and risk assessment of pharmaceutically active compounds in Donana Park (Spain). *J. Hazard. Mater.* **2010**, *183*, 602–608. [CrossRef]

11. Xu, M.; Huang, H.; Li, N.; Li, F.; Wang, D.; Luo, Q. Occurrence and ecological risk of pharmaceuticals and personal care products (PPCPs) and pesticides in typical surface watersheds, China. *Ecotox. Environ. Safe* **2019**, *175*, 289–298. [CrossRef] [PubMed]
12. Essadki, A.H.; Bennajah, M.; Gourich, B.; Vial, C.; Azzi, M.; Delmas, H. Electrocoagulation/electroflotation in an external-loop airlift reactor—Application to the decolorization of textile dye wastewater: A case study. *Chem. Eng. Process. Process Intensif.* **2008**, *47*, 1211–1223. [CrossRef]
13. Boroski, M.; Rodrigues, A.C.; Garcia, J.C.; Gerola, A.P.; Nozaki, J.; Hioka, N. The effect of operational parameters on electrocoagulation–flotation process followed by photocatalysis applied to the decontamination of water effluents from cellulose and paper factories. *J. Hazard. Mater.* **2008**, *160*, 135–141. [CrossRef] [PubMed]
14. Chou, W.-L.; Wang, C.-T.; Chang, S.-Y. Study of COD and turbidity removal from real oxide-CMP wastewater by iron electrocoagulation and the evaluation of specific energy consumption. *J. Hazard. Mater.* **2009**, *168*, 1200–1207. [CrossRef]
15. Ge, J.; Qu, J.; Lei, P.; Liu, H. New bipolar electrocoagulation–electroflotation process for the treatment of laundry wastewater. *Sep. Purif. Technol.* **2004**, *36*, 33–39. [CrossRef]
16. Emamjomeh, M.M.; Sivakumar, M. Fluoride removal by a continuous flow electrocoagulation reactor. *J. Environ. Manag.* **2009**, *90*, 1204–1212. [CrossRef]
17. Bansal, S.; Kushwaha, J.P.; Sangal, V.K. Electrochemical Treatment of Reactive Black 5 Textile Wastewater: Optimization, Kinetics, and Disposal Study. *Water Environ. Res* **2013**, *85*, 2294–2306. [CrossRef]
18. Sirés, I.; Brillas, E. Remediation of water pollution caused by pharmaceutical residues based on electrochemical separation and degradation technologies: A review. *Environ. Int.* **2012**, *40*, 212–229. [CrossRef]
19. Liu, Y.-J.; Lo, S.-L.; Liou, Y.-H.; Hu, C.-Y. Removal of nonsteroidal anti-inflammatory drugs (NSAIDs) by electrocoagulation–flotation with a cationic surfactant. *Sep. Purif. Technol.* **2015**, *152*, 148–154. [CrossRef]
20. Liu, Y.-J.; Hu, C.-Y.; Lo, S.-L. Direct and indirect electrochemical oxidation of amine-containing pharmaceuticals using graphite electrodes. *J. Hazard. Mater.* **2019**, *366*, 592–605. [CrossRef]
21. Mollah, M.Y.A.; Schennach, R.; Parga, J.R.; Cocke, D.L. Electrocoagulation (EC)—Science and applications. *J. Hazard. Mater.* **2001**, *84*, 29–41. [CrossRef]
22. Awual, M.R.; Hasan, M.M. A ligand based innovative composite material for selective lead(II) capturing from wastewater. *J. Mol. Liq.* **2019**, *294*, 111679. [CrossRef]
23. Awual, M.R.; Hasan, M.M.; Rahman, M.M.; Asiri, A.M. Novel composite material for selective copper(II) detection and removal from aqueous media. *J. Mol. Liq.* **2019**, *283*, 772–780. [CrossRef]
24. Awual, M.R. A novel facial composite adsorbent for enhanced copper(II) detection and removal from wastewater. *Chem. Eng. J.* **2015**, *266*, 368–375. [CrossRef]
25. Anglada, Á.; Urtiaga, A.; Ortiz, I. Contributions of electrochemical oxidation to waste-water treatment: Fundamentals and review of applications. *J. Chem. Technol. Biot.* **2009**, *84*, 1747–1755. [CrossRef]
26. Liu, H.; Zhao, X.; Qu, J. Electrocoagulation in Water Treatment. In *Electrochemistry for the Environment*; Comninellis, C., Chen, G., Eds.; Springer: New York, NY, USA, 2010; pp. 245–262. [CrossRef]
27. Rychen, P.; Provent, C.; Pupunat, L.; Hermant, N. Domestic and Industrial Water Disinfection Using Boron-Doped Diamond Electrodes. In *Electrochemistry for the Environment*; Comninellis, C., Chen, G., Eds.; Springer: New York, NY, USA, 2010; pp. 143–161. [CrossRef]
28. Pulkka, S.; Martikainen, M.; Bhatnagar, A.; Sillanpää, M. Electrochemical methods for the removal of anionic contaminants from water—A review. *Sep. Purif. Technol.* **2014**, *132*, 252–271. [CrossRef]
29. Griesbach, U.; Malkowsky, I.M.; Waldvogel, S.R. Green Electroorganic Synthesis Using BDD Electrodes. In *Electrochemistry for the Environment*; Comninellis, C., Chen, G., Eds.; Springer: New York, NY, USA, 2010; pp. 125–141. [CrossRef]
30. Steckhan, E.; Arns, T.; Heineman, W.R.; Hilt, G.; Hoormann, D.; Jörissen, J.; Kröner, L.; Lewall, B.; Pütter, H. Environmental protection and economization of resources by electroorganic and electroenzymatic syntheses. *Chemosphere* **2001**, *43*, 63–73. [CrossRef]
31. Dos Santos, A.J.; Cabot, P.L.; Brillas, E.; Sirés, I. A comprehensive study on the electrochemical advanced oxidation of antihypertensive captopril in different cells and aqueous matrices. *Appl. Catal. B Environ.* **2020**, *277*, 119240. [CrossRef]
32. Chen, X.; Chen, G.; Yue, P.L. Investigation on the electrolysis voltage of electrocoagulation. *Chem. Eng. Sci.* **2002**, *57*, 2449–2455. [CrossRef]

33. He, C.-C.; Hu, C.-Y.; Lo, S.-L. Integrating chloride addition and ultrasonic processing with electrocoagulation to remove passivation layers and enhance phosphate removal. *Sep. Purif. Technol.* **2018**, *201*, 148–155. [CrossRef]
34. Teixeira, S.; Delerue-Matos, C.; Alves, A.; Santos, L. Fast screening procedure for antibiotics in wastewaters by direct HPLC-DAD analysis. *J. Sep. Sci.* **2008**, *31*, 2924–2931. [CrossRef] [PubMed]
35. Chou, W.-L.; Wang, C.-T.; Huang, K.-Y.; Liu, T.-C. Electrochemical removal of salicylic acid from aqueous solutions using aluminum electrodes. *Desalination* **2011**, *271*, 55–61. [CrossRef]
36. Zhang, G.H.; Yin, L.L.; Zhang, S.T.; Li, X. Adsorption Behavior of Sulfamethoxazole as Inhibitor for Mild Steel in 3% HCl Solution. *Adv. Mater. Res.* **2011**, *194–196*, 8–15. [CrossRef]
37. Ren, M.; Song, Y.; Xiao, S.; Zeng, P.; Peng, J. Treatment of berberine hydrochloride wastewater by using pulse electro-coagulation process with Fe electrode. *Chem. Eng. J.* **2011**, *169*, 84–90. [CrossRef]
38. Indermuhle, C.; Martín de Vidales, M.J.; Sáez, C.; Robles, J.; Cañizares, P.; García-Reyes, J.F.; Molina-Díaz, A.; Comninellis, C.; Rodrigo, M.A. Degradation of caffeine by conductive diamond electrochemical oxidation. *Chemosphere* **2013**, *93*, 1720–1725. [CrossRef]
39. Gao, S.; Zhao, Z.; Xu, Y.; Tian, J.; Qi, H.; Lin, W.; Cui, F. Oxidation of sulfamethoxazole (SMX) by chlorine, ozone and permanganate—A comparative study. *J. Hazard. Mater.* **2014**, *274*, 258–269. [CrossRef]
40. Martins, A.F.; Mallmann, C.A.; Arsand, D.R.; Mayer, F.M.; Brenner, C.G.B. Occurrence of the Antimicrobials Sulfamethoxazole and Trimethoprim in Hospital Effluent and Study of Their Degradation Products after Electrocoagulation. *CLEAN Soil Air Water* **2011**, *39*, 21–27. [CrossRef]

 © 2020 by the authors. Licensee MDPI, Basel, Switzerland. This article is an open access article distributed under the terms and conditions of the Creative Commons Attribution (CC BY) license (http://creativecommons.org/licenses/by/4.0/).

Article

Removal Characteristics of Effluent Organic Matter (EfOM) in Pharmaceutical Tailwater by a Combined Coagulation and UV/O₃ Process

Jian Wang [1,2], **Yonghui Song** [2,*], **Feng Qian** [2,3,*], **Cong Du** [2,3], **Huibin Yu** [2,3] and **Liancheng Xiang** [2,3]

1. State Key Joint Laboratory of Environment Simulation and Pollution Control, School of Environment, Tsinghua University, Beijing 100084, China; jian-wan14@mails.tsinghua.edu.cn
2. State Key Laboratory of Environmental Criteria and Risk Assessment, Chinese Research Academy of Environmental Sciences, Beijing 100012, China; ducongducong@126.com (C.D.); yhbybx@163.com (H.Y.); xianglc@craes.org.cn (L.X.)
3. Department of Urban Water Environmental Research, Chinese Research Academy of Environmental Sciences, Beijing 100012, China
* Correspondence: songyh@craes.org.cn (Y.S.); qianfeng@craes.org.cn (F.Q.); Tel.: +86-10-8492-4787 (Y.S.); +86-10-8491-7906 (F.Q.)

Received: 21 August 2020; Accepted: 30 September 2020; Published: 5 October 2020

Abstract: A novel coagulation combined with UV/O$_3$ process was employed to remove the effluent organic matter (EfOM) from a biotreated pharmaceutical wastewater for harmlessness. The removal behavior of EfOM by UV/O$_3$ process was characterized by synchronous fluorescence spectroscopy (SFS) integrating two-dimensional correlation (2D-COS) and principal component analysis (PCA) technology. The highest dissolved organic carbon (DOC) and ratio of UV$_{254}$ and DOC (SUVA) removal efficiency reached 55.8% and 68.7% by coagulation-UV/O$_3$ process after 60 min oxidation, respectively. Five main components of pharmaceutical tail wastewater (PTW) were identified by SFS. Spectral analysis revealed that UV/O$_3$ was selective for the removal of different fluorescent components, especially fulvic acid-like fluorescent (FLF) component and humus-like fluorescent (HLF) component. Synchronous fluorescence/UV-visible two-dimensional correlation spectra analysis showed that the degradation of organic matter occurred sequentially in the order of HLF, FLF, microbial humus-like fluorescence component (MHLF), tryptophan-like fluorescent component (TRLF), tyrosine-like fluorescent component (TYLF). The UV/O$_3$ process removed 95.6% of HLF, 80.0% of FLF, 56.0% of TRLF, 50.8% of MHLF and 44.4% of TYLF. Therefore, the coagulation-UV/O$_3$ process was proven to be an attractive way to reduce the environmental risks of PTW.

Keywords: pharmaceutical wastewater; UV/O$_3$; coagulation; fluorescence spectrum; removal characteristics

---

1. Introduction

Pharmaceutical wastewater is one of the important sources for emerging pollutants such as hormones, antibiotics and non-biodegradable organic intermediates [1–3]. The pharmaceutical residue usually entered the aquatic environment via sewage, and even low concentration can impact the drinking water and human health [4]. Despite undergoing biological treatment, the pharmaceutical residue cannot be completely metabolized [2,5,6]. Some nonbiodegradable and toxic substances still exist in biologically treated effluent (pharmaceutical tail wastewater, PTW). Therefore, intensive treatment must be carried out to realize harmlessness and reduce environmental risks [7].

After biological treatment, the biodegradability of PTW was low and not suitable for continued biological treatment [3,7,8]. The BOD$_5$/COD (B/C) of PTW was only close to 0.1, meaning that the organic matter from PTW was difficult to biodegrade. Therefore, the effect of biological treatment would be very poor in the further processing, physical chemistry and some advanced oxidation techniques should be considered. Among these, coagulation and UV/O_3 were the most two common technologies to remove specific organic pollutants from wastewater, such as contaminated groundwater, drinking water, industrial wastewater and landfill leachate [9–11]. Combined application of coagulation and UV/O_3 are suitable to treat PTW, since it has a certain removal effect on most organic matter ranging from low to high molecular weight. Coagulation treatment can effectively remove suspended particles, colloidal particles and dissolved organic matter (DOM) in sewage [12]. UV radiation can stimulate ozone oxidation to produce highly reactive hydroxyl radicals, which have the ability to oxidize and remove almost any organic contaminants [13]. However, studies on using UV/O_3, coagulation or combined coagulation-UV/O_3 to remove organic components of PTW have rarely been reported. Understanding the removal characteristics of effluent organic matter (EfOM) of PTW was an important basis for examining the effect of coagulation-UV/O_3 treatment. Therefore, it was necessary to find alternative characterization methods to evaluate the removal of EfOM during the oxidation process.

DOM, as the most important part of EfOM, can determine the coagulant, disinfection by-products, membrane fouling, type of microbial activity, and removal effect of contaminants [14,15]. The fluorescent component of DOM can be analyzed by fluorescence spectroscopy technology. The outstanding advantage of fluorescence spectroscopy is that it can quickly obtain measured DOM characteristic information with high sensitivity and has the characteristics of non-destructiveness and low cost [16,17]. Many fluorescent components in DOM can be analyzed and determined by combining fluorescence excitation-emitter matrix with parallel factors, self-organizing mapping algorithms or area integral methods [18]. Among them, synchronous fluorescence spectroscopy (SFS) was constituted by the measured fluorescence intensity signal and the corresponding excitation wavelength (or emission wavelength), based on simultaneous wavelength scanning of excitation and emission monochromators [19]. It has been widely used in the simultaneous analysis of heterogeneous mixtures, due to its advantages of simplified spectrum, reduced light scattering and improved selectivity. However, partial overlap of SFS wavelength can reduce its selectivity in DOM analysis. This problem can be solved through the combination of SFS and two-dimensional correlation (2D-COS) by extending peaks in the second dimension. Additionally, 2D-COS can be used as strong evidence for detecting correlations between features of dynamic spectra at two different wavelengths. It showed that combined SFS with 2D-COS can effectively describe the removal characteristics of DOM in PTW. Dynamic spectral changes are triggered by external disturbances, including various physical, chemical, and biological phenomena. The use of 2D-COS can explain the fundamental mechanisms of complex and heterogeneous material changes by enhancing the resolution of the spectrum and identifying the sequence of any subtle spectral changes in response to external disturbances [20,21].

Although the above methods have been widely used, there are few studies on the application of SFS technology and 2D-COS in the field of DOM structure in biotreated pharmaceutical wastewater by using coagulation-UV/O_3 combination method. The purpose of this paper is to (1) study the dynamic spectral changes of DOM in PTW after UV/O_3 oxidation using SFS technology, and (2) characterize the dynamic spectra and their relationship with different wavelengths by using 2D-COS combined with SFS technology, so as to reveal the degradation characteristics of different organic components in PTW.

2. Material and Methods

2.1. Pharmaceutical Tail Wastewater

The PTW was collected from the secondary settling tank of a pharmaceutical company wastewater treatment plant in Northeast China. The pharmaceutical company is a comprehensive enterprise, focusing on the production of chemosynthetic drugs and bio-fermentation drugs. The average daily production of pharmaceutical wastewater is 30,000 m^3. After being pretreated in a regulation tank, the wastewater was hydrolyzed and acidified in a two-stage hydrolysis acidification tank, and then entered into a one-unit activated sludge reactor (UNITANK) for biological treatment. The water quality index of pharmaceutical wastewater before and after biotreatment was provided in Table 1. All samples were filtered through a 0.45 µm glass fiber membrane prior to measurement, except for turbidity.

Table 1. Characteristics of the pharmaceutical wastewater before and after biotreatment.

Value	Value	
	Before Biotreatment	After Biotreatment
pH	5.3–6.1	6.8–7.2
DOC (mg/L)	739–892	77–126
COD (mg/L)	3331–5183	201–343
SCOD	2861–4658	180–277
NH_4^+-N (mg/L)	61–73	13–15
B/C	0.25–0.32	0.09–0.13
UV_{254} (/cm)	-	0.906–1.31
Turbidity (NTU)	178–582	46–52
Conductivity (mS/cm)	-	14.95–15.57

2.2. Coagulation and UV/O_3 Treatment

The coagulation test was carried out in a six-unit agitator (ZR4-6, Zhongrun technology development Co., Ltd. Shenzhen, China). The 1000 mL pharmaceutical tail wastewater was placed in a cylindrical beaker with a volume of 1.5 L. The initial pH of the wastewater was adjusted by 0.1 mol/L of NaOH and HCl. In order to fully dissolve and react the added coagulant in the wastewater, the procedure of the agitator was set as follows: quick stirring for 2 min with a speed of 250 rpm, slow stirring for 10 min at a speed of 50 rpm. At last, the supernatant of treated wastewater was collected after 1 h of static treatment for the relevant water quality analysis and three-dimensional fluorescence analysis. Polyferric sulfate ($(Fe_2(OH)_n(SO_4)_{3-n/2})_m$, PFS) was chosen as the coagulant in this study. The coagulant dosage is 0.4 g/L (according to the results of preliminary laboratory research), and the initial solution pH is 7.0.

Individual UV irradiation, O_3 oxidation and UV/O_3 oxidation experiments were processed in a closed double-layer cylindrical glass reactor with diameter of 8 cm, height of 13.5 cm and effective volume of 750 mL. The low-pressure UV lamp with the power of 15 W was used in this study to emit monochromatic light at emission spectra $\lambda = 254$ nm. The UV lamp was placed in a 3.3 cm diameter quartz tube, which was placed in the center of the reactor. Both ozone generator (CFS20) and ozone concentration detector (UV_{300}) used in this study were purchased from Beijing Shanmei Shuimei Environmental Protection Technology Co., Ltd. (Beijing, China), and the ozone gas flow rate was controlled at 48–50 L/h to maintain the ozone concentration at around 10 mg/L (according to the results of preliminary laboratory research) during the O_3 oxidation and UV/O_3 treatment test. For comparison, the pH value of the biochemical treated pharmaceutical wastewater was adjusted to 7.0 by 0.1 mol/L H_2SO_4 and NaOH solution prior to individual UV irradiation, O_3 oxidation alone and UV/O_3 treatment, consistent with the pH value of the coagulation treatment. All the tests were carried out at room temperature (25 ± 2 °C).

2.3. Water Quality and Spectral Analysis

The turbidity was measured using a WGZ-1 turbidity meter from Xinrui Instrument Co., Ltd. (Shanghai, China). The dissolved organic carbon (DOC) was determined by Shimadzu TOC-VCPH analyzer (Shimadzu, Kyoto, Japan). Chemical oxygen demand (COD) was measured by spectrophotometry (DRB200, HACH, Loveland, CO, USA). OxiTop® system was used to determine BOD_5. The 254 nm UV absorbance (UV_{254}) was measured by UV-Vis spectrophotometer (UV-6100, METASH, Shanghai, China) with UV absorption spectrum in the range of 200–600 nm. The NH_4^+-N was determined by Nessler spectrophotometry. The analysis methods of the above conventional indicators are all standard analytical methods published by the Ministry of Environmental Protection

(HJ 535-2009). The SUVA value is defined as the ratio of the absorbance at 254 nm to the solution DOC concentration.

Synchronous fluorescence spectroscopy (SFS) was performed using a Hitachi fluorescence spectrophotometer (F-7000) from Hitachi, Tokyo, Japan. The response time of the spectrometer was set to 0.5 s and the photomultiplier tubes (PMT) voltage was 400 V. The wavelength range was from 260 nm to 550 nm by simultaneously scanning the excitation wavelength (ex) and emission wavelength (em). The passband was 0.2 nm and at the same time we kept the wavelength difference constant, $\Delta\lambda = \lambda_{em} - \lambda_{ex} = 18$ nm. The scanning speed was set to 240 nm/min. All spectral values were deducted from their respective program blanks. Fluorescence spectroscopy was usually performed within one day of sampling.

Two-dimensional correlation fluorescence spectroscopy (2D-COS) was analyzed by using the "2D-Shige" software developed by Kwansei-Gakuin University. Samples were taken at different time points from the UV/O$_3$ oxidation test, the sequence of time can be regarded as the specified external disturbance. Therefore, a matrix depending on sampling units at different points in time can be generated.

The synchronous correlation spectra can be calculated by the following formula [22]:

$$\phi(x_1, x_2) = \frac{1}{m-1}\Sigma_{j=1}^{m} I_j(x_1, t) I_j(x_2, t) \tag{1}$$

The asynchronous correlation spectrum is determined by the following formula [22]:

$$\varphi(x_1, x_2) = \frac{1}{m-1}\Sigma_{j=1}^{m} I_j(x_1, t)\Sigma_{k=1}^{m} N_{jk} I_j(x_2, t) \tag{2}$$

where x represented an index variable (number of sampling points) of the synchronous fluorescence spectrum caused by disturbance variable t. $I_{(x, t)}$ was the analytical spectrum of m evenly distributed on t (between T_{min} and T_{max}). N_{jk} including the j column and original element k was called the discrete Hilbert-Noda transformation matrix and was defined as follows [22]:

$$N_{jk} = \{^{0}_{\frac{1}{\pi(k-j)}} \quad if = k; \; otherwise \tag{3}$$

The fluorescence intensity ϕ (x_1, x_2) of the synchronous fluorescence spectrum can represent a simultaneous or consistent change in the two independent spectral intensities determined by the perturbation variable t between T_{min} and T_{max} at x_1 and x_2.

The spectral intensity of the asynchronous two-dimensional φ (x_1, x_2) represented a continuous or continuous but inconsistent change in the spectral intensity measured at x_1 and x_2, respectively.

3. Results and Discussion

3.1. Changes in DOC and SUVA by Coagulation and UV/O$_3$ Process

DOC and SUVA are two important indicators for characterizing the organic matter in water treatment. The lower the SUVA value, the lower the risk of producing disinfection byproducts. As showed in Figure 1, after the coagulation pretreatment stage, the removal rate of DOC and SUVA reached 37.5% and 24.4%, respectively. After the oxidation stage, the DOC and SUVA was further reduced, the highest DOC and SUVA removal efficiency reached 55.8% and 68.7% by coagulation-UV/O$_3$ process after 60 min oxidation.

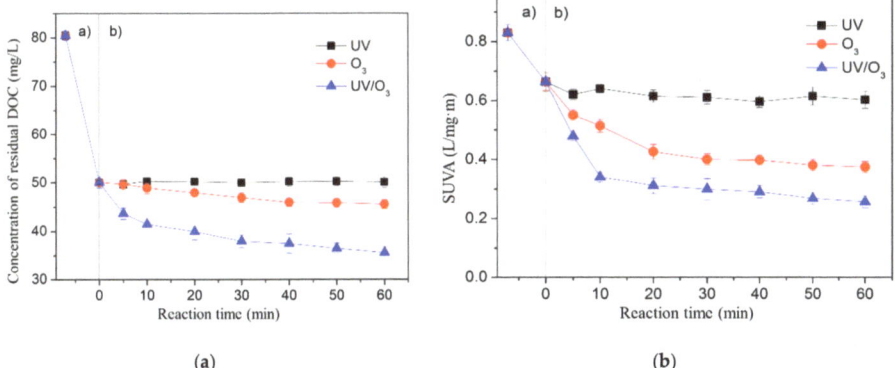

Figure 1. Effect of coagulation-UV, coagulation-O_3 and coagulation-UV/O_3 process on dissolved organic carbon (DOC) in pharmaceutical tail wastewater (PTW) (**a**: coagulation stage; **b**: oxidation stage).

The DOC concentration slightly changed during individual O_3 oxidation, indicating that the O_3 treatment mainly changed the structure of the organic matter by direct oxidation to form intermediate products instead of mineralizing organics into CO_2 and H_2O. Therefore, the 30.5% of SUVA was reduced by O_3 oxidation but only 5% of DOC was removed at the same time. In comparison, the ·OH formed by UV excitation in O_3 oxidation enhanced the degradation and mineralization of the active compound, thereby UV/O_3 achieved better DOC removal efficiency than O_3 oxidation.

Since the removal of organic matter was not achieved by UV irradiation alone, the effectiveness of UV/O_3 on PTW treatment demonstrated a synergistic effect of the combination of UV and O_3. In the UV/O_3 treatment process, the reduction in DOC content included direct and indirect effects. Direct effect was manifested in the photochemical mineralization of photosensitive substances (such as aromatic ring), and indirect effects represented the utilization of reactive oxygen species (OH), whose major production pathways require O_2 [23]. It can be seen in the first 20 min of the oxidation reaction that the SUVA value of the wastewater was rapidly decreased. Then, as time goes on, the decline rate of SUVA slowed down. At the end of the reaction (60 min), the removal rate of SUVA by O_3 or UV/O_3 reached 55% and 68.7%, respectively. These results indicate that organic components with UV absorption were easily oxidized by the O_3 or UV/O_3 process. The removal of SUVA (about 68.7%) was even higher than the removal rate of DOC. This indicates the preferential removal of aromatic luminescent chromophore or partially aromatic structure that can be converted to a non-UV absorbing compound by photochemical reaction. Many photochemical products were reported to have molecular weights as low as those of organic acids, alcohols, aldehydes, and inorganic carbons [24,25].

3.2. SFS Characteristics and Component Identification

3.2.1. SFS Characteristics and Fluorescent Component Removal

In order to reveal the removal mechanism of organics in PTW in the process of UV/O_3 treatment, the SFS characteristics of PTW taken at different time points during the oxidation treatment were analyzed. Three main peaks and two broad shoulders were shown in SFS (Figure 2a), including tyrosine-like fluorescent component (TYLF, λ = 265–300 nm), tryptophan-like fluorescent component (TRLF, λ = 300–360 nm) [26,27], microbial humus-like fluorescence component (MHLF, λ = 360–420 nm), fulvic acid-like fluorescent component (FLF, λ = 420–460 nm) and humus-like fluorescent component (HLF, λ = 460–520 nm) [28]. It can be seen that the fluorescence intensity in the whole wavelength range decreased as the oxidation reaction proceeded. Similar to the fluorescence spectrum, the absolute fluorescence loss of the solution during the reaction can be taken as a function of wavelength and compared over different irradiation oxidation times (Figure 2b).

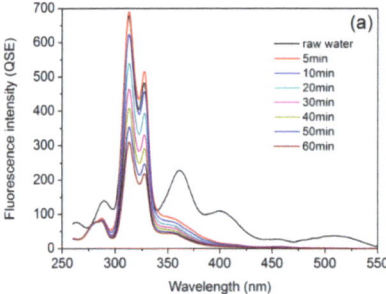
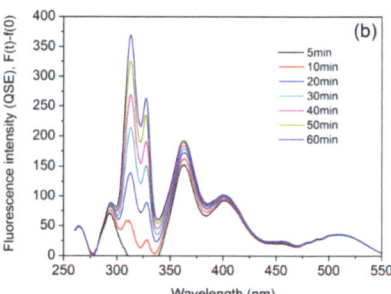

Figure 2. Changes in spectral responses of PTW with different irradiation times in UV/O$_3$ system: (a) synchronous fluorescence spectroscopy (SFS) ($\Delta\lambda = 18$ nm) and (b) absolute fluorescence losses of excitation wavelength function.

As expected, the initial loss did not occur in the TRLF region, but in the TYLF, MHLF and the FLF component. The TYLF, MHLF and FLF components decreased rapidly and TRLF increased slightly in the first 5 min of UV/O$_3$ irradiation. This was likely to be due to the conversion of MHLF and FLF into TRLF during the irradiation oxidation.

3.2.2. SFS Component Identification and Principle Component Analysis

Principle component analysis (PCA) was carried out in this study to identify the component based on SFS at different times. The independent fluorescent components in PTW can be distinguished by cluster scores at different spectroscopic wavelength through PCA analysis. The Kaiser-Meyer-Olkin (KMO) test and Bartlett's test of sphericity (P) were two indexes to test the correlation between variables in the relevant array. In this study, the KMO value was 0.843, $p < 0.001$, indicating that SFS was well suited for PCA application [29]. After performing PCA analysis on the SFS of eight time points, two principal components (PC) were generated with the cumulative variance contribution rate of 99.69%, which can reflect most of the characteristics of the SFS (Figure 3). The scores curve of different wavelengths can show the characteristics of the spectral waveform of each principal component. Therefore, fluorescent groups that contributed variance to SFS signal can be identified [30]. PC1 with a variance of 98.8% showed three main peaks and four shoulders (Figure 3a). Same as the SFS characteristics, five fluorescent components were identified by wavelength at PC1, including TYLF (at 289 nm), TRLF (at 312 nm, 329 nm, 359 nm), MHLF (at 380 nm), FLF (at 450 nm) and HLF (at 470 nm). PC2 (variance of 0.89%) showed four main peaks and two shoulders (Figure 3b).

The main peak of TYLF at 292 nm exhibited a red-shift of 3 nm and its wavelength migration was longer than that in PC1, indicating that the polarity of the organic component was enhanced and the hydrophobicity was reduced. The MHLF peak was transferred to a shorter wavelength (360 nm) compared to PC1. The peak of FLF component had a red-shift of 5 nm at the wavelength of 455 nm. The TRLF component was identified at the wavelengths of 312 nm and 329 nm, no red-shift and blue-shift occurred compared PC1.

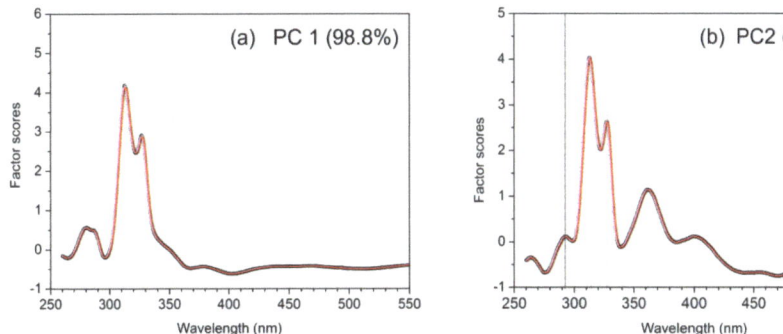

Figure 3. Scores plots of principle component 1 (PC1) (**a**) and principal component 2 (PC2) (**b**) for spectral wavelengths in UV/O$_3$ system.

3.2.3. Organic Component Spectral Area Integral

The SFS can be divided into five synchronous fluorescence regions based on excitation wavelength, which were corresponding to TYLF, TRLF, MHLF, FLF and HLF. The area integral of wavelength and fluorescence intensity can represent the relative abundance of homologous component. According to the area integral, the removal efficiency of fluorescence components can be studied. The distributions of the abundance of the DOM components were shown in Figure 4. It turned out that the DOC concentration showed a significant positive correlation to the variation of TYLF ($r = 0.940$, $p = 0.002$), TRLF ($r = 0.988$, $p = 0.0003$) and HLF ($r = 0.929$, $p = 0.003$), but a weak positive correlation to the MHLF ($r = 0.430$, $p = 0.335$) and FLF ($r = 0.499$, $p = 0.254$), which indicated that TYLF, TRLF and HLF could more accurately indicate DOC changes than MHLF and FLF. This was caused by the selective removal of fluorescent component in UV/O$_3$ irradiation oxidation process.

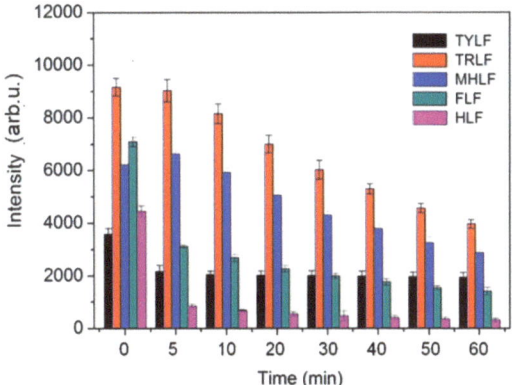

Figure 4. Distributions of the abundance of the dissolved organic matter (DOM) components in UV/O$_3$ system.

It can be seen in Figure 4 that the FLF, HLF and TYLF components decreased rapidly due to the UV/O$_3$ oxidation. Consequently, 82.2% of HLF, 40.4% of FLF and 40.0% of TYLF were removed in the first 5 min, indicating that the FLF, HLF and TYLF can be effectively removed by UV/O$_3$. This well explained the quick degradation of DOC and SUVA in the first reaction time, which was discussed before. However, in the subsequent reactions, the removal rate of FLF, HLF and TYLF gradually slowed down, and the TRLF and MHLF components began to decline. After 60 min of UV/O$_3$ oxidation, the maximum removal efficiency was obtained by HLF component of 95.6%, followed by FLF of 80.0%, TRLF of 56.0%, MHLF of 50.8% and TYLF of 44.4%. This indicated that the UV/O$_3$ oxidation system could effectively remove the different fluorescent component in PTW, especially the HLF and FLF, while it took more time to completely remove the TYLF, TRLF and MHLF components.

3.3. Synchronous Fluorescence/UV-Visible Two-Dimensional Correlation Spectra Analysis

To reduce spectral overlap and reveal the DOM transformation process, 2D-COS was used to analyze two-dimensional fluorescence spectra [26]. Two-dimensional synchronous fluorescence spectrum of PTW was a symmetric spectrum about diagonals, in which there were four main self-peaks at wavelengths of 313 nm, 326 nm, 360 nm, and 400 nm (Figure 5a). The results are consistent with the PCA analysis. The self-peak represented the overall sensitivity of the corresponding spectral region after the change in spectral intensity as an external disturbance. During the reaction, the intensity of the self-peak at 313 nm and 326 nm was higher than that at 360 nm and 400 nm, indicating that the fluorescence intensity of the TYLF and TRLF component was clearly higher than that of the MHLF. The solid cross peaks indicated a positive correlation of fluorescence components at 313 nm, 326 nm, 360 nm, and 400 nm. Indirectly this confirmed the fluorescence components of DOM in PTW were synchronously removed in the process of the coagulation-UV/O$_3$ process. In the asynchronous fluorescence spectrum (Figure 5b), two negative cross peaks appeared at 289/313 nm and 289/326 nm, while five positive cross peaks were found at: 313/360, 313/400, 313/501, 326/360 and 326/400 nm. The positive or negative asynchronous crossover peak could provide information on the disturbance order of spectral band along the external variables. According to Nado rules, the change order of the spectral bands was: 501, 400, 360, 326, 313, 289 nm. This indirectly confirmed the removal characteristics of each component in the tail water treatment process of the synthetic pharmaceutical park, which had been discussed before.

The 2D-COS was used to study, in more detail, the characteristics of different absorption wavelengths of organic matter in the PTW during the oxidation process, the synchronous and asynchronous 2D-COS analysis using UV-visible absorption spectra with different oxidation time was shown in Figure 5c. Strong self-peaks at 225 nm and 255 nm and weak self-peak at 350 nm were observed in the synchronous 2D-UV-visible correlation spectrum and the intensity of the first two absorption bands at 225 nm and 255 nm decreased significantly with the irradiation time. This result was consistent with our previous observations of the overall trend of higher absorption losses at shorter wavelengths. In the asynchronous 2D-UV-visible correlation spectrum, three absorption bands of the oxidative chemical reaction were observed, and the wavelength ranges were: 200–240, 240–270, and 270–320 nm, respectively (Figure 5d). The sequence relationship of spectral variation characteristics with oxidation time can be derived: 270–320, 200–240, 240–270 nm. This indicated that the oxidation reaction first removed the macromolecular organic matter with a conjugated bond in the wavelength range of 270 nm–320 nm, and then removed the organic matter at 240–270 nm.

In order to further reveal the characteristics of organic matter removal and degradation, synchronous fluorescence/UV-visible two-dimensional correlation spectra were obtained. From the synchronous two-dimensional correlation map (Figure 5e), it can be found that the aromatic group at the wavelength of 255 nm in UV-visible spectrum was corresponding to the variation of TRLF, MHLF, FLF and HLF components at 313 nm, 326 nm, 360 nm and 400 nm in synchronous fluorescence. This indicated that the TRLF, MHLF, FLF and HLF components containing aromatic groups were degraded during the oxidation process. At the same time, the wavelength in the range of 400–500 nm in the UV-visible spectrum was opposite to the variation trend of the TRLF and MHLF components, which represented a humus-like substance with a high degree of humification and low lignin content that was generated after the oxidation reaction. Through analysis of asynchronous two-dimensional correlation map (Figure 5f), UV-visible spectrum at 285 nm (organic matter containing unsaturated conjugated double-bond structure) has a negative peak with $v1/v2$ at 360 nm (FLF) and 400 nm (HLF) in synchronous fluorescence, respectively. It was shown that the organic compounds with unsaturated conjugated double-bond structure in HLF and FLF was first degraded by the oxidation process. At the same time, the UV-visible spectrum at 255 nm showed a positive peak with $v1/v2$ at 313 nm (TRLF) and 326 nm (MHLF) in the synchronous fluorescence, respectively, indicating the MHLF- and TRLF-containing aromatic groups were removed sequentially in subsequent oxidation. The spectral characteristics can explain that the degradation of organic matter occurred sequentially in the order of HLF→FLF→MHLF→TRLF.

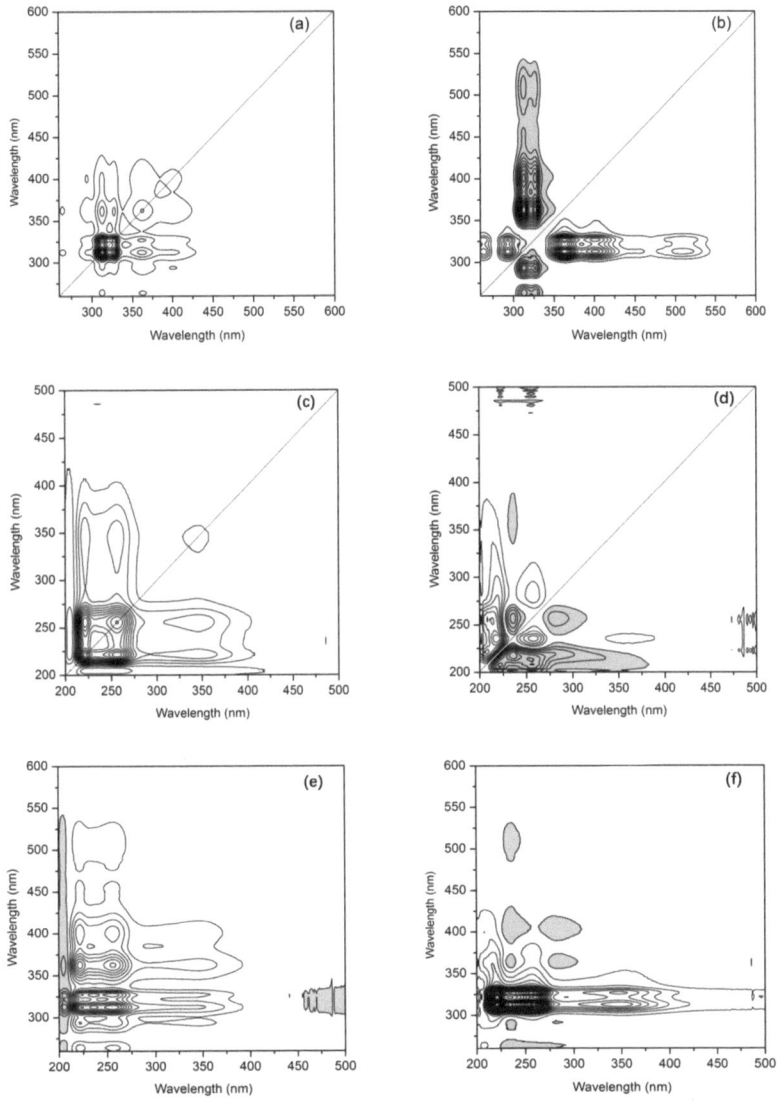

Figure 5. Two-dimensional- (2D)-correlation spectroscopy results of PTW: (**a**) synchronous fluorescence map, (**b**) asynchronous fluorescence map, (**c**) synchronous UV-visible map, (**d**) asynchronous UV-visible map, (**e**) synchronous fluorescence/UV-visible two-dimensional correlation map, (**f**) asynchronous fluorescence/UV-visible two-dimensional correlation map. The solid and the gray represent the positive and negative signs, respectively.

The organic matter in PTW can be effectively removed by coagulation-UV/O_3 pretreatment. After 60 min reaction, the removal rate of SUVA and DOC reached 68.7% and 55.8%, respectively. The UV can significantly enhance the mineralization of organic matters by O_3, therefore the treatment effect was significantly better than the O_3 oxidation alone. TRLF, TYLF, MHLF, FLF and HLF components were identified by the SFS combined with 2D-COS and PCA analysis. All the results consistently show that HLF and FLF were the main components of DOM in the tail water of synthetic pharmaceutical parks. UV/O_3 was selective for the removal of different fluorescent components, which can quickly remove FLF and HLF from the PTW and convert them into TRLF. The order of degradation of the different fluorescent components was HLF→FLF→MHLF→TRLF→TYLF. SFS combined with 2D-COS and PCA can quickly and effectively reveal the spectral dynamics of DOM in UV/O_3 treatment system and the removal and degradation characteristics of different organic components.

Author Contributions: Conceptualization, J.W. and F.Q.; methodology, J.W., H.Y. and C.D.; software, J.W., F.Q., C.D. and H.Y.; validation, Y.S. and L.X.; formal analysis, J.W., F.Q.; investigation, J.W., C.D., H.Y. and F.Q.; resources, Y.S.; data curation, Y.S.; writing—original draft preparation, J.W.; writing—review and editing, J.W. and F.Q.; visualization, J.W. and C.D.; supervision, Y.S. and L.X.; project administration, Y.S.; funding acquisition, Y.S. and F.Q. All authors have read and agreed to the published version of the manuscript.

Funding: This research was funded by National Major Program of Science and Technology for Water Pollution Control and Governance (Fund number, 2012ZX07202-005, PR China).

Conflicts of Interest: The authors declare no conflict of interest.

References

1. Qian, F.; He, M.; Song, Y.; Tysklind, M.; Wu, J. A bibliometric analysis of global research progress on pharmaceutical wastewater treatment during 1994–2013. *Environ. Earth Sci.* **2015**, *73*, 4995–5005. [CrossRef]
2. Balcıoğlu, I.A.; Ötker, M. Treatment of pharmaceutical wastewater containing antibiotics by O_3 and O_3/H_2O_2 processes. *Chemosphere* **2003**, *50*, 85–95. [CrossRef]
3. López-Fernández, R.; Martínez, L.; Villaverde, S. Membrane bioreactor for the treatment of pharmaceutical wastewater containing corticosteroids. *Desalination* **2012**, *300*, 19–23. [CrossRef]
4. Rivera-Utrilla, J.; Sánchez-Polo, M.; Ferro-García, M.; Prados-Joya, G.; Ocampo-Perez, R. Pharmaceuticals as emerging contaminants and their removal from water. A review. *Chemosphere* **2013**, *93*, 1268–1287. [CrossRef] [PubMed]
5. Sirés, I.; Brillas, E.; Sadornil, I.S. Remediation of water pollution caused by pharmaceutical residues based on electrochemical separation and degradation technologies: A review. *Environ. Int.* **2012**, *40*, 212–229. [CrossRef] [PubMed]
6. Ikehata, K.; Naghashkar, N.J.; El-Din, M.G. Degradation of Aqueous Pharmaceuticals by Ozonation and Advanced Oxidation Processes: A Review. *Ozone Sci. Eng.* **2006**, *28*, 353–414. [CrossRef]
7. Gadipelly, C.; Pérez-González, A.; Yadav, G.D.; Ortiz, I.; Ibáñez, R.; Rathod, V.K.; Marathe, K.V. Pharmaceutical Industry Wastewater: Review of the Technologies for Water Treatment and Reuse. *Ind. Eng. Chem. Res.* **2014**, *53*, 11571–11592. [CrossRef]
8. Tang, C.-J.; Zheng, P.; Chen, T.-T.; Zhang, J.-Q.; Mahmood, Q.; Ding, S.; Chen, X.-G.; Chen, J.-W.; Wu, D.-T. Enhanced nitrogen removal from pharmaceutical wastewater using SBA-ANAMMOX process. *Water Res.* **2011**, *45*, 201–210. [CrossRef]
9. Biń, A.K.; Sobera-Madej, S. Comparison of the Advanced Oxidation Processes (UV, UV/H_2O_2 and O_3) for the Removal of Antibiotic Substances during Wastewater Treatment. *Ozone Sci. Eng.* **2012**, *34*, 136–139. [CrossRef]
10. Huber, M.M.; Göbel, A.; Joss, A.; Hermann, N.; Löffler, D.; McArdell, C.S.; Ried, A.; Siegrist, H.; Ternes, T.; Von Gunten, U. Oxidation of Pharmaceuticals during Ozonation of Municipal Wastewater Effluents: A Pilot Study. *Environ. Sci. Technol.* **2005**, *39*, 4290–4299. [CrossRef]
11. Gong, J.; Liu, Y.; Sun, X. O_3 and UV/O_3 oxidation of organic constituents of biotreated municipal wastewater. *Water Res.* **2008**, *42*, 1238–1244. [CrossRef] [PubMed]

12. Qian, F.; Sun, X.; Lei, J. Removal characteristics of organics in bio-treated textile wastewater reclamation by a stepwise coagulation and intermediate GAC/O_3 oxidation process. *Chem. Eng. J.* **2013**, *214*, 112–118. [CrossRef]
13. Sirtori, C.; Zapata, A.; Oller, I.; Gernjak, W.; Agüera, A.; Malato, S. Decontamination industrial pharmaceutical wastewater by combining solar photo-Fenton and biological treatment. *Water Res.* **2009**, *43*, 661–668. [CrossRef] [PubMed]
14. Zhu, G.; Yin, J.; Zhang, P.; Wang, X.; Fan, G.; Hua, B.; Ren, B.; Zheng, H.; Deng, B. DOM removal by flocculation process: Fluorescence excitation–emission matrix spectroscopy (EEMs) characterization. *Desalination* **2014**, *346*, 38–45. [CrossRef]
15. Meng, F.; Huang, G.; Li, Z.; Li, S. Microbial Transformation of Structural and Functional Makeup of Human-Impacted Riverine Dissolved Organic Matter. *Ind. Eng. Chem. Res.* **2012**, *51*, 6212–6218. [CrossRef]
16. Carstea, E.; Bridgeman, J.; Baker, A.; Reynolds, D. Fluorescence spectroscopy for wastewater monitoring: A review. *Water Res.* **2016**, *95*, 205–219. [CrossRef]
17. Miano, T.; Senesi, N. Synchronous excitation fluorescence spectroscopy applied to soil humic substances chemistry. *Sci. Total Environ.* **1992**, *117*, 41–51. [CrossRef]
18. Henderson, R.; Baker, A.; Murphy, K.R.; Hambly, A.; Stuetz, R.M.; Khan, S.J. Fluorescence as a potential monitoring tool for recycled water systems: A review. *Water Res.* **2009**, *43*, 863–881. [CrossRef]
19. Baker, A.; Cumberland, S.; Bradley, C.; Buckley, C.; Bridgeman, J. To what extent can portable fluorescence spectroscopy be used in the real-time assessment of microbial water quality? *Sci. Total Environ.* **2015**, *532*, 14–19. [CrossRef]
20. Jung, Y.M.; Noda, I. New Approaches to Generalized Two-Dimensional Correlation Spectroscopy and Its Applications. *Appl. Spectrosc. Rev.* **2006**, *41*, 515–547. [CrossRef]
21. Ozaki, Y.; Czarnik-Matusewicz, B.; Sasic, S. Two-Dimensional Correlation Spectroscopy in Analytical Chemistry. *Anal. Sci.* **2002**, *17*, i663–i666.
22. Noda, I.; Liu, Y.; Ozaki, Y. Two-Dimensional Correlation Spectroscopy Study of Temperature-Dependent Spectral Variations of N-Methylacetamide in the Pure Liquid State. 2. Two-Dimensional Raman and Infrared—Raman Heterospectral Analysis. *J. Phys. Chem.* **1996**, *100*, 8674–8680. [CrossRef]
23. Lou, T.; Xie, H. Photochemical alteration of the molecular weight of dissolved organic matter. *Chemosphere* **2006**, *65*, 2333–2342. [CrossRef]
24. Pullin, M.J.; Bertilsson, S.; Goldstone, J.V.; Voelker, B.M. Effects of sunlight and hydroxyl radical on dissolved organic matter: Bacterial growth efficiency and production of carboxylic acids and other substrates. *Limnol. Oceanogr.* **2004**, *49*, 2011–2022. [CrossRef]
25. Vidali, R.; Remoundaki, E.; Tsezos, M. Humic Acids Copper Binding Following Their Photochemical Alteration by Simulated Solar Light. *Aquat. Geochem.* **2009**, *16*, 207–218. [CrossRef]
26. Hur, J. Microbial Changes in Selected Operational Descriptors of Dissolved Organic Matters From Various Sources in a Watershed. *Water Air Soil Pollut.* **2010**, *215*, 465–476. [CrossRef]
27. Yu, H.; Song, Y.; Tu, X.; Du, E.; Liu, R.; Peng, J. Assessing removal efficiency of dissolved organic matter in wastewater treatment using fluorescence excitation emission matrices with parallel factor analysis and second derivative synchronous fluorescence. *Bioresour. Technol.* **2013**, *144*, 595–601. [CrossRef]
28. Pan, H.; Yu, H.; Wang, Y.; Liu, R.; Lei, H. Investigating variations of fluorescent dissolved organic matter in wastewater treatment using synchronous fluorescence spectroscopy combined with principal component analysis and two-dimensional correlation. *Environ. Technol.* **2017**, *39*, 1–8. [CrossRef]
29. Yu, H.; Song, Y.; Liu, R.; Pan, H.; Xiang, L.; Qian, F. Identifying changes in dissolved organic matter content and characteristics by fluorescence spectroscopy coupled with self-organizing map and classification and regression tree analysis during wastewater treatment. *Chemosphere* **2014**, *113*, 79–86. [CrossRef]
30. Guo, X.-J.; Yuan, D.-H.; Jiang, J.-Y.; Zhang, H.; Deng, Y. Detection of dissolved organic matter in saline–alkali soils using synchronous fluorescence spectroscopy and principal component analysis. *Spectrochim. Acta Part A Mol. Biomol. Spectrosc.* **2013**, *104*, 280–286. [CrossRef]

© 2020 by the authors. Licensee MDPI, Basel, Switzerland. This article is an open access article distributed under the terms and conditions of the Creative Commons Attribution (CC BY) license (http://creativecommons.org/licenses/by/4.0/).

Heterogeneous Fenton-Like Catalytic Degradation of 2,4-Dichlorophenoxyacetic Acid by Nano-Scale Zero-Valent Iron Assembled on Magnetite Nanoparticles

Xiaofan Lv [1], Yiyang Ma [2], Yangyang Li [3] and Qi Yang [1,*]

1. Beijing Key Laboratory of Water Resources & Environmental Engineering, China University of Geosciences (Beijing), Beijing 100083, China; bh79858@sina.com
2. Institute of Environmental Engineering & Nano-Technology, Tsinghua Shenzhen International Graduate School, Tsinghua University, Shenzhen 518055, China; shelfunfun@yeah.net
3. Beijing Que Chen Architectural Design Consulting Co., Ltd., Beijing 100038, China; 3005160002@cugb.edu.cn
* Correspondence: yq@cugb.edu.cn; Tel.: +86-(010)-8232-3917

Received: 27 August 2020; Accepted: 15 October 2020; Published: 18 October 2020

Abstract: Fe^0@Fe_3O_4 nanoparticles with dispersibility and stability better than single nano zero-valent iron (nZVI) were synthesized and combined with hydrogen peroxide to constitute a heterogeneous Fenton-like system, which was creatively applied in the degradation of 2,4-dichlorophenoxyacetic acid (2,4-D). The effects of different reaction conditions like pH, hydrogen peroxide concentration, temperature, and catalyst dosage on the removal of 2,4-D were evaluated. The target pollutant was completely removed in 90 min; nearly 66% of them could be mineralized, and the main intermediate product was 2,4-dichlorophenol. Synergistic effects between nZVI and Fe_3O_4 made the 2,4-D degradation efficiency in the Fe^0@Fe_3O_4/H_2O_2 system greater than in either of them alone. More than a supporter, Fe_3O_4 could facilitate the degradation process by releasing ferrous and ferric ions from the inner structure. The reduction of 2,4-D was mainly attributed to hydroxyl radicals including surface-bound ·OH and free ·OH in solution and was dominated by the former. The possible mechanism of this Fe^0@Fe_3O_4 activated Fenton-like system was proposed.

Keywords: 2,4-dichlorophenoxyacetic acid; nano-scale zero-valent iron; magnetite; heterogeneous Fenton-like system; hydroxyl radicals

1. Introduction

2,4-dichlorophenoxyacetic acid (2,4-D) is a white powdery crystal; it is corrosive, slightly soluble in water, and easily soluble in most organic solvents [1]. Among phenoxyl herbicides, 2,4-D is one of the most widely used as a plant growth regulator, for lawn care, and for broadleaf weeding of corn, sugarcane, sorghum, and other crops [2]. As an endocrine disruptor with strong biological toxicity, it can enter the human body through inhalation, ingestion, and dermal contact [3]. Exposure to 2,4-D may cause the decline of insulin secretion and immunity, bringing damage to kidneys, liver, muscles, and the central nervous system. After accumulating to a certain concentration in the body, 2,4-D may induce abnormality, gene mutation, or cancer; its presence poses serious threat to humans and other life in nature [4]. At the same time, the low volatility and refractory properties of 2,4-D make it easy to migrate to groundwater by leaching during its production and use, causing persistent environmental pollution [5]. Therefore, remediation of 2,4-D polluted water has become an urgent problem of aquatic environment.

Advanced oxidation processes (AOPs) are some of the effective remediation methods for reducing organic pollutants in contaminated soils or groundwater. The contribution of AOPs is to introduce a high oxidation potential source to produce the primary oxidant species, hydroxyl radicals (·OH), which react rapidly and unselectively with most organic compounds. Many AOPs have been developed and investigated to mineralize 2,4-D in aqueous solutions [6,7].

As one of the most widely used AOPs, Fenton oxidation has shown significant effects on the remediation of organic pollutants in contaminated soils or groundwater including 2,4-D. Traditional Fenton technology has its effects by utilizing Fe^{2+} to catalyze the decomposition of hydrogen peroxide to generate radicals like the hydroxyl radical, which can react with most organic compounds unselectively and complete the reduction of target pollutants rapidly [8]. For the high efficiency, non-toxicity, and easy operation, it has been successfully applied to the treatment of many kinds of organic wastewater. Nevertheless, the high operating cost, limited optimum pH range, and difficulties in recycling ferrous catalysts restrict the large-scale application of Fenton technology [9,10]. To overcome these disadvantages, many researchers use heterogeneous or homogeneous catalysts like Fe^{3+}, Cu^{2+}, pyrite, and some zero-valent metals instead of Fe^{2+} to establish Fenton-like processes to enhance the remediation effect [11–13]. Among them, nano zero-valent iron (nZVI) has received great attention for its large specific surface area, high permeability, good reactivity, easy accessibility, and low cost [14–16]. However, the high interfacial energy and magnetic and gravitational forces make it easily agglomerate. Meanwhile, the formation of passive shells caused by the oxidation of water or oxygen in surrounding media will hinder the further reaction of inner Fe^0 and decrease electron transfer [17,18].

In our previous research, a complex catalyst with nZVI particles attached on the surface of Fe_3O_4 was prepared and successfully applied to the reductive degradation of carbon tetrachloride [19]. Introducing Fe_3O_4 nanoparticles (NPs) as the supporter could help to overcome the aggregation and passivation problems of nZVI; the magnetic properties make nZVI tightly attracted to the surface of Fe_3O_4, thereby optimizing the problems associated with separating the nanomaterial and stabilizing the reaction system. It is noteworthy that the characteristics of nZVI and magnetite determine that the $Fe^0@Fe_3O_4$ NPs can complete the degradation of chlorinated organic pollutants through both the reductive and oxidative pathway. To our knowledge, there is limited reporting on the application of $Fe^0@Fe_3O_4$ NPs in terms of AOPs, especially the heterogeneous Fenton-like reaction system for the rapid degradation of 2,4-D contaminated water.

In this study, a heterogeneous Fenton-like system was constructed with the combination of hydrogen peroxide and $Fe^0@Fe_3O_4$ NPs and was applied to the oxidative degradation of 2,4-D. Factors such as the initial pH of the solution, temperature, hydrogen peroxide concentration, and catalyst dosage were investigated to evaluate the effects of different reaction conditions. The degradation pathway of 2,4-D and the possible mechanism in the $Fe^0@Fe_3O_4/H_2O_2$ system were proposed by testing the changes of ferrous ions/chloride ions and determining the products during the degradation of the target pollutants. The main purpose of this study is to provide a more effective way to achieve the degradation of 2,4-D in the heterogeneous Fenton-like system motivated by the $Fe^0@Fe_3O_4$ NPs and hydrogen peroxide, thereby improving the adaptability of this modified catalyst. Meanwhile, it may also contribute to the remediation of other organochlorine pollutants.

2. Materials and Methods

2.1. Chemicals and Reagents

All chemicals were obtained from commercial sources and used as received: 2,4-D (>98.8% Merck, Darmstadt, Germany), Ferrous sulfate ($FeSO_4 \cdot 7H_2O$, 99.5%), Ferric sulfate ($Fe_2(SO_4)_3$, >99.5%), sodium borohydride ($NaBH_4$, >99.5%), hydrogen peroxide (H_2O_2, 30%), methanol (>99.9%), ethanol (>99.7%), n-butanol (>99.5%), and potassium iodide (KI) (>99.0%). Except 2,4-D, all the above chemicals were analytical grade (AR). The prepared 2,4-D stock solution (50 mg/L) was stored at 35 °C.

2.2. Preparation of Fe^0@Fe$_3$O$_4$ Nanoparticles

The Fe$_3$O$_4$ NPs were prepared by the co-precipitation method and are depicted in Equation (1) [20]. Under the atmosphere of N$_2$ injection (0.1 L/min), the Fe^{2+}/Fe^{3+} solution was prepared (4.99 g Fe$_2$(SO$_4$)$_3$, 6.95g FeSO$_4$·7H$_2$O) and added drop-wise to the heated NaOH solution (0.5 M) with electric stirring (150 r/min). When the reaction was completed, the gained Fe$_3$O$_4$ NPs were washed twice by ultra-pure water, collected by magnetic separation, and kept by freeze-drying preservation.

$$Fe^{2+} + 2Fe^{3+} + 8OH^- \rightarrow Fe_3O_4 + 4H_2O \tag{1}$$

With a continuous N$_2$ injection (0.1 L/min), FeSO$_4$·7H$_2$O (0.62 g) was dissolved in 250 mL of ultra-pure water and mixed with the prepared Fe$_3$O$_4$ NPs (0.5 g), then a drop-wise addition of 250 mL of NaBH$_4$ solution (0.018 M) into the mixed solution was conducted to realize the reduction of Fe^{2+} and form the nZVI particles (Equation (2)) [21].

$$nFe_3O_4 - Fe^{2+} + 2BH_4^- + 6H_2O \rightarrow nFe_3O_4 - nZVI + 2B(OH)_3 + 7H_2\uparrow \tag{2}$$

The Fe^0@Fe$_3$O$_4$ nanoparticles were collected by magnetic separation and washed by ultra-pure water to neutral pH. Finally, the Fe^0@Fe$_3$O$_4$ NPs were freeze-dried and preserved under nitrogen protection from oxidation for further use. For this prepared catalyst, the mass ratio of Fe$_3$O$_4$:Fe^0 was 4:1.

2.3. Characterization and Analysis Methods

The particle morphology was investigated by scanning electron microscope (SEM) (Carl-Zeiss Microscopy GmbH, 73447 Oberkochen, Germany), and X-ray photo-electron spectroscopy (XPS) (JPS-9010TR, JEOL, Japan) was applied for qualitative analysis and elemental composition.

High performance liquid chromatography (HPLC Waters 1525) with a reversed-phase C-18 column (4.6 × 150 mm, 5 µm particle diameter) and a UV detector (Waters 2487) was used for the detection of 2,4-D concentration. The mobile phase composition was methanol/water/acetic acid (75:23:2%, v/v/v), while the flow rate of the mobile phase was settled at 1 mL/min. The injection volume was 20 µL, and the analysis was conducted at λ = 285 nm via UV detection. The pH value was measured using a pH meter (Thermo star-A 211 pH meter).

Gas chromatography/mass spectrometry (GC–MS, Agilent 7890A/Agilent 5975C) equipped with a HP-5MS column (30 m × 0.25 µm × 0.25 mm) was chosen for intermediate products' analysis. High-purity helium (99.99%) was used as the carrier gas with a flow rate of 1.0 mL/min, and the split ratio was 5:1. After holding for 1 min, the original GC oven temperature (35 °C) was increased to 300 °C at a rate of 7.0 °C/min and held for 1 min. The injector and detector temperatures were both 280 °C, and the injection volume was 1 µL.

The 1,10-phenanthroline method was applied to measure the concentration of total dissolved iron and ferrous ion by a UV-Vis spectrophotometer (Agilent Cary 300), while the hydrogen peroxide concentration was determined by the spectrophotometric method [7,22].

Chloride and some short-chain organic acids were measured by ion chromatograph (IC, Dionex ICS-900) with a DS5 conductivity detector and an IonPac AS23 analytical column (Dionex 4 × 250 mm) with an IonPac AG 23 guard column (Dionex 4 × 50 mm). Total organic carbon was tested by the Multi TOC Analyzer (2100, Analytik Jena AG Corporation, Jena, Germany).

2.4. Batch Experiments

Batch experiments were conducted to investigate the impacts of different reaction conditions on 2,4-D removal efficiency by Fe^0@Fe$_3$O$_4$. A certain amount of Fe^0@Fe$_3$O$_4$ particles and H$_2$O$_2$ were added into 100 mL serum bottles with a given concentration of 2,4-D solution. The bottles, double sealed with Teflon butyl stopper and an aluminum cover, were placed in a water bath shaker with a rotation speed of 175 rpm under a constant temperature and run in triplicate. A powerful magnet was used to separate $Fe^0@Fe_3O_4$ particles from the solution to have an easier sampling and reduce errors. Aliquots of the samples were taken by a glass syringe at certain time intervals, filtered through 0.22 μm filter membranes, and collected into 2 mL vials containing 10 μL of n-butanol (1 M), which was to quench the reaction. To explore the optimum reaction conditions, the temperature, pH, $Fe^0@Fe_3O_4$ dosage, and H_2O_2 concentration were tested, with one parameter changing while the others remained unchanged. The initial temperature was set for 20 °C, 30 °C, 35 °C, 40 °C, and 45 °C, while the initial pH ranged from 3.0 to 8.0. Different $Fe^0@Fe_3O_4$ dosages of 0.1 g/L, 0.3 g/L, 0.5 g/L, 0.8 g/L, and 1.0 g/L were investigated, and the H_2O_2 concentration was studied from 0.5 mg/L to 5.0 mg/L. Under the optimum reaction conditions, contrast experiments of the 2,4-D removal efficiency in different reaction systems ($Fe^0@Fe_3O_4$, nZVI and Fe_3O_4 particles with or without hydrogen peroxide) were conducted. EtOH and TBA were employed as radical scavengers to distinguish the degradation contributions of the radicals formed in the reaction system.

3. Results and Discussion

3.1. Characterization of $Fe^0@Fe_3O_4$ NPs

As shown in Figure 1b, $Fe^0@Fe_3O_4$ NPs have a chainlike distribution because of the magnetic dipole interaction and their nano-size. The nanoparticle sizes were in the range of 30–200 nm. According to the results of our previous studies [23], these spherically shaped particles presented in Figure 1d had a structure consisting of nZVI attaching to the surface of Fe_3O_4 particles. Compared with pure nZVI in Figure 1a and some relevant studies [24,25], the prepared complex composite had a better dispersity, which might be owed to the introduction of Fe_3O_4. As the supporter, magnetite could not only help to optimize the agglomeration of nZVI, but also accelerate electrons' transfer from nZVI to targeted pollutants for the formation of numerous Fe^0-Fe_3O_4 batteries. It is demonstrated in Figure 1c that, after 90 min of reaction, particle aggregation occurred with the increasing of particle sizes, owing to the consumption of both Fe^0 and Fe_3O_4, which would change the morphology and components of the prepared catalyst.

The XPS spectra in Figure 2a revealed that the original $Fe^0@Fe_3O_4$ NPs consisted of elements Fe, O, and C. The C1s peak at 284.8 eV could be attributed to the adventitious carbon from the sample preparation and analysis.

The freshly gained catalyst had two peaks at 710.4 eV and 724.0 eV, which were assigned to Fe_3O_4 (Fe(II)/Fe(III)), whose positions were consistent with the related literature values of 710.6 and 724.1 eV, respectively. The binding energies of Fe 2p3/2 and Fe 2p1/2 detected at 712.2 eV and 725.9 eV were similar to those of Fe_2O_3 (Fe(III)) ($Fe2p_{3/2}$, 711.0 eV; $Fe2p_{1/2}$, 726.4 eV). Meanwhile, it was reported that the Fe 2p3/2 peak of Fe_2O_3 has associated satellite peaks, which are located approximately 8 eV higher than the main Fe 2p3/2 peak [26]. It is the fact to which the peak with lower binding energy of 719.4 eV could be ascribed. The existence of Fe_2O_3 in the originally prepared sample might be due to the unavoidable oxidation of superficial nZVI during preparation and storage.

The composite comprised of small amounts of nZVI (Fe(0)) showed a peak at a low binding energy of 707.4 eV, which was attributed to $Fe2p_{3/2}$ in Fe^0 (706.7 eV). Moreover, the low-intensity peaks of nZVI might be ascribed to the thickness of the iron-oxide shell because XPS can only detect the photoelectrons from within 10 nm of the outer surface of a target.

With increasing reaction time, peaks of Fe(0) disappeared for the consumption of nZVI and the deposition of iron oxides, while peaks of Fe_2O_3 and Fe_3O_4 changed in different degrees. The formation of more iron oxide would indeed lead to the increase of Fe_2O_3, but similar to Fe_3O_4, those newly formed ones could also react with hydrogen peroxide to help with pollutant degradation, causing the decrease of Fe_2O_3 and Fe_3O_4 in the composite.

As time passed, the peak of lepidocrocite (γ-FeOOH) was detected at 713.3 eV in the sample after the reaction. Sampling of the reaction solution by syringe inevitably led to air egress, and the produced Fe^{2+} would react with oxygen and water, which might lead to the production of FeOOH (Equation (3)).

$$4Fe^{2+} + O_2 + 6H_2O \rightarrow 4FeOOH + 8H^+ \tag{3}$$

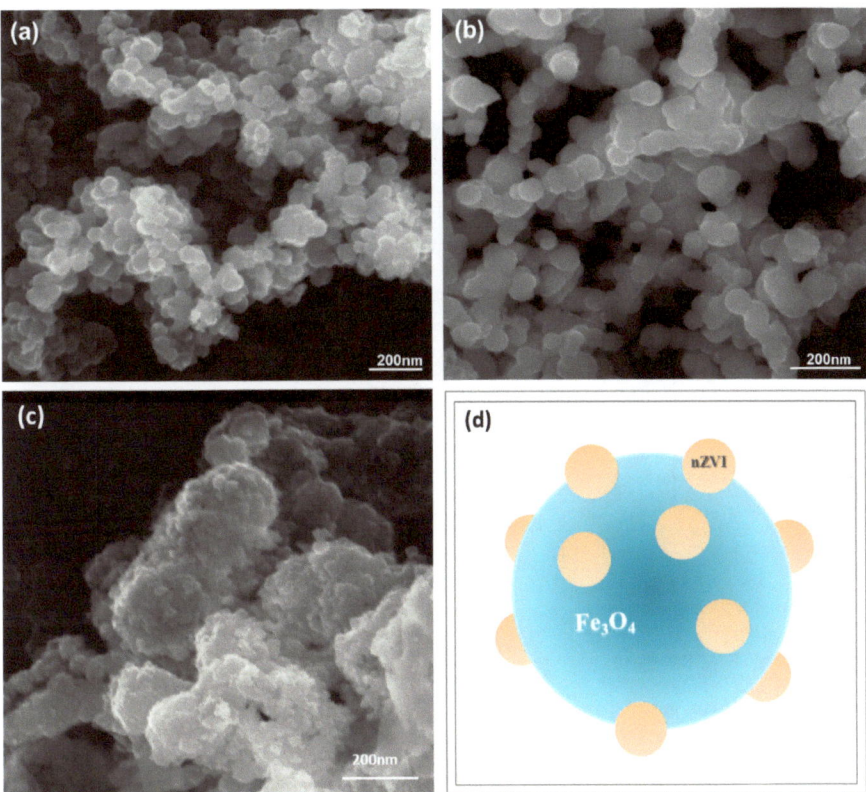

Figure 1. SEM images of (**a**) nano zero-valent iron (nZVI), (**b**) the original $Fe^0@Fe_3O_4$ NPs, and (**c**) the $Fe^0@Fe_3O_4$ NPs after reaction, and (**d**) the conceptual model of the $Fe^0@Fe_3O_4$ NPs.

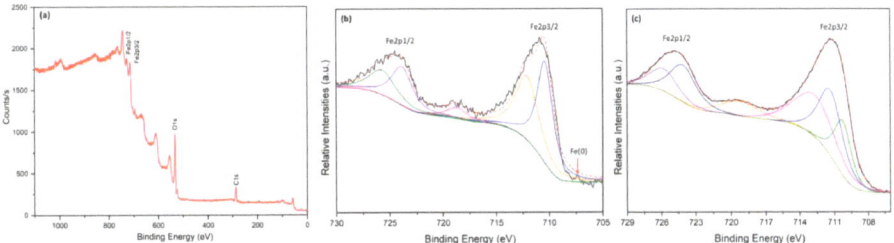

Figure 2. (**a**) Full spectrum X-ray photoelectron patterns of the $Fe^0@Fe_3O_4$ NPs; narrow region X-ray photoelectron patterns of Fe2p for the $Fe^0@Fe_3O_4$ NPs: (**b**) original (**c**) after reaction.

3.2. Batch Experiments

3.2.1. Effect of Initial pH

The pH value is an important factor that can have significant effects on the degradation of organic pollutants, as well as the form of iron and hydrogen peroxide in Fenton and Fenton-like systems. Removal of 2,4-D under different pH values varying from 3.0 to 8.0 were investigated, and the results were shown in Figure 3. Under the conditions of pH = 3.0, 4.5, and 5.0, no 2,4-D was left in the solution after 90 min, but when it continued to increase to 5.5 and above, the pollutant removal efficiency decreased rapidly; the final reduction of 2,4-D was only 11% at pH = 8.0.

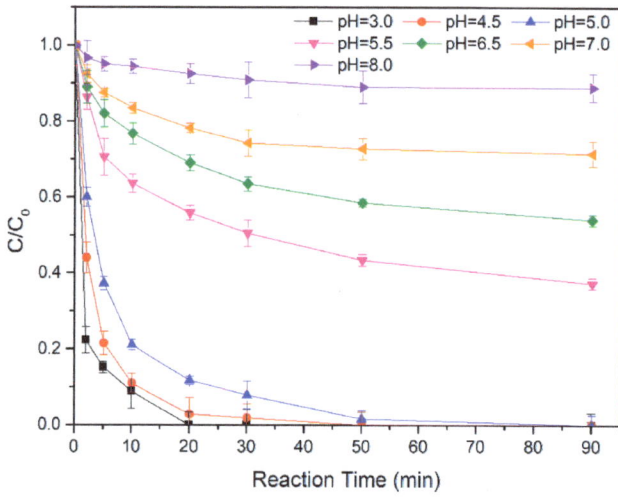

Figure 3. Effect of pH on 2,4-dichlorophenoxyacetic acid (2,4-D) degradation ($[2,4\text{-}D]_0$ = 5.0 mg/L, $[Fe^0@Fe_3O_4]_0$ = 0.5 g/L, H_2O_2 = 1 mM, T = 30 °C).

Decreasing pH can gradually accelerate the removal of target pollutants, while a higher pH value may lead to the formation of iron oxyhydroxide and the precipitation of iron hydroxide, which will reduce the activity of the Fenton reagent. Under alkaline conditions, the reduction of ferrous/ferric ions will in turn restrict the generation of hydroxyl radicals; the oxidation potential of the hydroxyl radicals will also decrease with the increasing pH value. In addition, a higher pH will accelerate the decomposition of hydrogen peroxide into oxygen and water. All the above reasons are unfavorable for the efficient degradation of 2,4-D.

The optimal pH for the traditional Fenton process was found to be around 3 as a pH above 3 could significantly reduce the degradation efficiency of the target pollutants [27]. Meanwhile, if the pH value of the solution is too low, there will be a large amount of $[Fe(H_2O)_6]^{2+}$ existing in the system, which has more difficulty reacting with hydrogen peroxide when compared with the other iron substances [28]. At the same time, the reaction rate between hydrogen ions and hydroxyl radicals will also be greatly accelerated (Equation (4)). In addition, a high concentration of hydrogen ions will cause the dissolution of peroxide to form the hydronium ion, which is much more stable and will make it very difficult for hydrogen peroxide to react with Fe^{2+}/Fe^{3+} and then influence the release of free radicals. Therefore, the appropriate range of pH is an important prerequisite to ensure efficient degradation of pollutants since the Fenton reaction may not proceed effectively when the solution pH is too high or too low.

$$OH + H^+ + e^- \rightarrow H_2O \tag{4}$$

However, in the range of pH = 3.0–5.5, removal of 2,4-D was maintained above 80%, and even under the condition of pH = 6.5, still nearly 50% of 2,4-D could be removed in 90 min. Compared with other studies [29,30], the inhibition along with the change of pH were not that obvious, indicating that the Fe^0@Fe_3O_4/H_2O_2 system in this study could adapt to a wider pH range.

Generally speaking, the acidic condition was more favorable to the reaction; but there are still research works pointing out that the degradation efficiency of pollutants was higher under neutral or even weakly alkaline conditions. For a different catalyst, the same pH value might lead to different results due to differences in the mechanism of their reaction. The effects of catalysts by releasing metal ions was greatly affected by pH; the increase of pH would slow down or even stop the dissolution of metal ions and make it inactivated through hydrolysis and precipitation; but those catalysts participating in the reaction through the active sites on their surface were more adaptable to pH change. Anyway, regardless of any catalytic system, the reaction pH needs to be controlled within the appropriate range. Considering the high efficiency of the Fe^0@Fe_3O_4/H_2O_2 system without pH adjusting, the follow-up experiments were conducted with no adjustment of pH (pH = 5.0 ± 0.2).

3.2.2. Effect of Hydrogen Peroxide Concentration

Hydrogen peroxide is the primary source of hydroxyl radicals and plays a vital role in the Fenton-like catalytic oxidation process. For different systems, its dosage changes according to the types and properties of the target pollutants and the catalyst. Experiments with different ranges of hydrogen peroxide were undertaken to determine the optimal concentration for Fe^0@Fe_3O_4. As shown in Figure 4, without the addition of the catalyst, hydrogen peroxide alone could hardly achieve 2,4-D degradation. Little hydrogen peroxide would result in insufficient release of the hydroxyl radical, which directly affects the degradation efficiency of organic contaminants, but in this study, a small amount of hydrogen peroxide could achieve a good removal of 2,4-D. At a concentration of 0.5 mM, eighty-two percent of 2,4-D was removed in 90 min; when the concentration of hydrogen peroxide increased to 1 mM, no 2,4-D remained in the solution at the end of the reaction. Further increasing the concentration to 2 mM and 3 mM could accelerate the removal of 2,4-D and shorten the reaction time to 50 min, but it is worth noting that although the time consumption was reduced, at the beginning of the reaction (especially within the first 10 min), for the condition of 2 mM and 3 mM, the decline of 2,4-D was slower than that of 1 mM, then gradually accelerated with time and came to an early end.

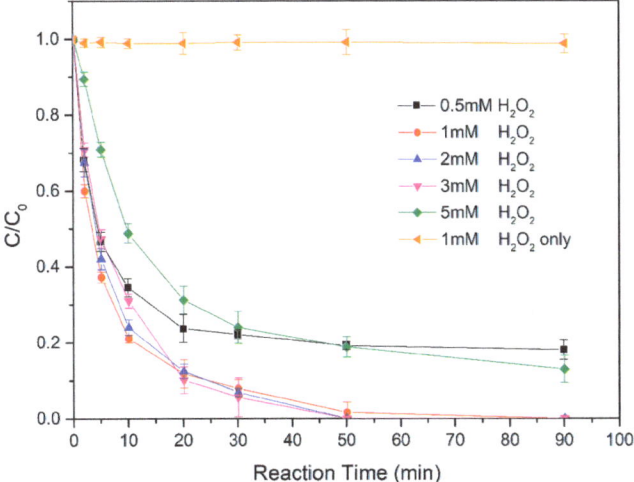

Figure 4. Effect of H_2O_2 concentration on 2,4-D degradation ([2,4-D]$_0$ = 5.0 mg/L, [Fe^0@Fe_3O_4]$_0$ = 0.5 g/L, pH = 5.0 ± 0.2, T = 30 °C).

The addition of too much hydrogen peroxide would cause a sharp increase of its concentration in solution; part of the hydrogen peroxide quickly decomposed into water and oxygen, while some of it could react with the generated free radicals [31,32]. As a result, the quantity of free radicals was reduced, and in turn, the degradation of 2,4-D was inhibited. When this excessive part of hydrogen peroxide was depleted, more free radicals could be released so that the removal rate of the pollutants increased.

However, when the dosage of hydrogen peroxide continued to increase to 5 mM, both the reaction rate and the final removal efficiency deteriorated accordingly. After 90 min, twelve percent of the contaminants were left in the solution. This may be due to the fact that when there was too much hydrogen peroxide existing in the solution, excess H_2O_2 would quickly eliminate the generated hydroxyl radicals through the following reactions (Equations (5) and (6)) and convert them to less active $HO_2\cdot$ or $O_2^-\cdot$ and greatly influence the treatment of 2,4-D.

In practical applications, excessive hydrogen peroxide will increase the operating cost and the unconsumed H_2O_2 will interfere with the Chemical Oxygen Demand (COD) measurement [33]. Besides, since Fenton/Fenton-like processes can be used as a pretreatment and are usually combined with other processes, the biotoxicity of hydrogen peroxide will significantly reduce the degradation efficiency of pollutants in some Fenton-biological combined oxidation treatment process.

$$H_2O_2 + OH \rightarrow H_2O + HO_2 \quad (5)$$

$$HO_2 \leftrightarrow O_2^- + H^+ \quad (6)$$

3.2.3. Effects of Temperature

The effects of temperature on the oxidative degradation of 2,4-D in the Fe^0@Fe_3O_4 heterogeneous Fenton system were investigated under 20 °C, 30 °C, 35 °C, 40 °C, and 45 °C. It can be seen from Figure 5a that in the range of 20 °C to 45 °C, as temperature rises, the degradation rate of 2,4-D increases continuously, while the reaction time consumed for the complete removal of target pollutant is gradually shortened; the reaction rate constant k_a increased from 4.58×10^{-2} min^{-1} to 3.65×10^{-1} min^{-1}, respectively. At the maximum temperature of 45 °C, it only took 30 min for all 5.0 mg/L 2,4-D in the solution to disappear. The rise of temperature could not only contribute to the corrosion of zero-valent iron and the generation of hydroxyl radicals, but also accelerate the movement of reactant molecules, increase the collision probability of 2,4-D molecules with free radicals, then enhance removal efficiency.

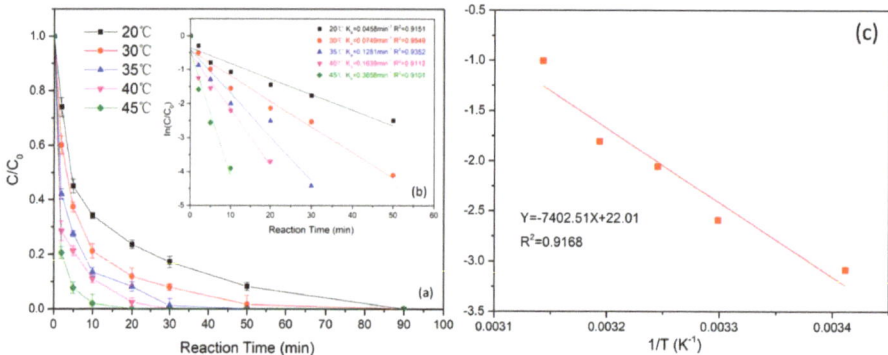

Figure 5. (a) Effect of temperature on 2,4-D degradation. (b) Dynamic simulation of 2,4-D degradation under different temperatures. (c) Fitting curve of ln(K)/T ([2,4-D]$_0$ = 5.0 mg/L, [Fe^0@Fe$_3$O$_4$]$_0$ = 0.5 g/L, H$_2$O$_2$ = 1 mM, pH = 5.0 ± 0.2).

The thermodynamic analysis was conducted for a further study of the temperature effects on the oxidative degradation process of 2,4-D. According to the Arrhenius equation, the functional relationship between temperature and rate constants could be described as follows [34,35]:

$$k = A \times e^{-\frac{E_a}{RT}} \tag{7}$$

where k is the measured pseudo-first-order rate constant, E_a is the activation energy, A is a frequency factor, R is the universal gas constant (8.314 Jmol^{-1}K^{-1}), and T is the temperature (K). Integrating Equation (7) resulted in:

$$\ln k = -\frac{E_a}{R} \times \frac{1}{T} + \ln A \tag{8}$$

Therefore, defining the k_a obtained from Figure 5b as k, a plot of lnk versus 1/T resulted in a linear relationship with the slope and the intercept equal to $-E_a/R$ and lnA as illustrated in Figure 5c. In this study, we investigated the activation energy for the removal of 2,4-D by Fe^0@Fe$_3$O$_4$/H$_2$O$_2$ at different temperatures ranging from 20 °C to 45 °C, and the value was calculated to be 61.55 KJ/mol, which was greater than that of diffusion-controlled reactions (~29 kJ/mol). It could be inferred that this process was a surface-mediated reaction; the rate-limiting step of 2,4-D degradation was not diffusion, but surface-chemical reaction [36]. Since the optimal temperature of the Fenton-like process is usually controlled at about 25–30 °C, excessive temperature will accelerate the decomposition of hydrogen peroxide into water and oxygen, thereby reducing removal efficiency. Considering the cost and operation convenience, thirty degrees Celsius was chosen as the experimental temperature in this study.

3.2.4. Effect of Fe^0@Fe$_3$O$_4$ Dosage

Without adding hydrogen peroxide, about 20% of 2,4-D could be removed in 90 min, which was due to the adsorption and the reductive degradation by Fe^0@Fe$_3$O$_4$ particles. Dechlorination was the main approach for 2,4-D reducing, when there was no hydrogen peroxide added in solution, but the removal rate of pollutants by Fe^0@Fe$_3$O$_4$ was much slower than that of the Fenton system. The results in Figure 6 show that when the concentration of hydrogen peroxide was 1 mM, the removal rate of 2,4-D gradually increased with the increase of catalyst dosage, and even a small amount of Fe^0@Fe$_3$O$_4$ nanoparticles could achieve a good removal efficiency. Under the condition of 0.1 g/L and 0.3 g/L, the 2,4-D removal efficiencies were greater than 70%, but there were still contaminants remaining in the solution at the end (29% and 16%, respectively). nZVI in the catalytic particles was the main source of Fe^{2+} in this reaction system, and insufficient ferrous ion would lead to less production of free radicals, which directly affected the degradation of 2,4-D. In addition, a proper dosage of Fe^0@Fe$_3$O$_4$ was the guarantee of sufficient active sites for reaction.

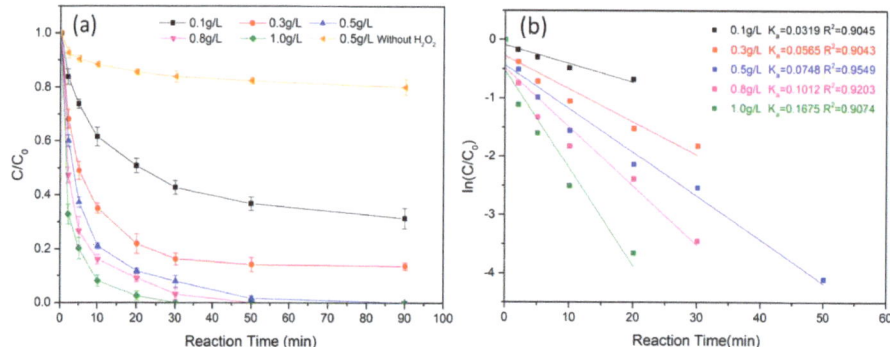

Figure 6. (a) Effect of catalyst dosage on 2,4-D degradation. (b) Dynamic simulation of 2,4-D degradation under different catalyst dosages ([2,4-D]$_0$ = 5.0 mg/L, H$_2$O$_2$ = 1 mM, pH = 5.0 ± 0.2, T = 30 °C).

When the dosage of catalyst continuously increased to 0.5 g/L and above, the pollutants could be completely removed, and the required reaction time was gradually shortened. By calculating the reaction rate constant, it could be seen that as the dosage increased from 0.1 g/L to 0.5 g/L, K$_a$ increased by 4.29 × 10^{-2} min^{-1}, while the increase from 0.5 g/L to 0.8 g/L and 1.0 g/L led to K$_a$ changes of 2.64 × 10^{-2} min^{-1} and 6.53 × 10^{-2} min^{-1}, respectively. It was inferred that, when the dosage of Fe^0@Fe$_3$O$_4$ exceeded 0.5 g/L, the increasing of the reaction rate would gradually slow down. Within a proper range, increasing the amount of catalyst could accelerate the removal of pollutants, but a large amount of Fe^0@Fe$_3$O$_4$ might cause over release of Fe^{2+}/Fe^{3+} to react with hydroxyl radicals, leading to the consumption of active ingredients (Equations (9) and (10)) [37]. At the same time, excessive dosing would cause the waste of catalysts and increase the operating costs in practical applications.

$$Fe^{2+} + OH \rightarrow Fe^{3+} + OH^- \tag{9}$$

$$Fe^{3+} + H_2O_2 \rightarrow FeOOH^{2+} + H^+ \tag{10}$$

3.3. Comparison between Different Catalysts on the Degradation of 2,4-D

To investigate the role of each component in the Fe^0@Fe$_3$O$_4$/H$_2$O$_2$ heterogeneous Fenton system and have a better understanding of its promotion on 2,4-D removal, the performances of different systems were compared. From high to low, the removal efficiency of 2,4-D in different reaction systems was Fe^0@Fe$_3$O$_4$/H$_2$O$_2$ > nZVI/H$_2$O$_2$ > Fe$_3$O$_4$/H$_2$O$_2$ > Fe^0@Fe$_3$O$_4$ > H$_2$O$_2$. As shown in Figure 7, one-hundred percent of 2,4-D could be removed by the combination of Fe^0@Fe$_3$O$_4$ and H$_2$O$_2$ within 90min. Without the catalysis of Fe^0@Fe$_3$O$_4$, few hydroxyl radicals could be generated from the natural decomposition of hydrogen peroxide; simply adding hydrogen peroxide almost had no ability to degrade the pollutants. A shortage of free radicals led to the poor degradation effect of 2,4-D in the H$_2$O$_2$-only system. The Fe$_3$O$_4$/H$_2$O$_2$ system achieved a removal efficiency of 52.8% in 90 min, and without the addition of hydrogen peroxide, Fe^0@Fe$_3$O$_4$ itself reduced about 20% of 2,4-D through reduction and adsorption.

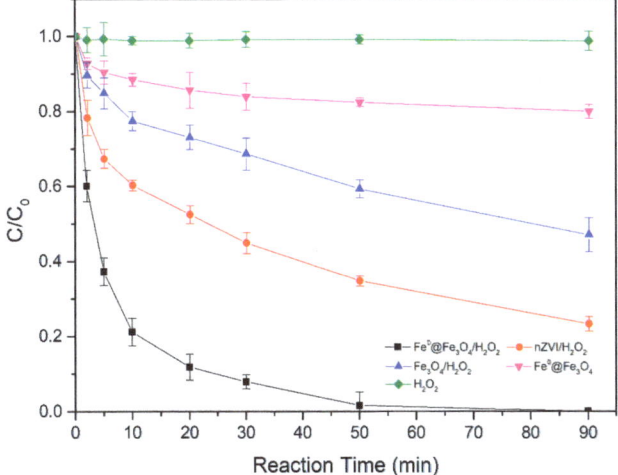

Figure 7. Comparison between different catalysts for the degradation of 2,4-D ([2,4-D]$_0$ = 5.0 mg/L, [catalyst]$_0$ = 0.5 g/L, H$_2$O$_2$ = 1 mM, pH = 5.0 ± 0.2, T = 30 °C).

Although the reductive ability of the nZVI/H$_2$O$_2$ system was not as good as that of the Fe^0@Fe$_3$O$_4$/H$_2$O$_2$ system, the removal of 2,4-D still reached 76.7% in the same duration, and there was still a trend for further reduction. In this system, the degradation of 2,4-D was mainly attributed to hydroxyl radicals released from the reaction of hydrogen peroxide and ferrous ions produced by the oxidation of nZVI. However, as mentioned above, pure nZVI particles were prone to agglomeration, and that would in turn reduce their reactivity. On the one hand, the introduction of Fe$_3$O$_4$ as a supporter could avoid the agglomeration of nZVI and increase the specific surface area of catalyst particles. On the other hand, as an active substance, Fe$_3$O$_4$ itself would participate in the Fenton reaction by releasing structural Fe^{2+}/Fe^{3+}, increasing the amount of free radicals, and then, improving the degradation effect. Nevertheless, because of the slow release rate of iron ions from the internal structure, its contribution to improving the degradation efficiency of target pollutants was relatively limited. Fe^0@Fe$_3$O$_4$ could reduce the concentration of 2,4-D through reductive dechlorination. However, as a highly chlorinated compound, the complex structure of 2,4-D made it more difficult to be degraded, and its reduction process would be much slower.

In summary, the high speed removal of 2,4-D in the Fe^0@Fe$_3$O$_4$/H$_2$O$_2$ system was dominated by nZVI, which could release ferrous ions to promote the decomposition of hydrogen peroxide to generate free radicals; Fe$_3$O$_4$ not only could increase the surface area and improve the dispersion and stability of particles, but also could take part in the degradation processes of pollutants by releasing ferrous and ferric ions from the inner structure. The synergistic effect of nZVI and Fe$_3$O$_4$ made the 2,4-D degradation efficiency in the Fe^0@Fe$_3$O$_4$/H$_2$O$_2$ system greater than that either of them alone.

3.4. Degradation Products and Mechanism of 2,4-D in the Fe^0@Fe$_3$O$_4$/H$_2$O$_2$ System

3.4.1. Changes of Ferrous/Ferric Ions Concentration

As demonstrated in Figure 8, the concentration of Fe^{2+} in the Fe^0@Fe$_3$O$_4$/H$_2$O$_2$ system increased from 0 to 0.31 mg/L in the first 10 min and then gradually decreased; a second increase appeared at about 30 min followed by a tendency to decline. Throughout, the ferrous curve floated in a concentration range; even the top one reached only 0.171 mg/L. This might be because the existence of the oxide layer on the surface of the catalyst, which would limit the dissolution of nZVI and slow down the release of Fe^{2+} at the beginning of the reaction. As the oxide layer gradually dissolved and fell off, fresh nZVI was exposed in the solution, and the generation of ferrous iron could also be accelerated. However, the released Fe^{2+} would quickly react with hydrogen peroxide to convert to Fe^{3+}, resulting in the low ferrous concentration in solution.

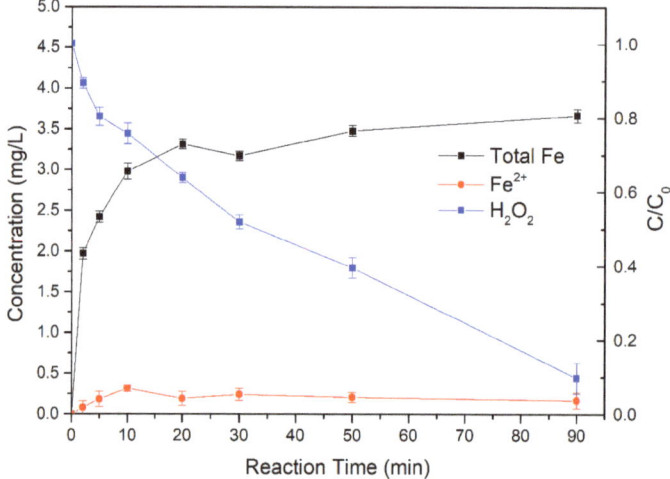

Figure 8. Changes of Fe^{2+} and Fe^{3+} concentrations in the Fe^0@Fe_3O_4/H_2O_2 system ([2,4-D]$_0$ = 5.0 mg/L, [Fe^0@Fe_3O_4]$_0$ = 0.5 g/L, H_2O_2 = 1 mM, pH = 5.0 ± 0.2, T = 30 °C).

The concentration of total iron rapidly increased to 3.3 mg/L in the first 20 min and then decreased slightly; finally, it remained at 3.66 mg/L at the end of the reaction. Ferrous ions released during the reaction of nZVI and hydrogen peroxide were quickly consumed and converted into ferric iron, leading to the rapid rise of the total iron concentration. Meanwhile, the resultant Fe^{3+} could react with Fe^0 and be transferred into Fe^{2+} again, realizing the cycle of Fe^{2+}/Fe^{3+}. In addition, the precipitation of Fe^{3+} would cause a decrease in the total iron concentration. According to the concentration curve of hydrogen peroxide, it could be seen that the decomposition rate of hydrogen peroxide was very fast in the first 30 min and then gradually slowed down, which reflected that the reaction between Fe^0 and Fe^{2+} with hydrogen peroxide was also changed from rapid to slow, which was consistent with the concentration change of total iron and the degradation of 2,4-D. After 90 min, the residual hydrogen peroxide in the solution was less than 10%, indicating that Fe^0@Fe_3O_4 nanoparticles could efficiently catalyze the decomposition of the hydrogen peroxide.

3.4.2. Identification of the Predominant Radical Species Generated in the Fe^0@Fe_3O_4/H_2O_2 System

In order to find the major active species responsible for the degradation of 2,4-D and have a better understanding of the mechanism in the Fe^0@Fe_3O_4/H_2O_2 heterogeneous Fenton system, KI and n-butanol were selected as free radical quenching agents in this study to eliminate the hydroxyl radicals. In the heterogeneous Fenton process, there were surface-bound hydroxyl radicals and free hydroxyl radicals in solution. KI mainly reacted with the former ones, while excessive n-butanol could quench both of them [38,39]. It can be seen from Figure 9 that the addition of KI and n-butanol had a significant inhibitory effect on the degradation of 2,4-D. Compared with the condition without quenching agents, the removal efficiency of the target pollutants was reduced from 100% to 23% and 10%, respectively, confirming the fact that the oxidative degradation of 2,4-D mainly depends on hydroxyl radicals' oxidation. It was calculated that only about 13% of the target pollutant removal could be attributed to the free hydroxyl radicals in solution, and the rest of 2,4-D (about 77%) was reduced by surface-bound hydroxyl radicals. Meanwhile, without the existence of two types of hydroxyl radicals, there was still a 23% removal of the target pollutant be achieved, indicating that apart from the adsorption of the

Fe⁰@Fe₃O₄ particles, other active ingredients with high redox capacity like active [H] might exist in the Fe⁰@Fe₃O₄/H₂O₂ system, which could also degrade a small amount of 2,4-D.

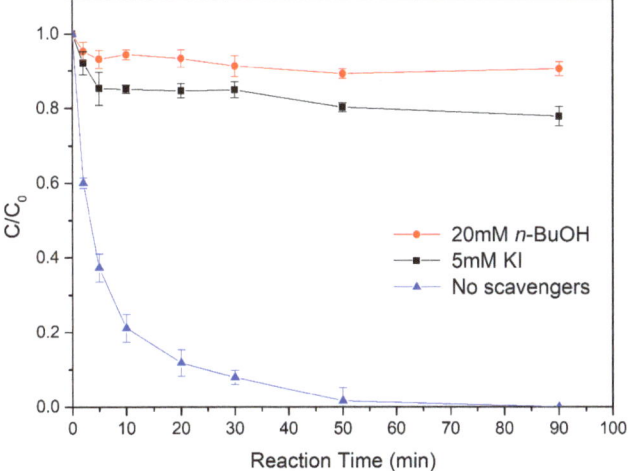

Figure 9. Effect of radical scavengers on the degradation of 2,4-D in the Fe⁰@Fe₃O₄/H₂O₂ system ([2,4-D]$_0$ = 5.0 mg/L, [Fe⁰@Fe₃O₄]$_0$ = 0.5 g/L, H$_2$O$_2$ = 1 mM, pH = 5.0 ± 0.2, T = 30 °C).

3.4.3. Degradation Products and Mineralization of 2,4-D

Changes of total organic carbon (TOC) and chloride ion concentration during the degradation of 2,4-D in the Fe⁰@Fe₃O₄/H₂O₂ system were detected to study the mineralization and dechlorination of pollutants and shown in Figure 10. Unlike the rapid reduction of 2,4-D, the decreasing of TOC tended to be slower. In the end, no 2,4-D was detected in the solution, while 34% of TOC remained. It can be inferred that although 2,4-D could be quickly degraded and removed in this system, it had not been completely mineralized. This might be due to the formation and remaining refractory organic substances in the solution. At the same time, the final concentration of Cl⁻ (0.038 mM) was lower than the theoretical calculated value of complete dechlorination (0.046 mM), suggesting that there were other chlorine-containing intermediate products generated in the degradation process of the pollutants, which were difficult to be effectively degraded or required a longer reaction time. Although this heterogeneous Fenton system did not completely mineralize 2,4-D within 90 min, the elimination of TOC also exceeded 66%, and it was observed that the TOC decreasing curve had a downward trend, indicating the potential for further mineralization in this system.

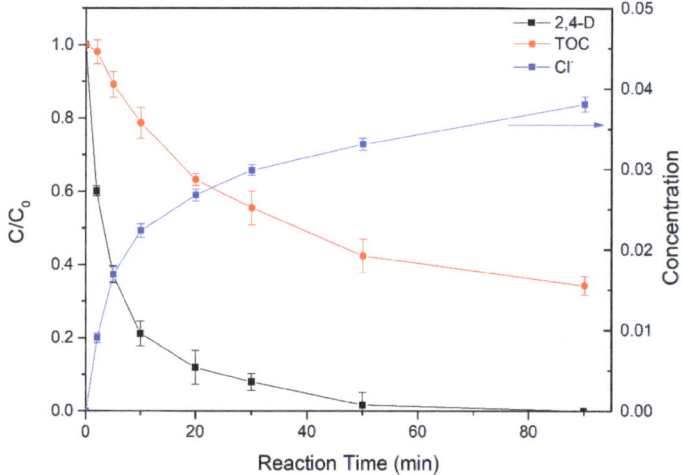

Figure 10. Changes of TOC and Cl⁻ concentrations with the degradation of 2,4-D ([2,4-D]$_0$ = 5.0 mg/L, [Fe^0@Fe$_3$O$_4$]$_0$ = 0.5 g/L, H$_2$O$_2$ = 1 mM, pH = 5.0 ± 0.2, T = 30 °C).

GC-MS was used to investigate the intermediate products formed during the degradation of 2,4-D in the Fe^0@Fe$_3$O$_4$/H$_2$O$_2$ system. Figure 11 depicts the main degradation products in the solution after 20 min. By comparing the peak value and response intensity of the products, it could be seen that 2,4-dichlorophenol (2,4-DCP) was one of the most important degradation products of 2,4-D. In addition, a certain concentration of 2,6-dichlorophenol (2,6-DCP) and some small-molecule organic acids such as formic acid, acetic acid, glycolic acid, and glyoxylic acid were generated during the reaction. Meanwhile, a small amount of 4,6-dichlororesorcinol (4,6-DCR), 2-chlorohydroquinone (2-CHQ), and 2-chloro-1,4-dichlorobenzoquinone (2-CBQ) was generated.

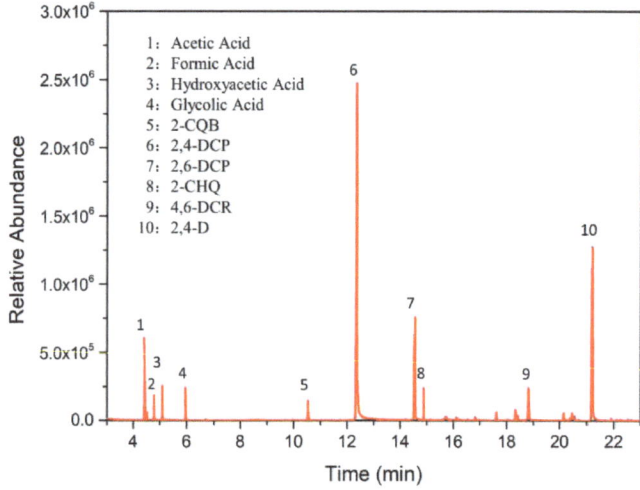

Figure 11. GC-MS spectra of intermediate products formed in 2,4-D degradation ([2,4-D]$_0$ = 5.0 mg/L, [Fe^0@Fe$_3$O$_4$]$_0$ = 0.5 g/L, H$_2$O$_2$ = 1 mM, pH = 5.0 ± 0.2, T = 30 °C). 2-CBQ, 2-chloro-1,4-dichlorobenzoquinone; 2,4-DCP, 2,4-dichlorophenol; 4,6-DCR, 4,6-dichlororesorcinol.

With the degradation of 2,4-D, 2,4-DCP gradually formed and accumulated in the solution. This could be further degraded into 4,6-DCR and 2-CHQ through the electrophilic attack and dechlorinative substitution of the hydroxyl radical. 4,6-DCR was subsequently decomposed into small molecule organic acids through oxidation or dechlorination, while 2-CHQ would be hydroxylated and converted into 2-CBQ, then become small molecule organic acid in the same way as 4,6-DCR. Furthermore, through a series of reactions, oxalic acid and formic acid could be generated from the degradation of glycolic acid. All the small molecule organic acids formed above could be decomposed into H_2O and CO_2, completing the mineralization of the target pollutant.

3.4.4. Mechanism of 2,4-D Degradation in the $Fe^0@Fe_3O_4/H_2O_2$ System

In the $Fe^0@Fe_3O_4/H_2O_2$ system, it was the generated hydroxyl radicals (surface-bound hydroxyl radicals and free hydroxyl radicals) that were mainly attributed to the degradation of 2,4-D, and their formation could be realized by the decomposition of hydrogen peroxide, which was motivated by the surface-bound $\equiv Fe(II)$ and the free Fe^{2+} ions released from the oxidation of zero-valent iron. As shown in Equations (11)–(17), it was possible that there were two types of interface mechanisms in this system. One was the homogeneous Fenton reaction initiated by dissolved Fe^{2+}/Fe^{3+} in solution, and the other one was the heterogeneous Fenton reaction involving the $\equiv Fe(II)$ on the surface of the catalyst particle [12,13,40].

$$Fe^0 + 2H^+ \rightarrow Fe^{2+} / \equiv Fe(II) + H_2 \qquad (11)$$

$$Fe^0 + O_2 + 2H^+ \rightarrow Fe^{2+} / \equiv Fe(II) + H_2O_2 \qquad (12)$$

$$Fe^0 + H_2O_2 \rightarrow Fe^{2+} / \equiv Fe(II) + OH + OH^- \qquad (13)$$

$$\equiv Fe(II) + H_2O_2 \rightarrow \equiv Fe(III) - OH + OH \qquad (14)$$

$$Fe^{2+} + H_2O_2 \rightarrow Fe^{3+} + OH^- + OH \qquad (15)$$

$$\equiv Fe(III) + H_2O_2 \rightarrow \equiv Fe(II) + H^+ + HO_2 \qquad (16)$$

$$Fe^{3+} + H_2O_2 \rightarrow Fe^{2+} + H^+ + HO_2 \qquad (17)$$

Both pathways could generate hydroxyl radicals for the effective degradation of 2,4-D. However, in terms of the free radical inhibition experiments, the heterogeneous way accounted for a much larger proportion of the removal than the homogeneous one, and surface-bound hydroxyl radicals were the major active species in the system to remove 2,4-D. Due to the stability of most iron-based catalytic materials, the releasing of ferrous ions was limited. Free radicals involved in the oxidation of organic compounds were mainly generated through the heterogeneous Fenton reaction process. In summary, the two pathways of degrading 2,4-D in the $Fe^0@Fe_3O_4/H_2O_2$ system are shown in Figure 12.

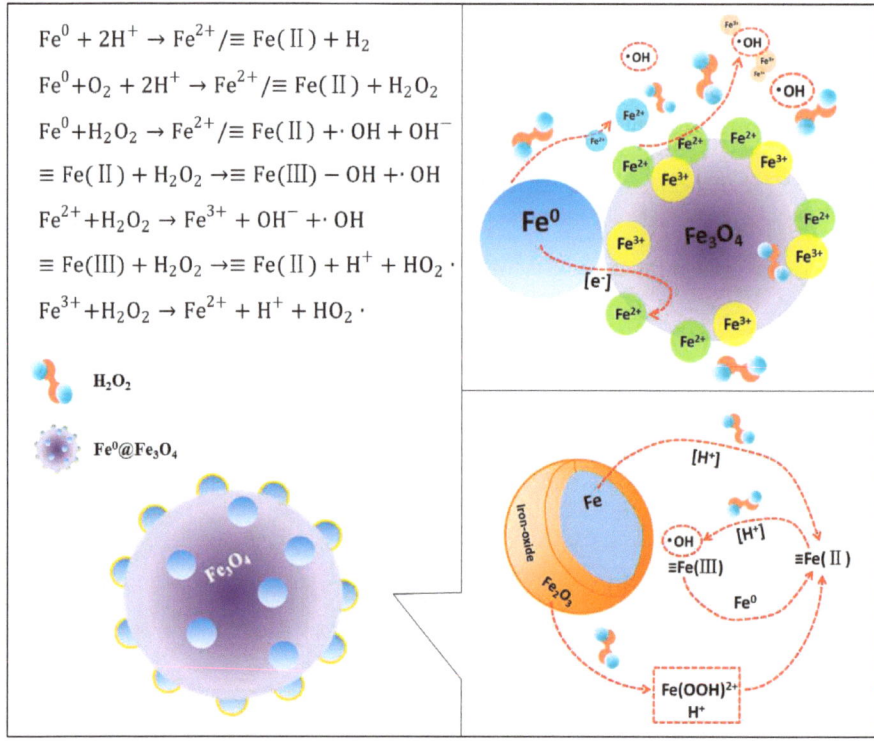

Figure 12. Possible mechanism for 2,4-D degradation in the Fe^0@Fe$_3$O$_4$/H$_2$O$_2$ system.

In the homogeneous Fenton process, nZVI oxidized and released Fe^{2+} into the solution, catalyzing the decomposition of hydrogen peroxide to generate hydroxyl radicals for 2,4-D degradation; in the heterogeneous Fenton-like process, nZVI particles were oxidized to form ≡Fe(II) through electron transfer. Hydrogen peroxide diffusing from the solution to the surface of the Fe^0@Fe$_3$O$_4$ particles would be catalyzed by ≡Fe(II) to generate surface-bound hydroxyl radicals. Simultaneously, the adsorption of the catalyst made 2,4-D accumulate on its surface and react with the generated hydroxyl radicals at the active sites to complete the degradation, then the degradation products again diffused back into the solution from the solid phase interface. It is worth noting that the Fe$_3$O$_4$ molecule has a special reverse spinel structure, which makes it possible for its structural Fe^{2+} and Fe^{3+} at the octahedral site to transpose, so it is conducive to the internal oxidation-reduction reaction of Fe^{2+}/Fe^{3+}. The electron transfer between ferric ions and ferrous ions endowed Fe$_3$O$_4$ with unique electromagnetic properties, and Fe$_3$O$_4$ could participate in the Fenton-like reaction through releasing of structural Fe^{2+}/Fe^{3+}, motivating the decomposition of hydrogen peroxide and the formation of free radicals, thereby enhancing the 2,4-D removal effect. However, compared with that of nZVI particles, the dissolution of magnetite was much slower and usually consumed more hydrogen peroxide and more time when Fe$_3$O$_4$ participated in the Fenton reaction as a catalyst alone. At the same time, under acidic conditions, there was a redox of Fe^{2+} (≡Fe(II)) and Fe^{3+} (≡Fe(III)), which could accelerate the regeneration of ferrous ions and was beneficial to the decomposition of hydrogen peroxide and the generation of free radical [41].

4. Conclusions

Fe^0@Fe_3O_4 NPs was synthesized and successfully applied in the reduction of 2,4-D by constructing a heterogeneous Fenton-like system with hydrogen peroxide. It could adapt to a pH range from 3.0 to 6.5, which was wider than the traditional Fenton process. The introduction of magnetite overcame the aggregation of nZVI and accelerated electron transfer. The modified catalyst with a better dispersity than single nZVI could remove 5 mg/L of 2,4-D effectively in 90 min. Two possible interface mechanisms might exist in the Fe^0@Fe_3O_4/H_2O_2 system: one is the homogeneous Fenton reaction initiated by dissolved Fe^{2+}/Fe^{3+} in solution, and the other one is the heterogeneous Fenton reaction involving the ≡Fe (II) on the surface of the catalyst particle. The hydroxyl radicals (especially the surface-bound one) generated during both of them could achieve the oxidative degradation and mineralization of 2,4-D effectively. This research put forward an easily available modified catalyst with strong adaptability for the rapid remediation of highly-chlorinated organic compounds including 2,4-D.

Author Contributions: Conceptualization, X.L.; methodology, X.L. and Y.M.; validation, X.L. and Y.M.; formal analysis, X.L.; investigation, X.L. and Y.L.; data curation, X.L.; writing, original draft preparation, X.L.; supervision, Q.Y. All authors read and agreed to the published version of the manuscript.

Funding: This work was funded by the Major Science and Technology Program for Water Pollution Control and Treatment of China (2015ZX07406005-001) and the Fundamental Research Funds for the Central Universities (2652018205).

Conflicts of Interest: The authors declare no conflict of interest. The funders had no role in the design of the study; in the collection, analyses, or interpretation of data; in the writing of the manuscript; nor in the decision to publish the results.

References

1. Boivin, A.; Amellal, S.; Schiavon, M.; Van Genuchten, M.T. 2,4-Dichlorophenoxyacetic acid (2,4-D) sorption and degradation dynamics in three agricultural soils. *Environ. Pollut.* **2005**, *138*, 92–99. [CrossRef] [PubMed]
2. Song, Y. Insight into the mode of action of 2, 4-dechlorophenoxyacetic acid (2,4-D) as an herbicide. *J. Integr. Plant Biol.* **2014**, *56*, 106–113. [CrossRef]
3. Garabrant, D.H.; Philbert, M.A. Review of 2,4-Dichlorophenoxyacetic Acid (2,4-D) Epidemiology and Toxicology. *Crit. Rev. Toxicol.* **2002**, *32*, 233–257. [CrossRef]
4. Chu, W.; Kwan, C.; Chan, K.; Chong, C. An unconventional approach to studying the reaction kinetics of the Fenton's oxidation of 2,4-dichlorophenoxyacetic acid. *Chemosphere* **2004**, *57*, 1165–1171. [CrossRef] [PubMed]
5. Atamaniuk, T.M.; Kubrak, O.I.; Storey, K.B.; Lushchak, V.I. Oxidative stress as a mechanism for toxicity of 2,4-dichlorophenoxyacetic acid (2,4-D): Studies with goldfish gills. *Ecotoxicology* **2013**, *22*, 1498–1508. [CrossRef] [PubMed]
6. Brillas, E. Mineralization of 2,4-D by advanced electrochemical oxidation processes. *Water Res.* **2000**, *34*, 2253–2262. [CrossRef]
7. Lee, C.; Keenan, C.R.; Sedlak, D.L. Polyoxometalate-Enhanced Oxidation of Organic Compounds by Nanoparticulate Zero-Valent Iron and Ferrous Ion in the Presence of Oxygen. *Environ. Sci. Technol.* **2008**, *42*, 4921–4926. [CrossRef]
8. Babuponnusami, A.; Muthukumar, K. A review on Fenton and improvements to the Fenton process for wastewater treatment. *J. Environ. Chem. Eng.* **2014**, *2*, 557–572. [CrossRef]
9. Pignatello, J.J.; Oliveros, E.; Mackay, A. Advanced Oxidation Processes for Organic Contaminant Destruction Based on the Fenton Reaction and Related Chemistry. *Crit. Rev. Environ. Sci. Technol.* **2006**, *36*, 1–84. [CrossRef]
10. Benatti, C.T.; da Costa, A.C.S.; Tavares, C.R.G. Characterization of solids originating from the Fenton's process. *J. Hazard. Mater.* **2009**, *163*, 1246–1253.
11. Chu, L.; Wang, J.; Dong, J.; Liu, H.; Sun, X. Treatment of coking wastewater by an advanced Fenton oxidation process using iron powder and hydrogen peroxide. *Chemosphere* **2012**, *86*, 409–414. [CrossRef] [PubMed]
12. Joo, S.H.; Feitz, A.J.; Sedlak, D.L.; Waite, T.D. Quantification of the Oxidizing Capacity of Nanoparticulate Zero-Valent Iron. *Environ. Sci. Technol.* **2005**, *39*, 1263–1268. [CrossRef] [PubMed]

13. Xu, L.; Wang, J.L. A heterogeneous Fenton-like system with nanoparticulate zero-valent iron for removal of 4-chloro-3-methyl phenol. *J. Hazard. Mater.* **2011**, *186*, 256–264. [CrossRef]
14. Kim, Y.-H.; Carraway, E.R. Dechlorination of Pentachlorophenol by Zero Valent Iron and Modified Zero Valent Irons. *Environ. Sci. Technol.* **2000**, *34*, 2014–2017. [CrossRef]
15. Noubactep, C.; Carè, S. On nanoscale metallic iron for groundwater remediation. *J. Hazard. Mater.* **2010**, *182*, 923–927. [CrossRef]
16. Garbou, A.M.; Liu, M.; Zou, S.; Yestrebsky, C. Degradation kinetics of hexachlorobenzene over zero-valent magnesium/graphite in protic solvent system and modeling of degradation pathways using density functional theory. *Chemosphere* **2019**, *222*, 195–204. [CrossRef] [PubMed]
17. Kim, J.S.; Shea, P.J.; Yang, J.E.; Kim, J.-E. Halide salts accelerate degradation of high explosives by zerovalent iron. *Environ. Pollut.* **2007**, *147*, 634–641. [CrossRef]
18. Bae, S.; Lee, W. Influence of Riboflavin on Nanoscale Zero-Valent Iron Reactivity during the Degradation of Carbon Tetrachloride. *Environ. Sci. Technol.* **2014**, *48*, 2368–2376. [CrossRef]
19. Lv, X.; Li, H.; Ma, Y.; Yang, H.; Yang, Q. Degradation of Carbon Tetrachloride by nanoscale Zero-Valent Iron@ magnetic Fe3O4: Impact of reaction condition, Kinetics, Thermodynamics and Mechanism. *Appl. Organomet. Chem.* **2018**, *32*, 4139. [CrossRef]
20. Huang, R.; Fang, Z.; Yan, X.; Cheng, W. Heterogeneous sono-Fenton catalytic degradation of bisphenol A by Fe3O4 magnetic nanoparticles under neutral condition. *Chem. Eng. J.* **2012**, *197*, 242–249. [CrossRef]
21. Tan, L.; Lu, S.; Fang, Z.; Cheng, W.; Tsang, E.P. Enhanced reductive debromination and subsequent oxidative ring-opening of decabromodiphenyl ether by integrated catalyst of nZVI supported on magnetic Fe3O4 nanoparticles. *Appl. Catal. B Environ.* **2017**, *200*, 200–210. [CrossRef]
22. Chen, H.; Zhang, Z.; Feng, M.; Liu, W.; Wang, W.; Yang, Q.; Hu, Y. Degradation of 2,4-dichlorophenoxyacetic acid in water by persulfate activated with FeS (mackinawite). *Chem. Eng. J.* **2017**, *313*, 498–507. [CrossRef]
23. Lv, X.; Prastistho, W.; Yang, Q.; Tokoro, C. Application of nano-scale zero-valent iron adsorbed on magnetite nanoparticles for removal of carbon tetrachloride: Products and degradation pathway. *Appl. Organomet. Chem.* **2020**, *34*, e5592. [CrossRef]
24. Zhu, H.; Jia, Y.; Wu, X.; Wang, H. Removal of arsenic from water by supported nano zero-valent iron on activated carbon. *J. Hazard. Mater.* **2009**, *172*, 1591–1596. [CrossRef]
25. Fan, L.; Luo, C.; Sun, M.; Li, X.; Lu, F.; Qiu, H. Preparation of novel magnetic chitosan/graphene oxide composite as effective adsorbents toward methylene blue. *Bioresour. Technol.* **2012**, *114*, 703–706. [CrossRef] [PubMed]
26. Yamashita, T.; Hayes, P. Analysis of XPS spectra of Fe2+ and Fe3+ ions in oxide materials. *Appl. Surf. Sci.* **2008**, *254*, 2441–2449. [CrossRef]
27. Rivas, F.J.; Beltrán, F.J.; Frades, J.; Buxeda, P. Oxidation of p-hydroxybenzoic acid by Fenton's reagent. *Water Res.* **2001**, *35*, 387–396. [CrossRef]
28. Xu, X.R.; Li, X.Y.; Li, X.Z.; Li, H.B. Degradation of melatonin by UV, UV/H2O2, Fe2+/H2O2 and UV/Fe2+/H2O2 processes. *Sep. Purif. Technol.* **2009**, *68*, 261–266. [CrossRef]
29. Ma, Y.S.; Huang, S.T.; Lin, J.G. Degradation of 4-nitro phenol using the Fenton process. *Water Sci. Technol.* **2000**, *42*, 155–160.
30. Babuponnusami, A.; Muthukumar, K. Degradation of Phenol in Aqueous Solution by Fenton, Sono-Fenton and Sono-photo-Fenton Methods. *Clean-Soil Air Water* **2011**, *39*, 142–147. [CrossRef]
31. Kwan, W.P.; Voelker, B.M. Decomposition of Hydrogen Peroxide and Organic Compounds in the Presence of Dissolved Iron and Ferrihydrite. *Environ. Sci. Technol.* **2002**, *36*, 1467–1476. [CrossRef]
32. De La Plata, G.B.O.; Alfano, O.M.; Cassano, A.E. Decomposition of 2-chlorophenol employing goethite as Fenton catalyst II: Reaction kinetics of the heterogeneous Fenton and photo-Fenton mechanisms. *Appl. Catal. B Environ.* **2010**, *95*, 14–25. [CrossRef]
33. Lin, S.H.; Lo, C.C. Fenton process for treatment of desizing wastewater. *Water Res.* **1997**, *31*, 2050–2056. [CrossRef]
34. Liu, C.-C.; Tseng, D.-H.; Wang, C.-Y. Effects of ferrous ions on the reductive dechlorination of trichloroethylene by zero-valent iron. *J. Hazard. Mater.* **2006**, *136*, 706–713. [CrossRef] [PubMed]
35. Lookman, R.; Bastiaens, L.; Borremans, B.; Maesen, M.; Gemoets, J.; Diels, L. Batch-test study on the dechlorination of 1,1,1-trichloroethane in contaminated aquifer material by zero-valent iron. *J. Contam. Hydrol.* **2004**, *74*, 133–144. [CrossRef]

36. Zhou, T.; Li, Y.; Lim, T.-T. Catalytic hydrodechlorination of chlorophenols by Pd/Fe nanoparticles: Comparisons with other bimetallic systems, kinetics and mechanism. *Sep. Purif. Technol.* **2010**, *76*, 206–214. [CrossRef]
37. Wang, N.; Zheng, T.; Zhang, G.; Wang, P. A review on Fenton-like processes for organic wastewater treatment. *J. Environ. Chem. Eng.* **2016**, *4*, 762–787. [CrossRef]
38. Pham, A.L.-T.; Lee, C.; Doyle, F.M.; Sedlak, D.L. A Silica-Supported Iron Oxide Catalyst Capable of Activating Hydrogen Peroxide at Neutral pH Values. *Environ. Sci. Technol.* **2009**, *43*, 8930–8935. [CrossRef] [PubMed]
39. Navalón, S.; Alvaro, M.; Garcia, H. Heterogeneous Fenton catalysts based on clays, silicas and zeolites. *Appl. Catal. B Environ.* **2010**, *99*, 1–26. [CrossRef]
40. Sychev, A.Y.; Isak, V.G. Iron compounds and the mechanisms of the homogeneous catalysis of the activation of O2 and H2O2 and of the oxidation of organic substrates. *Russ. Chem. Rev.* **1995**, *64*, 1105–1129. [CrossRef]
41. Wang, H.; Liang, H.S.; Chang, M.B. Chlorobenzene oxidation using ozone over iron oxide and manganese oxide catalysts. *J. Hazard. Mater.* **2011**, *186*, 1781–1787. [CrossRef]

Publisher's Note: MDPI stays neutral with regard to jurisdictional claims in published maps and institutional affiliations.

© 2020 by the authors. Licensee MDPI, Basel, Switzerland. This article is an open access article distributed under the terms and conditions of the Creative Commons Attribution (CC BY) license (http://creativecommons.org/licenses/by/4.0/).

Article

Degradation of Ketamine and Methamphetamine by the UV/H₂O₂ System: Kinetics, Mechanisms and Comparison

De-Ming Gu [1,2], Chang-Sheng Guo [1], Qi-Yan Feng [2], Heng Zhang [1] and Jian Xu [1,*]

1. Center for Environmental Health Risk Assessment and Research, Chinese Research Academy of Environmental Sciences, Beijing 100012, China; goodmingaust@163.com (D.-M.G.); guocs@craes.org.cn (C.-S.G.); zhangheng_craes@163.com (H.Z.)
2. School of Environment Science and Spatial Informatics, China University of Mining and Technology, Xuzhou 221116, China; fqycumt@126.com
* Correspondence: xujian@craes.org.cn

Received: 31 August 2020; Accepted: 22 October 2020; Published: 26 October 2020

Abstract: The illegal use and low biodegradability of psychoactive substances has led to their introduction to the natural water environment, causing potential harm to ecosystems and human health. This paper compared the reaction kinetics and degradation mechanisms of ketamine (KET) and methamphetamine (METH) by UV/H₂O₂. Results indicated that the degradation of KET and METH using UV or H₂O₂ alone was negligible. UV/H₂O₂ had a strong synergizing effect, which could effectively remove 99% of KET and METH (100 μg/L) within 120 and 60 min, respectively. Their degradation was fully consistent with pseudo-first-order reaction kinetics ($R^2 > 0.99$). Based on competition kinetics, the rate constants of the hydroxyl radical with KET and METH were calculated to be 4.43×10^9 and 7.91×10^9 $M^{-1} \cdot s^{-1}$, respectively. The apparent rate constants of KET and METH increased respectively from 0.001 to 0.027 and 0.049 min^{-1} with the initial H₂O₂ dosage ranging from 0 to 1000 μM at pH 7. Their degradation was significantly inhibited by HCO_3^-, Cl^-, NO_3^- and humic acid, with Cl^- having relatively little effect on the degradation of KET. Ultraperformance liquid chromatography with tandem mass spectrometry was used to identify the reaction intermediates, based on which the possible degradation pathways were proposed. These promising results clearly demonstrated the potential of the UV/H₂O₂ process for the effective removal of KET and METH from contaminated wastewater.

Keywords: ketamine; methamphetamine; UV/H₂O₂; degradation kinetics; reaction intermediates

1. Introduction

Illicit drugs are nonprescribed or psychostimulant substances which cannot be completely removed by conventional wastewater treatment, resulting in their widespread occurrence in aquatic environments [1,2]. Ketamine (KET) and methamphetamine (METH) were detected most frequently, with concentration levels up to 275 ng/L for KET and 239 ng/L for METH, in surface waters in China [3]. METH removal at most wastewater treatment plants was more than 80%, while the elimination of KET was less than 50% or even negative [4]. It was confirmed that chronic environmental concentrations of METH can lead to health issues in aquatic organisms [5]. Liao et al. [6] also reported that blood circulation and incubation time in medaka fish embryos could be significantly delayed at environmental concentration levels (0.004–40 μM) of KET and METH, which altered the swimming behavior of medaka fish larvae. Thus, there is an urgent need to explore new, efficient methods for eliminating these emerging contaminants in water.

Advanced oxidation processes (AOPs) have been employed to destroy illicit drugs due to their high efficiency and lower environmental impact [7,8]. The UV/H_2O_2 process is one of the AOPs and generates the strong, oxidizing hydroxyl radical (•OH, E_0 = 2.72 V), which attacks the organic compounds with rate constants ranging from 10^8 to 10^{10} M^{-1} s^{-1} [9]. Benzoylecgonine (BE), a metabolite of cocaine, was effectively removed by UV/H_2O_2 from different matrices [10]. The degradation of KET and METH was investigated using various AOPs, but no available report, so far, has addressed •OH assisted by UV/H_2O_2 treatment. After 3 min, 100 µg/L of METH that had been added to deionized water was completely eliminated by TiO_2 photocatalysis under UV_{365nm} irradiation [11]. Wei et al. [12] studied the synthesis of a novel sonocatalyst Er^{3+}:$YAlO_3$/Nb_2O_5 and its application for METH degradation. Gu et al. [13] observed that complete removal of KET was achieved by UV/persulfate, and possible transformation pathways were proposed.

To the best of our knowledge, there is little information about the theoretical calculation of the reactivity of KET and METH by radical attack using the UV/H_2O_2 process. Water constituents in actual wastewater could affect the degradation efficacy; therefore, a comprehensive understanding of the degradation of KET and METH using the UV/H_2O_2 system is needed. The aim of this study was to investigate the degradation kinetics and mechanisms of KET and METH during the UV/H_2O_2 process. The influence of various parameters on KET and METH removal was evaluated, including initial H_2O_2 dosage, pH and water background components. The degradation products were analyzed by ultraperformance liquid chromatography with tandem mass spectrometry (UPLC-MS/MS), and possible transformation paths were proposed.

2. Materials and Methods

2.1. Materials

The KET and METH were obtained from Cerilliant Corporation (Round Rock, TX); detailed information is listed in Table 1. HPLC grade acetonitrile (ACN) and methanol (MeOH) were purchased from Fisher Scientific (Poole, UK). Formic acid (FA, ≥98%) and benzoic acid (BA) were purchased from Sigma-Aldrich (Bellefonte, USA). Analytical grade H_2O_2 (30%, v/v), $NaHCO_3$ (≥99.7%), NaCl (≥99.0%), $NaNO_3$ (≥99.5%), NaOH (≥99.5%), humic acid (HA) and H_2SO_4 (≥98%) were obtained from Sinopharm Chemical Reagent Co., Ltd. (Beijing, China). All reaction solutions were configured with Milli-Q water produced by an ultrapure water system (Millipore, MA, USA).

Table 1. Chemical structures and properties of ketamine and methamphetamine.

Compound	Chemical Formula	Structure	CAS Number	pKa	Log K_{ow}
Ketamine	$C_{13}H_{16}ClNO$		6740-88-1	7.5	2.18
Methamphetamine	$C_{10}H_{15}N$		4846-07-5	9.9	2.07

2.2. Experimental Section

The experiments were operated in the quartz tubes (25 mm in diameter and 175 mm in length), which were placed in a photochemical reactor (Figure 1, XPA-7, Xujiang Machinery Factory, Nanjing, China). A low-pressure mercury lamp (11 W, emission at 254 nm, Philips Co., Zhuhai, China) was placed in the quartz sleeve. The UV lamp was preheated for 30 min to ensure irradiation stability. The UV fluence rate of 0.1 mW cm^{-2} was determined using three different methods [14]. The newly configured KET/METH and H_2O_2 stock solutions were supplemented with appropriate volumes to achieve a

50 mL reaction solution, which was then stirred thoroughly at 300 rpm with electromagnetic stirrers. Upon UV irradiation, the reaction started at pH 7.0 and room temperature. Specific samples were immediately quenched using a catalase and passed through 0.22 µm nylon filter before further analysis.

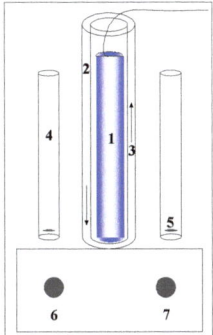

Figure 1. The schematic diagram of the experiment setup: (1) low-pressure Hg UV lamp, (2) quartz tube, (3) cooling water, (4) photoreactor, (5) magnetic stirrer, (6) magnetic stirrer apparatus, (7) thermostat.

2.3. Analytical Methods

The concentrations of KET and METH were quantified by UPLC-MS/MS equipped with a Waters Acquity liquid chromatography system and an Xevo T_QS triple quadrupole mass spectrometer (Waters Co., Milford, MA, USA). The analytes were separated by a reverse phase column (Acquity UPLC BEH C18, 1.7 µm, 50 × 2.1 mm, Waters, MA, USA). The mobile phases A and B, with a flow rate of 450 µL min^{-1}, were 0.1% FA in Milli-Q water and ACN, respectively. Ten percent of phase B was kept for 0.5 min at the initial proportion, linearly increased to 45% at 1.8 min, then increased to 95% within 0.1 min, held for 1.0 min, reverted to 10% at 3.0 min and held for 1.5 min. The injection volume was 5 µL with the column temperature at 40 °C. The chromatograms were recorded in the positive ion multiple reaction monitoring (MRM) mode. Nitrogen was used as the desolvation and nebulizing gas. The capillary voltage was set at 0.5 kV, and the desolvation temperature was 400 °C. Optimized UPLC-MS/MS parameters are given in Table 2.

Table 2. Detailed ultraperformance liquid chromatography with tandem mass spectrometry (UPLC-MS/MS) parameters for ketamine and methamphetamine.

Compound	Parent Ion (m/z)	Retention Time (min)	Production (m/z)	Cone Voltage (V)	Collision Voltage (V)
Ketamine	238	1.31	125	16	24
			179	16	16
Methamphetamine	150	1.11	91	22	16
			119	22	10

3. Results and Discussion

3.1. Degradation Kinetics of KET and METH

Figure 2 shows the degradation of KET and METH under different treatment processes. UV or H_2O_2 alone exhibited negligible effects on their degradation, suggesting that treatment by UV or H_2O_2 alone was unable to destroy KET and METH. However, nearly complete removal of KET and METH was achieved within 120 and 60 min, respectively, when treated with the combination of UV/H_2O_2. Similar results were reported regarding ofloxacin degradation, which was drastically increased due to the large amount of hydroxyl radicals (•OH) generated via the breakage of the H_2O_2

bond (Equation (1)) [15]. The degradation of KET and METH was consistent with the pseudo-first-order reaction kinetics. The apparent degradation rate constants (k_{obs}) of KET and METH by UV/H_2O_2 were 0.027 and 0.049 min^{-1}, respectively.

$$H - O - O - H + h\upsilon \rightarrow 2^\bullet OH \tag{1}$$

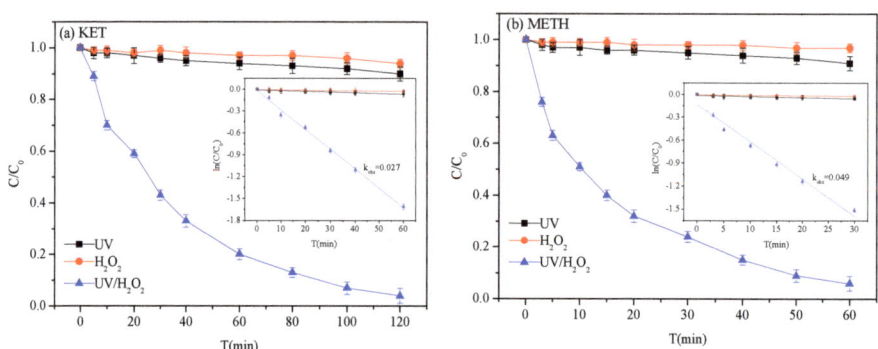

Figure 2. Degradation kinetics of ketamine (KET) (**a**) and methamphetamine (METH) (**b**) by different treatments. Conditions: Initial concentrations of KET and METH = 100 μg/L, Initial concentration of hydrogen peroxide ($H_2O_2)_0$ = 500 μM, pH_0 = 7.0, Temperature (T) = 25 ± 1 °C.

3.2. Determination of Bimolecular Reaction Rate

The generation of $^\bullet$OH in the UV/H_2O_2 system was proved by the photoluminescence (PL) technique using a probe molecule with terephthalic acid, which tends to react with $^\bullet$OH to form 2-hydroxyterephthalic acid, a highly fluorescent product [16]. The PL intensity of 2-hydroxyterephtalic acid is proportional to the amount of $^\bullet$OH radicals produced in water [17]. Figure 3 shows the PL spectral changes in the 5×10^{-4} M terephthalic acid solution with a concentration of 2×10^{-3} M NaOH (excitation at 315 nm), as described by Yu et al. [17]. Similar fluorescence intensity was found in the reaction systems with initial concentrations of 100 and 1000 μM of H_2O_2, suggesting a constant concentration of $^\bullet$OH with the initial H_2O_2 dosage ranging from 100 to 1000 μM. The PL signal at 425 nm increased with the irradiation time, which was attributed to the reaction of terephthalic acid with $^\bullet$OH generated in the UV/H_2O_2 system.

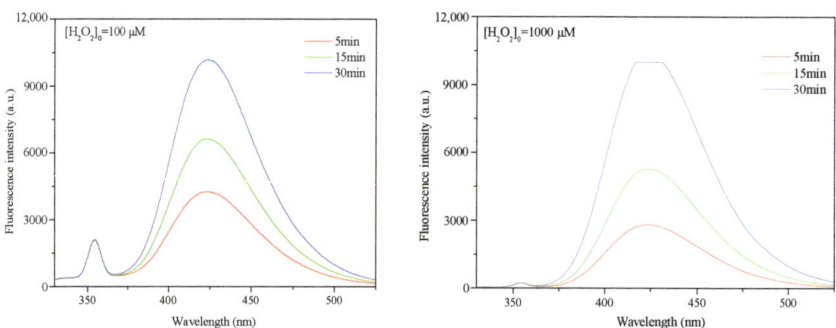

Figure 3. Photoluminescence (PL) spectral changes observed in the UV/H_2O_2 system in a 5×10^{-4} M basic solution of terephthalic acid (excitation at 315 nm).

The bimolecular reaction rates of KET and METH reacting with •OH were determined through the competition experiments at pH 7 (phosphate buffer solution, 5 mM). BA was used as the reference compound, with which the constant reaction rate of •OH is known to be 5.9×10^9 M^{-1} s^{-1} [18]. It is important to note that the degradation of KET, METH and BA using UV alone was negligible at less than 9%. Equations (2) and (3) describe the competing kinetics of KET and METH with •OH in the UV/H$_2$O$_2$ oxidation process, through which the bimolecular reaction rates of KET and METH reacting with •OH were 4.43×10^9 and 7.91×10^9 M^{-1} s^{-1}, respectively (Figure 4).

$$\ln \frac{(KET)_0}{(KET)_t} = \frac{k_{\bullet OH-KET}}{k_{\bullet OH-BA}} \ln \frac{(BA)_0}{(BA)_t} \qquad (2)$$

$$\ln \frac{(METH)_0}{(METH)_t} = \frac{k_{\bullet OH-METH}}{k_{\bullet OH-BA}} \ln \frac{(BA)_0}{(BA)_t} \qquad (3)$$

where (KET)$_0$, (METH)$_0$ and (BA)$_0$ are the initial concentrations (μmol/L) of target compounds. (KET)$_t$, (METH)$_t$ and (BA)$_t$ are the concentrations (μmol/L) at time t (min). $k_{\bullet OH-KET}$, $k_{\bullet OH-METH}$ and $k_{\bullet OH-BA}$ are the bimolecular reaction rates of KET, METH and BA reacting with •OH, respectively.

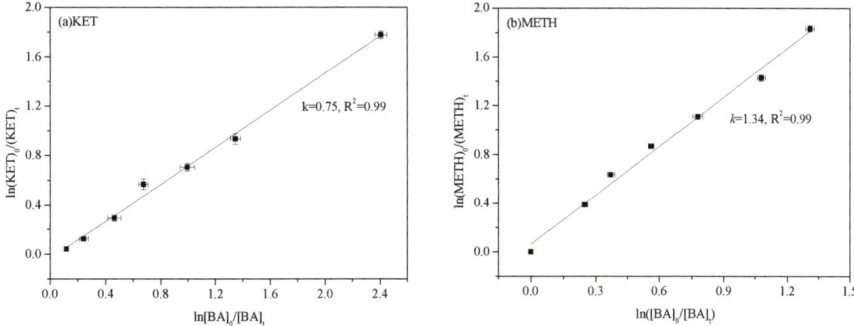

Figure 4. (a) The reaction rate constant of KET with •OH. Conditions: (KET)$_0$ = (BA)$_0$ = 0.42 μM, (H$_2$O$_2$)$_0$ = 1 mM, pH = 7, T = 25 ± 1 °C. (b) The reaction rate constant of METH with •OH. Conditions: (METH)$_0$ = (BA)$_0$ = 0.67 μM, (H$_2$O$_2$)$_0$ = 1 mM, pH = 7, temperature = 25 ± 1 °C.

3.3. Effect of H$_2$O$_2$ Dosage

The KET and METH degradation under different initial H$_2$O$_2$ dosages were consistent with the pseudo-first-order reaction model ($R^2 > 0.99$, Figure 5). The k_{obs} of KET and METH increased dramatically from 0.001 min^{-1} to 0.027 and 0.049 min^{-1} with the initial H$_2$O$_2$ dosage ranging from 0 to 1000 μM. The reason for this phenomenon is that the production of •OH increased with the initial H$_2$O$_2$ dosage ranging from 0 to 1000 μM, thus accelerating the degradation rate of target compounds [19]. However, the k_{obs} of METH decreased slightly with the initial concentration of H$_2$O$_2$ increased to 2000 μM. A similar phenomenon was observed in a previous report that indicated that the degradation rates of cyclophosphamide and 5-fluorouracil were proportional to the H$_2$O$_2$ dosage and slightly decreased with excess H$_2$O$_2$ [20]. An excessive amount of H$_2$O$_2$ would cause the self-scavenging effect of •OH to form HO$_2$• and O$_2^-$• (Equations (4) and (5)) [21], the low reactivity of which could reduce the degradation rate. Similar results were obtained concerning the degradation of ofloxacin [15] and chloramphenicol [22]. Moreover, large amounts of •OH were dimerized to H$_2$O$_2$, and the generated HO$_2$• and O$_2^-$• subsequently participated in other reactions (Equations (6)–(9)) [23]. This negative effect was not observed in this study, probably because the maximum H$_2$O$_2$ dosage (2000 μM) was not high enough to inhibit the KET degradation.

$$H_2O_2 + {}^{\bullet}OH \rightarrow HO_2^{\bullet} + H_2O \qquad (4)$$

$$H_2O_2 + {}^{\bullet}OH \rightarrow O_2^{-\bullet} + H^+ + H_2O \tag{5}$$

$${}^{\bullet}OH + {}^{\bullet}OH \rightarrow H_2O_2 \tag{6}$$

$$HO_2^{\bullet} + H_2O_2 \rightarrow {}^{\bullet}OH + H_2O + O_2 \tag{7}$$

$$HO_2^{\bullet} + {}^{\bullet}OH \rightarrow H_2O + O_2 \tag{8}$$

$$O_2^{-\bullet} + H_2O_2 \rightarrow {}^{\bullet}OH + OH^- + O_2 \tag{9}$$

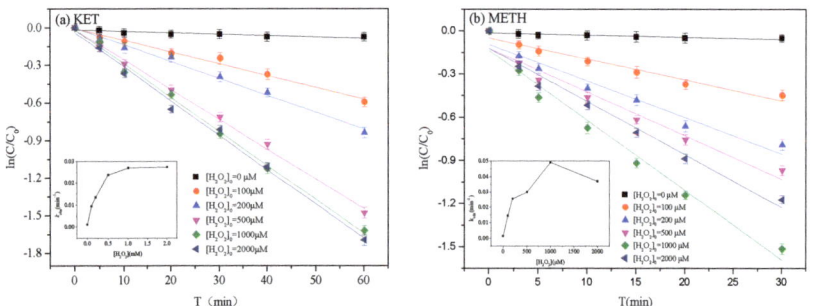

Figure 5. Effect of H_2O_2 dosage on KET (**a**) and METH (**b**) degradation in the UV/H_2O_2 system. Conditions: $(KET)_0 = (METH)_0 = 100$ µg/L, $(H_2O_2)_0 = 0$–2000 µM, $pH_0 = 7.0$, T = 25 ± 1 °C.

3.4. Effect of Initial pH

Figure 6 illustrates the KET and METH destruction at different initial pHs, which were adjusted with an H_2SO_4 or NaOH solution (0.1 M). No buffer was used due to its inhibiting effect on the decomposition of organics [24]. The KET and METH degradation at different initial pHs followed the pseudo-first-order reaction model well. The k_{obs} of KET and METH reached the highest levels in a neutral environment at 0.027 and 0.085 min^{-1}, respectively. Due to the greater stability of H_2O_2 at pH 5 and 7, the degradation rates of KET and METH under acidic and neutral conditions were obviously better than those under alkaline conditions. Under alkaline conditions, ${}^{\bullet}OH$ could be quenched by the HO_2^- produced by H_2O_2 dissociation, thus reducing the yield of ${}^{\bullet}OH$ in the system.

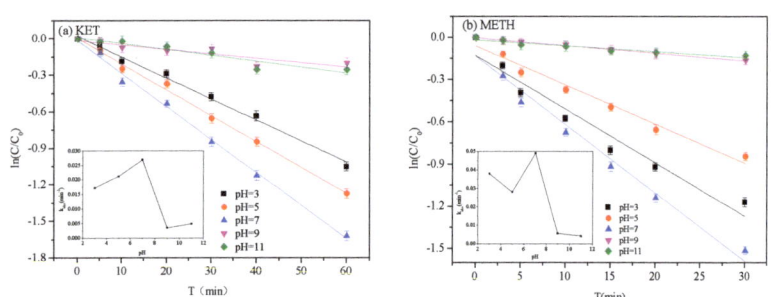

Figure 6. Effects of different initial pHs on the degradation of KET (**a**) and METH (**b**) in the UV/H_2O_2 system. Conditions: $(KET)_0 = (METH)_0 = 100$ µg/L, $(H_2O_2)_0 = 500$ µM, $pH_0 = 3$–11, T = 25 ± 1 °C.

3.5. Effect of Water Background Components on Degradation Efficiency of Target Compounds

There are many different substrates in natural water, including different kinds of anions, cations and organic matter. These ions could react with free radicals in advanced oxidation processes, thus inhibiting or promoting the reaction and affecting the overall oxidation effect. Therefore, it is of great significance to study the influence of different ion types and contents on the practical application of advanced oxidation technology.

3.5.1. Effect of HCO_3^-

The decomposition of KET and METH was significantly inhibited with the addition of HCO_3^- at different initial dosages in the UV/H_2O_2 oxidation process (Figure 7). When the initial dosage of HCO_3^- ranged from 0 to 10 mM, the reaction rate of KET and METH decreased from 0.027 and 0.049 min^{-1} to 0.008 and 0.011 min^{-1}, respectively. The reason for this experimental phenomenon was that HCO_3^- was the quenching agent for $^\bullet OH$ which was also consumed by the competing reaction of ionized CO_3^{2-} (Equations (10)–(13)). Therefore, the inhibitory effect of KET and METH degradation was more obvious with the increase of the HCO_3^- concentration.

$$CO_3^{2-} + {}^\bullet OH \rightarrow CO_3^{-\bullet} + OH^- \tag{10}$$

$$HCO_3^- + {}^\bullet OH \rightarrow HCO_3^\bullet + OH^- \tag{11}$$

$$HCO_3^\bullet \rightarrow CO_3^\bullet + H^+ \tag{12}$$

$$CO_3^{-\bullet} + H_2O_2 \rightarrow HCO_3^- + HO_2^\bullet \tag{13}$$

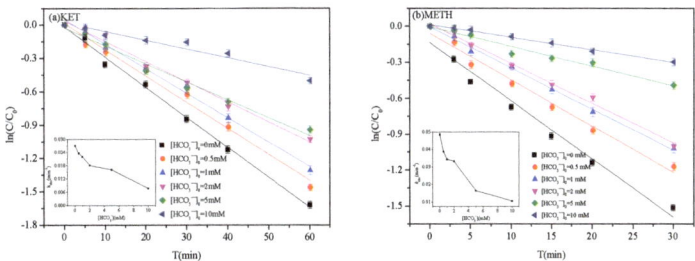

Figure 7. Effect of HCO_3^- on KET (**a**) and METH (**b**) degradation in UV/H_2O_2 system. Conditions: $(KET)_0 = (METH)_0 = 100$ µg/L, $(H_2O_2)_0 = 500$ µM, $pH_0 = 7.0$, T = 25 ± 1 °C.

3.5.2. Effect of Cl^-

With the initial concentration of Cl^- ranging from 0 to 10 mM, the destruction of KET was dramatically inhibited with the rate constant of KET decreased from 0.027 to 0.018 min^{-1} (Figure 8), which could be due to the elimination of $^\bullet OH$ by Cl^- according to Equations (14)–(16) [25]. The degradation reaction rate changed slightly as more Cl^- was added. However, the METH degradation was less affected by Cl^-, with the reaction rate remaining basically unchanged (0.0446–0.0485 min^{-1}).

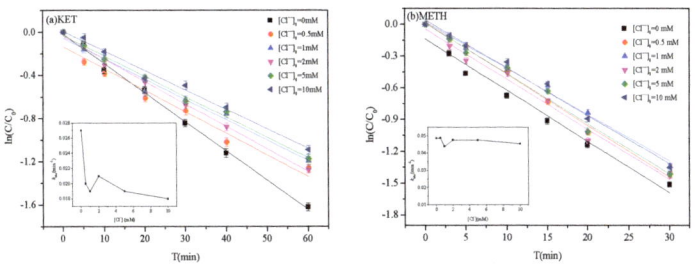

Figure 8. Effect of Cl^- on KET (**a**) and METH (**b**) degradation in the UV/H_2O_2 system. Conditions: $(KET)_0 = (METH)_0 = 100$ µg/L, $(H_2O_2)_0 = 500$ µM, $pH_0 = 7.0$, T = 25 ± 1 °C.

3.5.3. Effect of NO_3^-

$$^\bullet OH + Cl^- \rightarrow Cl^\bullet + OH^- \qquad (14)$$

$$Cl^\bullet + Cl^- \rightarrow Cl_2^{-\bullet} \qquad (15)$$

$$Cl^\bullet + Cl^\bullet \rightarrow Cl_2 \qquad (16)$$

The influence of NO_3^- on the decomposition of KET and METH is illustrated in Figure 9. With the initial concentration of NO_3^- ranging from 0 to 10 mM, the degradation of both target compounds was obviously inhibited. The reaction rate of KET and METH decreased from 0.027 and 0.049 min^{-1} to 0.007 and 0.012 min^{-1}, respectively. The above experimental phenomena were attributed to the following: First, a large amount of $^\bullet OH$ could be produced from NO_3^- under UV irradiation (Equations (17)–(18)), which is an important source of $^\bullet OH$ in natural water [26]. Second, as a photosensitizer, NO_3^- has a strong absorption in the ultraviolet range, which results in the formation of an internal filter that prevents the effective light transmittance and leads to the decline of $^\bullet OH$ production in the UV/H_2O_2 system [27]. The latter was found to be dominant after the degradation effect of the reaction was analyzed.

$$NO_3^- + h\nu \rightarrow NO_2^{-\bullet} + O^{-\bullet} \qquad (17)$$

$$O^{-\bullet} + H_2O \rightarrow {}^\bullet OH + OH^- \qquad (18)$$

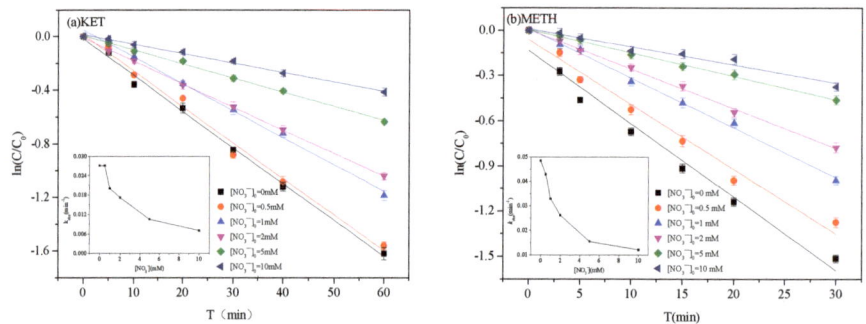

Figure 9. Effect of NO_3^- on KET (**a**) and METH (**b**) degradation in the UV/H_2O_2 system. Conditions: (KET)$_0$ = (METH)$_0$ = 100 µg/L, (H_2O_2)$_0$ = 500 µM, pH$_0$ = 7.0, T = 25 ± 1 °C.

3.5.4. Effect of HA

Due to its complex structure, HA may have uncontrollable effects on the destruction of target compounds. As illustrated in Figure 10, KET and METH degradation was dramatically inhibited once HA was added with different dosages in the UV/H_2O_2 system. As more HA (0–0.1 mM) was added, the reaction rate of KET and METH declined from 0.027 and 0.049 min^{-1} to 0.001 and 0.008 min^{-1}, respectively, while the degradation reaction rate changed slightly with the continued addition of the HA. UV irradiation was absorbed by HA, creating an inner filter (Figure 11) and significantly inhibiting the UV transmittance for UV photons, thus limiting the generation of $^\bullet OH$ in the UV/H_2O_2 process [28]. Moreover, the degradation of target compounds can be inhibited by the competing reaction of HA with the active radicals [29].

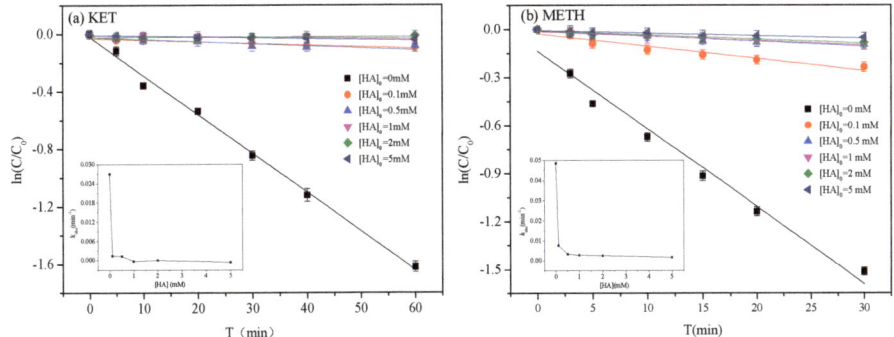

Figure 10. Effect of HA on KET (**a**) and METH (**b**) degradation in the UV/H$_2$O$_2$ system. Conditions: (KET)$_0$ = (METH)$_0$ = 100 µg/L, (H$_2$O$_2$)$_0$ = 500 µM, pH$_0$ = 7.0, T = 25 ± 1 °C.

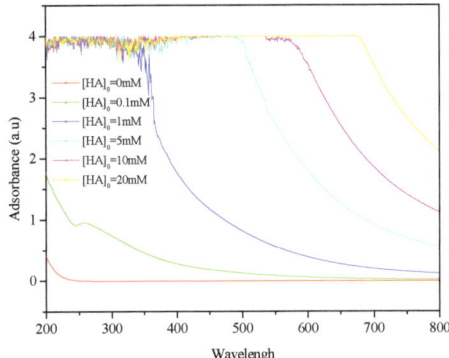

Figure 11. The ultraviolet–visible spectroscopy of reaction solutions at different concentrations of HA.

3.6. Degradation Products and Mechanism

Degradation intermediates and products of METH produced in the UV/H$_2$O$_2$ oxidation process were determined by using UPLC/MS/MS under full scans and product ion scans. During the whole METH degradation process, the mass spectra were compared to identify the intermediates. The structure of the transformation products was analyzed with the specific molecular ions and fragmentation patterns rather than direct comparison with corresponding standards. Figure 12 illustrates the mass spectra and possible structures of the degradation intermediates, based on which the possible transformation pathways of METH during UV/H$_2$O$_2$ are shown in Figure 13. The proposed degradation mechanisms of METH degradation involved in the UV/H$_2$O$_2$ system include hydrogenation, hydroxylation and electrophilic substitution.

With the molecular weight of 149, intermediate product 2 (P2, m/z = 150) was formed as a result of hydrogenation of METH. P1 (m/z = 91) with a stable structure was generated from the fracture of the C-C bond of the branched chain. Intermediates P3 (m/z = 110) and P4 (m/z = 73) were formed by electrophilic substitution of hydroxyl. METH was hydroxylated to form ephedrine (m/z = 165), of which the C-C bond of branched chain was fractured to form intermediate product P5 (m/z = 57). The hydroxylation of ephedrine induced the formation of intermediate P6 (m/z = 181) which was then achieved to form intermediate P7 (m/z = 89) after further hydroxylation. The mineralization of KET and METH was characterized by removal of total organic carbon (TOC), which achieved 41% and 57% within 60 min under UV/H$_2$O$_2$ treatment (Figure 14). The intermediate products were further degraded as the reaction continued.

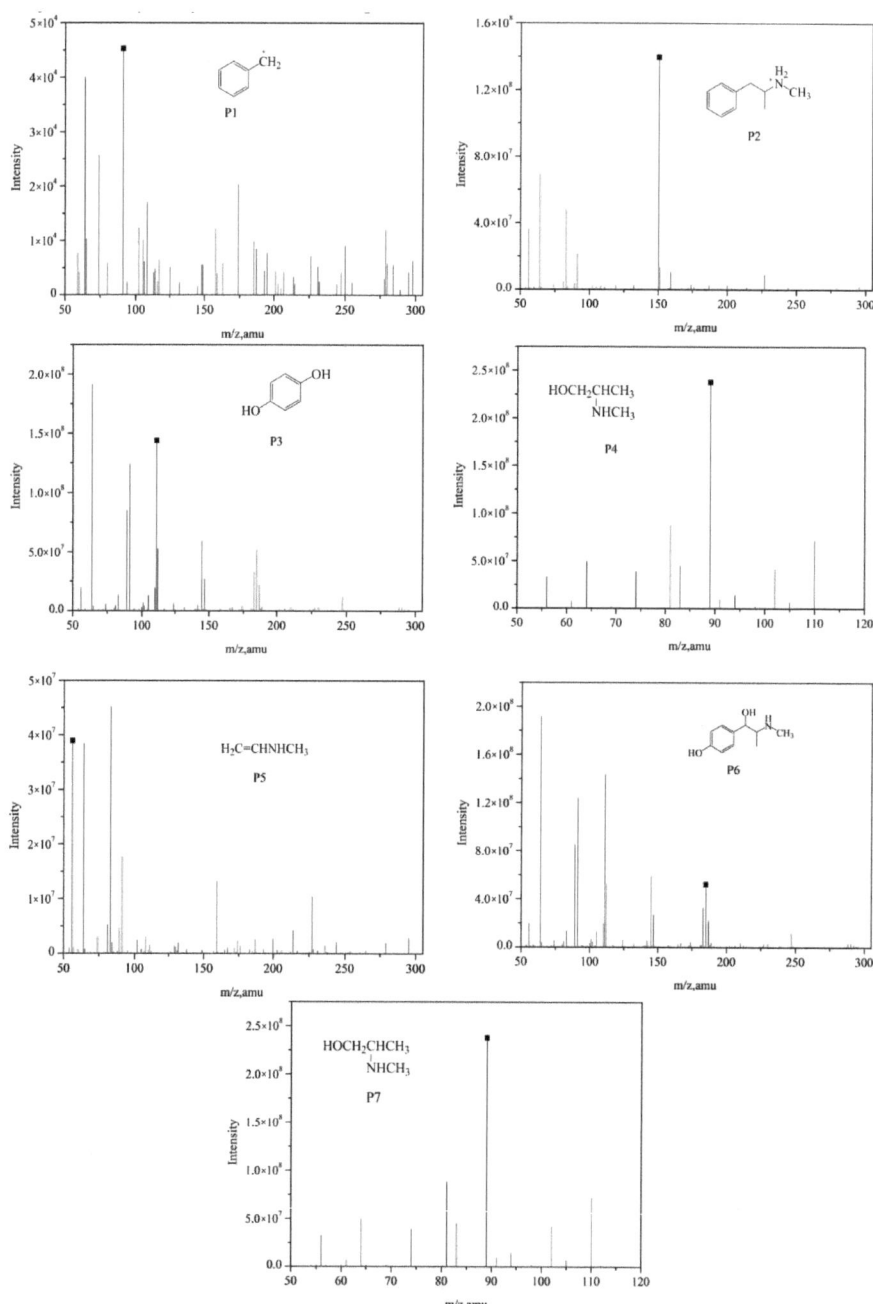

Figure 12. Mass spectra of the intermediate products of METH in the UV/PS system.

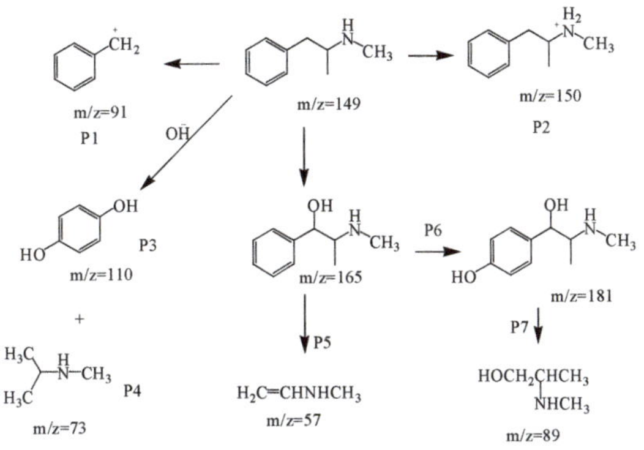

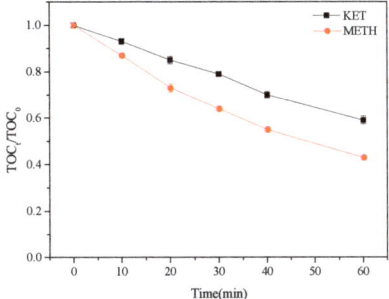

Figure 13. Tentative transformation of METH pathways in the UV/H$_2$O$_2$ system.

Figure 14. The mineralization of KET and METH during the UV/H$_2$O$_2$ system. Conditions: (KET)$_0$ = (METH)$_0$ = 100 µg/L, (H$_2$O$_2$)$_0$ = 500 µM, pH$_0$ = 7, T = 25 ± 1 °C.

4. Conclusions

The degradation kinetics and mechanisms of KET and METH using the UV/H$_2$O$_2$ process were investigated in this study. Their degradation in UV photolysis or H$_2$O$_2$ oxidation alone was negligible. However, 99% of KET and METH (100 µg/L) were effectively eliminated by the combination of UV and H$_2$O$_2$ within 120 and 60 min, respectively. According to the competition kinetics, the rate constants of •OH with KET and METH at pH 7 were calculated to be 4.43 × 10^9 and 7.91 × 10^9 M^{-1}·s^{-1}, respectively. The apparent rate constants of KET and METH reached the highest levels in a neutral environment. The degradation of KET and METH was significantly inhibited by HCO$_3^-$, Cl$^-$, NO$_3^-$ and HA; however, Cl$^-$ had little influence on the METH degradation. Seven reaction intermediates of METH in the UV/H$_2$O$_2$ system were identified by UPLC-MS/MS. Possible transformation mechanisms involved in the KET and METH degradation by UV/H$_2$O$_2$ oxidation system included hydrogenation, hydroxylation and electrophilic substitution. Results demonstrated that UV/H$_2$O$_2$ was an effective technique to remove KET and METH, providing a promising application for the decomposition of trace organic pollutants in natural water.

Author Contributions: Conceptualization, D.-M.G. and C.-S.G.; methodology, D.-M.G., C.-S.G., Q.-Y.F. and J.X.; software, D.-M.G., C.-S.G. and H.Z.; validation, D.-M.G., Q.-Y.F. and J.X.; formal analysis, D.-M.G. and H.Z.; resources, C.-S.G., Q.-Y.F. and J.X.; data curation, D.-M.G. and H.Z.; writing—original draft preparation, D.-M.G.; writing—review and editing, D.-M.G., C.-S.G., Q.-Y.F., H.Z. and J.X.; visualization, D.-M.G. and H.Z.; supervision, D.-M.G., Q.-Y.F. and J.X.; funding acquisition, C.-S.G. and J.X. All authors have read and agreed to the published version of the manuscript.

Funding: This work was funded by the National Natural Science Foundation of China (NSFC, 41673120) and Beijing Natural Science Foundation (8173058).

Acknowledgments: This study was carried out as part of the NSFC project, managed by the Jian Xu and supported by Center for Environmental Health Risk Assessment and Research, Chinese Research Academy of Environmental Sciences. We thank Wenli Qiu for her help in operating HPLC-MS. Reviewers are also thanked for the time dedicated and their comments.

Conflicts of Interest: The authors declare no conflict of interest.

References

1. Baker, D.R.; Kasprzyk-Hordern, B. Spatial and temporal occurrence of pharmaceuticals and illicit drugs in the aqueous environment and during wastewater treatment: New developments. *Sci. Total Environ.* **2013**, *454–455*, 442–456. [CrossRef]
2. Bijlsma, L.; Serrano, R.; Ferrer, C.; Tormos, I.; Hernández, F. Occurrence and behavior of illicit drugs and metabolites in sewage water from the Spanish Mediterranean coast (Valencia region). *Sci. Total Environ.* **2014**, *487*, 703–709. [CrossRef] [PubMed]
3. Wang, Z.; Xu, Z.; Li, X. Biodegradation of methamphetamine and ketamine in aquatic ecosystem and associated shift in bacterial community. *J. Hazard. Mater.* **2018**, *359*, 356–364. [CrossRef] [PubMed]
4. Du, P.; Li, K.; Li, J.; Xu, Z.; Fu, X.; Yang, J.; Zhang, H.; Li, X. Methamphetamine and ketamine use in major Chinese cities, a nationwide reconnaissance through sewage-based epidemiology. *Water Res.* **2015**, *84*, 76–84. [CrossRef] [PubMed]
5. Santos, M.E.S.; Grabicová, K.; Steinbach, C.; Schmidt-Posthaus, H.; Randák, T. Environmental concentration of methamphetamine induces pathological changes in brown trout (Salmo trutta fario). *Chemosphere* **2020**, *254*, 126882. [CrossRef] [PubMed]
6. Liao, P.H.; Hwang, C.C.; Chen, T.H.; Chen, P.J. Developmental exposures to waterborne abused drugs alter physiological function and larval locomotion in early life stages of medaka fish. *Aquat. Toxicol.* **2015**, *165*, 84–92. [CrossRef] [PubMed]
7. Awual, M.R.; Hasan, M.M. A ligand based innovative composite material for selective lead(II) capturing from wastewater. *J. Mol. Liq.* **2019**, *294*, 111679. [CrossRef]
8. Awual Rabiul, M. A novel facial composite adsorbent for enhanced copper(II) detection and removal from wastewater. *Chem. Eng. J.* **2015**, *266*, 368–375. [CrossRef]
9. Neta, P.; Huie, R.E.; Ross, A.B. Rate Constants for Reactions of Inorganic Radicals in Aqueous Solution. *J. Phys. Chem. Ref. Data* **1988**, *17*, 1027–1284. [CrossRef]
10. Russo, D.; Spasiano, D.; Vaccaro, M.; Cochran, K.H.; Richardson, S.D.; Andreozzi, R.; Puma, G.L.; Reis, N.M.; Marotta, R. Investigation on the removal of the major cocaine metabolite (benzoylecgonine) in water matrices by UV_{254}/H_2O_2 process by using a flow microcapillary film array photoreactor as an efficient experimental tool. *Water Res.* **2015**, *89*, 375–383. [CrossRef]
11. Kuo, C.; Lin, C.; Hong, P.A.K. Photocatalytic degradation of methamphetamine by UV/TiO_2—Kinetics, intermediates, and products. *Water Res.* **2015**, *74*, 1–9. [CrossRef] [PubMed]
12. Wei, C.; Yi, K.; Sun, G.; Wang, J. Synthesis of novel sonocatalyst Er3+:YAlO3/Nb2O5 and its application for sonocatalytic degradation of methamphetamine hydrochloride. *Ultrason. Sonochem.* **2018**, *42*, 57–67. [CrossRef]
13. Gu, D.; Guo, C.; Hou, S.; Lv, J.; Zhang, Y.; Feng, Q.; Zhang, Y.; Xu, J. Kinetic and mechanistic investigation on the decomposition of ketamine by UV-254 nm activated persulfate. *Chem. Eng. J.* **2019**, *370*, 19–26. [CrossRef]
14. He, X.; Pelaez, M.; Westrick, J.A.; O'Shea, K.E.; Hiskia, A.; Triantis, T.; Kaloudis, T.; Stefan, M.I.; Armah, A.; Dionysiou, D.D. Efficient removal of microcystin-LR by $UV-C/H_2O_2$ in synthetic and natural water samples. *Water Res.* **2012**, *46*, 1501–1510. [CrossRef] [PubMed]
15. Lin, C.C.; Lin, H.Y.; Hsu, L.J. Degradation of ofloxacin using UV/H_2O_2 process in a large photoreactor. *Sep. Purif. Technol.* **2016**, *168*, 57–61. [CrossRef]
16. Cheng, B.; Le, Y.; Yu, J. Preparation and enhanced photocatalytic activity of $Ag@TiO_2$ core-shell nanocomposite nanowires. *J. Hazard. Mater.* **2010**, *177*, 971–977. [CrossRef]
17. Yu, X.; Liu, S.; Yu, J. Superparamagnetic $\gamma\text{-}Fe_2O_3@SiO_2@TiO_2$ composite microspheres with superior photocatalytic properties. *Appl. Catal. B Environ.* **2011**, *104*, 12–20. [CrossRef]

18. Ismail, L.; Ferronato, C.; Fine, L.; Jaber, F.; Chovelon, J.M. Elimination of sulfaclozine from water with SO_4^- radicals: Evaluation of different persulfate activation methods. *Appl. Catal. B Environ.* **2016**, *201*, 573–581. [CrossRef]
19. Znad, H.; Abbas, K.; Hena, S.; Awual, M.R. Synthesis a novel multilamellar mesoporous TiO_2/ZSM-5 for photo-catalytic degradation of methyl orange dye in aqueous media. *J. Environ. Chem. Eng.* **2018**, *6*, 218–227. [CrossRef]
20. Lutterbeck, C.A.; Wilde, M.L.; Baginska, E.; Leder, C.; Machado, Ê.L.; Kümmerer, K. Degradation of cyclophosphamide and 5-fluorouracil by UV and simulated sunlight treatments: Assessment of the enhancement of the biodegradability and toxicity. *Environ. Pollut.* **2016**, *208 Pt B*, 467–476. [CrossRef]
21. Kwon, M.; Kim, S.; Yoon, Y.; Jung, Y.; Hwang, T.M.; Lee, J.; Kang, J.W. Comparative evaluation of ibuprofen removal by UV/H_2O_2 and $UV/S_2O_8^{2-}$ processes for wastewater treatment. *Chem. Eng. J.* **2015**, *269*, 379–390. [CrossRef]
22. Zuorro, A.; Fidaleo, M.; Fidaleo, M.; Lavecchia, R. Degradation and antibiotic activity reduction of chloramphenicol in aqueous solution by UV/H_2O_2 process. *J. Environ. Manag.* **2014**, *133*, 302–308. [CrossRef] [PubMed]
23. Qiu, W.; Zheng, M.; Sun, J.; Tian, Y.; Fang, M.; Zheng, Y.; Zhang, T.; Zheng, C. Photolysis of enrofloxacin, pefloxacin and sulfaquinoxaline in aqueous solution by UV/H_2O_2, UV/Fe(II), and UV/H_2O_2/Fe(II) and the toxicity of the final reaction solutions on zebrafish embryos. *Sci. Total Environ.* **2019**, *651*, 1457–1468. [CrossRef] [PubMed]
24. Sánchez-Polo, M.; Daiem, M.M.A.; Ocampo-Pérez, R.; Rivera-Utrilla, J.; Mota, A.J. Comparative study of the photodegradation of bisphenol A by HO·, $SO_4^{·-}$ and $CO_3^{·-}/HCO_3·$ radicals in aqueous phase. *Sci. Total Environ.* **2013**, *463–464*, 423–431.
25. Zhang, Y.; Xiao, Y.; Zhong, Y.; Lim, T. Comparison of amoxicillin photodegradation in the UV/H_2O_2 and UV/persulfate systems: Reaction kinetics, degradation pathways, and antibacterial activity. *Chem. Eng. J.* **2019**, *372*, 420–428. [CrossRef]
26. Yin, K.; Deng, L.; Luo, J.; Crittenden, J.; Liu, C.; Wei, Y.; Wang, L. Destruction of phenicol antibiotics using the UV/H_2O_2 process: Kinetics, byproducts, toxicity evaluation and trichloromethane formation potential. *Chem. Eng. J.* **2018**, *351*, 867–877. [CrossRef]
27. Moon, B.R.; Kim, T.K.; Kim, M.K.; Choi, J.; Zoh, K.D. Degradation mechanisms of Microcystin-LR during UV-B photolysis and UV/H_2O_2 processes: Byproducts and pathways. *Chemosphere* **2017**, *185*, 1039. [CrossRef]
28. Oh, B.T.; Seo, Y.S.; Sudhakar, D.; Choe, J.H.; Lee, S.M.; Park, Y.J.; Cho, M. Oxidative degradation of endotoxin by advanced oxidation process (O_3/H_2O_2 & UV/H_2O_2). *J. Hazard. Mater.* **2014**, *279*, 105–110.
29. Lutze, H.V.; Bircher, S.; Rapp, I.; Kerlin, N.; Bakkour, R.; Geisler, M.; von Sonntag, C.; Schmidt, T.C. Degradation of chlorotriazine pesticides by sulfate radicals and the influence of organic matter. *Environ. Sci. Technol.* **2015**, *49*, 1673–1680. [CrossRef]

Publisher's Note: MDPI stays neutral with regard to jurisdictional claims in published maps and institutional affiliations.

© 2020 by the authors. Licensee MDPI, Basel, Switzerland. This article is an open access article distributed under the terms and conditions of the Creative Commons Attribution (CC BY) license (http://creativecommons.org/licenses/by/4.0/).

Article

Oxidation of Selected Trace Organic Compounds through the Combination of Inline Electro-Chlorination with UV Radiation (UV/ECl₂) as Alternative AOP for Decentralized Drinking Water Treatment

Philipp Otter [1], Katharina Mette [2], Robert Wesch [2], Tobias Gerhardt [2], Frank-Marc Krüger [2], Alexander Goldmaier [1], Florian Benz [1], Pradyut Malakar [3] and Thomas Grischek [4],*

1. AUTARCON GmbH, D-34117 Kassel, Germany; otter@autarcon.com (P.O.); goldmaier@autarcon.com (A.G.); benz@autarcon.com (F.B.)
2. GNF e.V. Volmerstr. 7 B, 12489 Berlin, Germany; k.mette@gnf-berlin.de (K.M.); r.wesch@gnf-berlin.de (R.W.); t.gerhardt@gnf-berlin.de (T.G.); f.krueger@gnf-berlin.de (F.-M.K.)
3. International Centre for Ecological Engineering, University of Kalyani, Kalyani, West Bengal 741235, India; pradyutmalakar2@gmail.com
4. Division of Water Sciences, University of Applied Sciences Dresden, Friedrich-List-Platz 1, 01069 Dresden, Germany
* Correspondence: thomas.grischek@htw-dresden.de; Tel.: +49-0351 4623350

Received: 6 October 2020; Accepted: 17 November 2020; Published: 21 November 2020

Abstract: A large variety of Advanced Oxidation Processes (AOPs) to degrade trace organic compounds during water treatment have been studied on a lab scale in the past. This paper presents the combination of inline electrolytic chlorine generation (ECl₂) with low pressure UV reactors (UV/ECl₂) in order to allow the operation of a chlorine-based AOP without the need for any chlorine dosing. Lab studies showed that from a Free Available Chlorine (FAC) concentration range between 1 and 18 mg/L produced by ECl₂ up to 84% can be photolyzed to form, among others, hydroxyl radicals (·OH) with an UV energy input of 0.48 kWh/m³. This ratio could be increased to 97% by doubling the UV energy input to 0.96 kWh/m³ and was constant throughout the tested FAC range. Also the achieved radical yield of 64% did not change along the given FAC concentration range and no dependence between pH 6 and pH 8 could be found, largely simplifying the operation of a pilot scale system in drinking water treatment. Whereas with ECl₂ alone only 5% of benzotriazoles could be degraded, the combination with UV improved the degradation to 89%. Similar results were achieved for 4-methylbenzotriazole, 5-methylbenzotriazole and iomeprol. Oxipurinol and gabapentin were readily degraded by ECl₂ alone. The trihalomethanes values were maintained below the Germany drinking water standard of 50 μg/L, provided residual chlorine concentrations are kept within the permissible limits. The here presented treatment approach is promising for decentralized treatment application but requires further optimization in order to reduce its energy requirements.

Keywords: trace organic compounds; emerging pollutants; rural regions; electrochlorination; UV; AOP; energy per order

---

1. Introduction

Annually on average 15 g of pharmaceuticals are consumed per capita, but human bodies are unable to fully metabolize pharmaceuticals, which are then excreted as parental components or metabolites [1,2]. They are discharged into wastewater together with Personal Care Products

(PCPs), sweeteners, illicit and non-controlled drugs, complexing agents, nanoparticles, perfluorinated compounds, pesticides, flame retardants, fuel additives and endocrine disrupting chemicals and detergents [3,4]. Conventional wastewater treatment such as activated sludge processes, exhibits limitations in the removal of Trace Organic Compounds (TOrCs) [5,6]. TOrCs are additionally released into the environment from irrigation with treated or untreated wastewater [7], disposal of animal waste on agricultural sites [8] and artificial groundwater recharge [9]. World-wide studies have confirmed the occurrence of pharmaceutical residues in the effluents of wastewater treatment plants (WWTPs) as well as in surface and groundwaters [10–14].

The problem of water contamination with TOrCs is even more prevalent in some developing areas. E.g., in India, the amount of ciprofloxacin, sulfamethoxazole, amoxicillin, norfloxacin, and ofloxacin in treated wastewater was up to 40 times higher compared to other countries in Europe, Australia, Asia, and North America [5,8]. Here even remote rural areas are affected.

Conventional drinking water treatment processes such as coagulation and flocculation are not designed to effectively remove TOrCs [8]. Powdered activated carbon (PAC), membrane filtration technologies [15–17] or advanced oxidation processes (AOP) have been studied in the past to evaluate their efficiency on the removal of selected TOrCs [18]. Most studied AOPs are based on ozonation, Fenton oxidation, or UV based AOPs such as UV/H_2O_2. For the application of AOPs in water treatment, the supply of chemicals such H_2O_2 in UV/H_2O_2 AOP has been identified as major challenge and cost factor [19].

A relatively new AOP is the combination of chlorination with UV radiation (UV/Cl_2) [20–24]. When aqueous chlorine solutions are exposed to UV, ·OH and ·Cl radicals are also formed (Figure 1). The reduction potential of ·OH radicals is with 2.8 V vs. Standard Hydrogen Electrode (SHE) substantially higher compared to the potential of ozone (2.07 V) or chlorine (1.37 V) [25]. UV/Cl_2 has proven to produce higher amounts of (·OH) radicals compared to UV/H_2O_2 mainly due to the low absorbance of UV light by H_2O_2 [20,26]. Whereas absorption coefficients (e) for UV/H_2O_2 of 19.6 $M^{-1}·cm^{-1}$ have been identified for 254 nm [27] they have reached 59 $M^{-1}·cm^{-1}$ and 66 $M^{-1}·cm^{-1}$ for HOCl and OCl^- [28]. Higher quantum yields with regards to radical generation for UV/Cl_2 where reported by [26]. When radiated with UV light hypochlorous acid and its anion hypochlorite react in water not only to OH, but also to the reactive chlorine species (RCS) ·Cl and Cl_2 as well as to oxygen radicals. At higher pH, hydroxyl radicals might be consumed by chlorine itself also forming RCS [29].

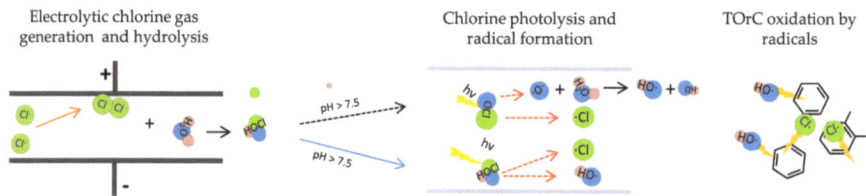

Figure 1. Pathway of chloride ions used for radical production (excerpt) in UV/ECl_2 process.

The RCS coexist with ·OH radicals and complement each other in degrading a wider variety of contaminants compared to e.g., UV/H_2O_2 AOP [29,30]. Depending on the chlorine species present two reaction pathways (Equations (1)–(3)) for the production of ·OH and ·Cl exist [26].

$$HOCl + h\nu \rightarrow ·OH + ·Cl, \tag{1}$$

$$OCl^- + h\nu \rightarrow ·O^- + ·Cl, \tag{2}$$

$$·O^- + H_2O \rightarrow ·OH + OH^- \tag{3}$$

However, the need to supply and dose chlorine reagents also persists for UV/Cl_2-based AOPs. Here the production of chlorine through inline electrolysis (ECl_2) could offer an alternative to transportation, storage and handling of chlorine reagents. During ECl$_2$ elementary chlorine is produced at the anode of an electrolytic cell from the natural chloride content of the water itself (Equation (4)). The chlorine immediately hydrolyses to form hypochlorous acid (Equation (5)) or hypochlorite (Equation (6)). Anodic side reactions and cathodic reactions are given in Equations (7) and (8):

Anodic reaction chlorine:
$$2Cl^- \rightarrow Cl_2 + 2e^-, \tag{4}$$

Hydrolysis of chlorine gas:
$$Cl_2 + 2H_2O \leftrightarrow HClO + Cl^- + H_3O^+, \tag{5}$$

Equilibrium at pH = 7.5:
$$HClO \leftrightarrow OCl^- + H^+, \tag{6}$$

Anodic side reaction oxygen:
$$2H_2O \rightarrow O_2 + 4H^+ + 4e^-, \tag{7}$$

Cathodic reaction:
$$2H_3O^+ + 2e^- \rightarrow H_2 + 2H_2O \tag{8}$$

In past studies this approach has already proven feasible to meet the disinfection requirements for remote drinking water supply [31–33]. In the presented study ECl$_2$ was combined with UV radiation (Figure 1).

The novelty of this approach is the operation of a chlorine based AOP for the degradation of TOrCs that is completely independent from any external chemical supply. On a small scale, ECl$_2$ as well as UV could be easily operated by photovoltaic (PV) which makes the process also independent of any external energy supply.

Available, studies to evaluate the degradation potential of AOPs using chlorine have been generally carried out for single compounds investigating concentrations much higher than those found in real waters. Further, deionized water free of other organic substances was used as solvent and sophisticated equipment was applied for e.g., determining reaction constants (k'). Therefore, uncertainties exist with regards to real case scenarios [24].

In this study a UV/ECl$_2$ setting was tested under lab conditions in order to evaluate chlorine production and radical formation in dependence of chloride concentrations, pH, cell currents and UV energy input applied. Following the lab test a UV/ECl$_2$ pilot setting was tested for the first time under real case conditions treating Elbe river water. In two short term sampling campaigns the removal efficiency of selected TOrCs, the energy consumption and the formation of disinfection by products (DBPs) by analyzing trihalomethanes (THM) was evaluated. The pilot system was operated long term to observe technical challenges that may occur under real case scenarios. The hypothesis of the here conducted work is, that the combination of ECl$_2$ with UV poses a technically feasible alternative to reduce TOrCs without the need for any external chemicals and electricity supply, which can be applied in decentralized water treatment. The conducted trials hereby allow a first insight into the application of UV/ECl$_2$ as part of an actual drinking water treatment system.

2. Methodology

To produce chlorine by means of ECl$_2$ mixed oxide electrodes (MOX) (GNF, Berlin, Germany) coated with ruthenium (Ru) and iridium (Ir) oxides. Ru- and Ir-mixed oxide electrodes have been selected due to their low overpotential for the oxidation reaction of chloride to chlorine and therefore offer a higher current efficiency for chlorine evolution compared to e.g., platinum coated electrodes [34].

2.1. Lab Test Setting

In order to assure sufficient chlorine generated by inline electrolysis as precursors for the chlorine photolysis and disinfection, lab experiments were conducted. Those included the variation of chloride ion concentration, flow rate, current density, pH and electrical conductivity.

During lab tests water was pumped through an array of two MOX electrodes and up to two UV lamps (Figure 2). The distance between the cell plates was 5 mm. Currents of up to 8 A were applied on the electrodes. With a total surface area of 959 cm^2, the current density accounted for 16.7 mA/cm^2. These currents were chosen because previous (unpublished) studies showed that this is the maximum applicable current density for long term chlorine evolution without damaging the coating. The MOXs were powered using a BaseTech BT-305 power supply unit (Hirschau, Germany).

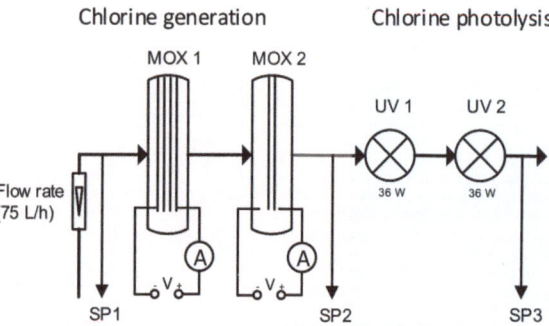

Figure 2. Lab test setting for performance evaluation of chlorine and radical formation.

Distilled water was pH adjusted with 10% HCl (Carl Roth, Karlsruhe, Germany) and 10% NaOH (Carl Roth). NaCl (Carl Roth) was added to the water to achieve concentrations of 25, 50 and 100 mg/L. The electric conductivity was adjusted to the desired value using NaHCO$_3$ (Carl Roth, Germany) and Na$_2$SO$_4$ (Carl Roth). Samples for chlorine measurement were taken directly behind the MOXs at Sampling Point (SP) 2 and SP 3 (Figure 2). The chlorine was then photolyzed by up to two PURION® 2500 36 W low pressure UV-C reactors with a volume of 0.75 L which were equipped with a calibrated silicon semiconductor-based UV-irradiance sensor (SUV-13A1Y2C, Purion, Zella-Mehlis, Germany) on each UV unit to assure constant irradiation of the UV lamps. The radical yield was determined at SP3.

The input of the electrical energy for the UV photolysis was 0.48 kWh/m^3 for one and 0.96 kWh/m^3 for two lamps, respectively. Samples to evaluate radical formation were taken prior (SP2) and after the UV reactor(s) (SP3).

The amount of HOCl and OCl$^-$ decomposed (Equations (1) and (2)) during FAC photodegradation is directly related to the amount of ·OH radicals [35] and the radical yield factor η can thus be quantified following Equation (9):

$$\eta = \frac{\Delta n_{\cdot OH} + \Delta n_{\cdot Cl}}{\Delta n_{FAC}} \qquad (9)$$

The quantification of radical formation was carried out following [36] as described below. The yield factors were calculated for pH values of 6, 7 and 8 in order to determine its dependence of chlorine species present (Equation (6)).

2.2. Field Test Settings

The field test was carried out for 10 months and was conducted with Elbe river water, which was filtrated by a UF system (150 kDa, 0.01 μm, Pall, Port Washington, NY, USA) before the UV/ECl$_2$ AOP in order to provide turbidity free water.

As natural water matrices contain a variety of radical scavengers the here tested setting will generate only site-specific results. Better pre-treatment hereby is expected to generate higher degradation as shown in [37].

The water was pumped by the pilot system in a flow through setting through three MOX with a total area of 1918 cm^2 operated in series followed by up to two Purion® 2500 36 W low pressure UV-C lamps (Figure 3).

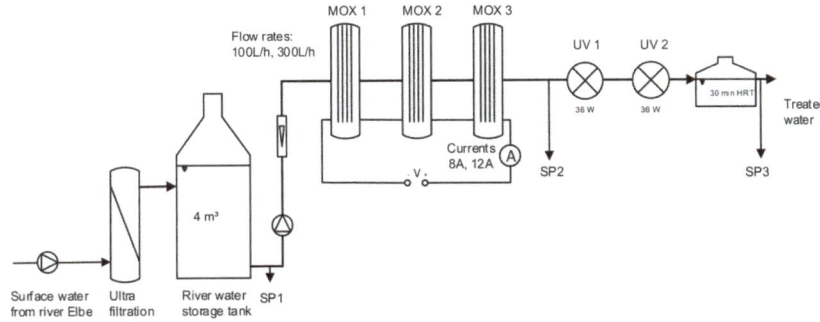

Figure 3. Pilot system tested with Elbe river water.

To release potentially formed calcareous deposits the polarity was inverted once every three hours. In own past studies, these comparable large intervals have proven to have a negligible effect on the lifetime of the electrolytic cells.

For the evaluation of TOrC degradation during drinking water treatment two short term sampling campaigns were conducted.

In order to evaluate the influence of the chlorine concentration and the UV irradiation on the degradation of TOrCs the ECl$_2$ current and resulting current densities, the number of UV lamps and the flow rate were varied during the field test. This tested pilot settings are documented in Table 1 including the specific energy demand for each of the tested settings.

A current of 8 A was hereby selected to meet a minimum total chlorine concentration of 2 mg/L after the electrolytic cell. Sampling was done before treatment (SP1), behind the ECl$_2$ (SP2) and after a hydraulic retention time (HRT) of 30 min behind the respective number of UV reactors. Samples for TOrCs, DOC, and THM analysis were quenched after 30 min using sulfite.

The MOX cell voltage was recorded in order to calculate energy demand and the "electric energy per order" (E_{EO}) of the ECl$_2$ (Equation (10)). As handling of contaminants at a waterworks site was prohibited, no spiking of the water with TOrCs was performed. Only selected substances, that were regularly present in the Elbe river water at sufficient concentrations (Table 2) were used for the evaluation of TOrC degradation. Due to the high costs of TOrC analysis the number of analyses and test settings was very limited in the here presented work and the results can only give a tendency of the degradation behavior.

Table 1. Test matrix and energy requirements for the different field test settings of UV/ECl$_2$.

Test Setting	Pump				ECl$_2$				UV		ECl$_2$/UV
	Flow Rate (L/h)	Energy Input (kWh/m^3)	Linear Flow Velocity (m/h)	Current (A)	Voltage (V)	Current Density (mA/cm^2)	Energy Input (kWh/m^3)	No. of UV Lamps	Energy Input (kWh/m^3)	Total Energy Input (kWh/m^3)
A	100	0.07	42	8	7.1	8.4	0.57	0	0.00	0.64
								1	0.36	0.99
								2	0.72	1.35
B	100	0.07	42	12	9.5	12.6	1.14	0	0.00	1.21
								2	0.72	1.93
C	300	0.20	126	12	11.7	12.6	0.47	0	0.00	0.67
								1	0.12	0.79
								2	0.24	0.91

Table 2. Analyzed TOrCs concentrations of UF Filtrate and relevant water quality parameters.

Parameter	Initial Concentrations/Values	Description
TOrC	1st and 2nd trial	
Benzotriazole (ng/L) CAS 95-14-7	410 and 440	Corrosion inhibitor
4-Methylbenzotriazole (ng/L) CAS 29878-31-7	190 and 170	Benzotriazole derivative used as a corrosion inhibitor
5-Methylbenzotriazole (ng/L) CAS 136-85-6	82 and 80	Antifreeze agent, corrosion inhibitor
Gabapentin (ng/L) CAS 60142-96-3	230 and 250	Anticonvulsant (Antiepileptic)
Iomeprol (ng/L) CAS 78649-41-9	380 and 520	X-ray contrast media
Oxipurinol (ng/L) CAS 2465-59-0	850 and 930	Active metabolite of Allopurinol (uricostatic)
Standard water quality parameters		
pH	8.1 and 7.9	
Temperature (°C)	18 and 25	
DOC (mg/L)	5.5 and 6.7	
Nitrate (mg/L NO_3^-) *	13 ± 2.7	
Nitrite (mg/L NO_2^-)*	0.043 ± 0.02	
Ammonium (mg/L NH_4) *	0.082 ± 0.07	
Total Hardness (mg/L $CaCO_3$)	67.5 ± 3.8	
Chloride (mg/L)	22.6 ± 3.7	

* Elbe river data derived from [38] and calculated by authors.

2.3. Energy Demand and Energy Efficiency

In order to estimate the energetic efficiency of the UV/ECl$_2$ AOP, the "electric energy per order" (E_{EO}), representing the electrical energy necessary for the degradation of one log unit of a contaminant, was calculated following Equation (10) [39]:

$$E_{EO} = \frac{P}{q \log\left(\frac{c_0}{c}\right)}, \tag{10}$$

where P: Electrical power applied to run the process (W), q: Flow rate (m^3/h), c_0: Initial contaminant concentration (µg/L) and c: Final contaminant concentration (µg/L)

The E_{EO} of the here generated data allows a comparison with alternative AOPs and may serve as requisite data for the evaluation of their economic feasibility and sustainability. It should be considered that for the generation of E_{EO} values of UV based AOPs in literature often only the energy required to power the UV lamp is considered. The energy required to produce, transport and dose e.g., H_2O_2 or chlorine are often neglected. Therefore, the E_{EO} calculated for the here presented approach distinguishes between the energy required for the UV and UV/ECl$_2$. Further deviations are to be expected when varying process capacity and source water quality [40].

2.4. Water Analysis

The quantification of hydroxyl radicals was carried out following [36] by adding methanol (Carl Roth, Karlsruhe, Germany) in excess as radical scavenger. According to [41] methanol reacts with OH and Cl radicals to formaldehyde following the Equations (11)–(14):

$$CH_3OH + \cdot OH \rightarrow CH_2OH \cdot + H_2O, \tag{11}$$

$$CH_3OH + \cdot Cl \rightarrow CH_2OH \cdot + HCl, \tag{12}$$

$$CH_2OH \cdot + O_2 \rightarrow O_2CH_2OH \cdot, \tag{13}$$

$$O_2CH_2OH \cdot \rightarrow CH_2O + HO_2 \cdot \tag{14}$$

Formaldehyde concentrations were analyzed following [42], were the extinction coefficient E_{DDL} caused by diacetyldihydrolutidin (DDL) formed during Hantzsch reaction was determined with a calibrated UV-VIS spectrometer (UV-1602, Shimadzu, Jena, Germany) using a 1 cm cuvette. Since the efficiency of ·OH radicals reacting with methanol is 93% [36], the OH radical production can be determined by the amount of DDL produced. As ·Cl is also an active species in this reaction [43], the calculated values only represent an upper limit of the ·OH production. Still, the obtained value is seen as useful to determine the capability of the formed radical species to react even with relatively inert C-H bonds like those of methanol.

Quantitative analysis of oxipurinol, gabapentin and iomeprol was done using HPLC-MS/MS following DIN 38407-47 using Agilent 1260 HPLC-System (Agilent Technologies, Waldbronn, Germany) und API 6500 MS/MS-System (AB Sciex, Darmstadt, Germany). Benzotriazoles were determined after automatic solid phase extraction with ASPEC (Gilson, Middleton, WI, USA) with HPLC-MS/MS Agilent 1200 HPLC-System (Agilent Technologies) and API 5500 MS/MS-System (AB Sciex) following TZW lab method. All substances had a Limit of Detection (LoD) of 0.01 µg/L except oxipurinol with a LoD of 0.025 µg/L. For analytic results below the LoD contaminant reduction rates and E_{EO} values were calculated by using the respective LoD.

The dissolved organic carbon (DOC) was analyzed with a TOC-V-CPH (Shimadzu, Kyoto, Japan) with integrated auto-sampler ASI-V (Shimadzu, Japan) following method DIN EN 1484:2019-04. The electrolytically produced Free Available Chlorine (FAC) and total chlorine was determined with AL410 photometer (Aqualytic, Dortmund, Germany) using Aqualytic DPD1 and DPD3 reagents with a measurement range between 0.01 and 6.0 mg/L. Electric conductivity (CDC401, Hach Düsseldorf, Germany) and pH (CDC401, Hach) were determined using a Hach HQ40d multimeter. THMs were analyzed following DIN EN ISO 10,301 using a 7890A GC/MS by Agilent Technologies (Santa Clara, CA, USA) with a detection limit of 0.1 µg/L.

3. Results and Discussion

3.1. Lab Test

Figure 4 shows the linear relation between current or current density applied at the MOX electrode and the formation of FAC at the given input chloride concentrations considering chloride concentration of 25, 50 and 100 mg/L in the synthetic water.

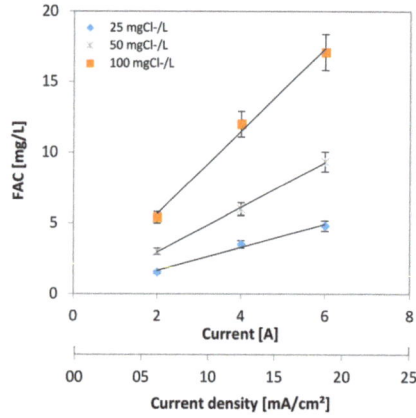

Figure 4. Relation between current and FAC concentration (n = 9) (Q = 75 L/h, EC = 400 µS/cm, T = 19 ± 1 °C).

Figure 5 shows the linear relation between the chloride concentration and the FAC as well as the FAC production rate achieved at the applied currents of 2, 4 and 6 A.

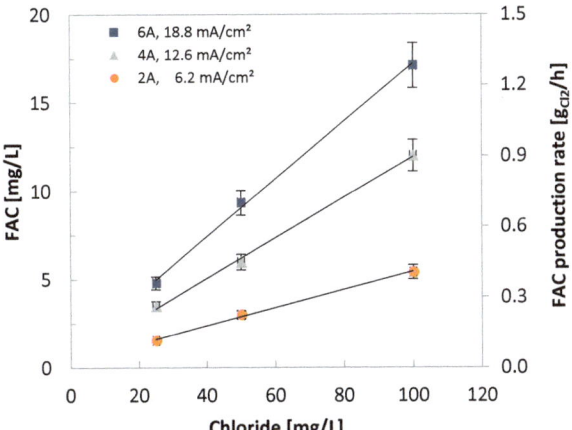

Figure 5. Relation between chloride concentration and FAC concentrations ($n = 9$) ($Q = 75$ L/h, EC = 400 µS/cm, T = 19 ± 1 °C).

Such relations were also derived in former studies as shown in [44]. The linearity largely simplifies the control of FAC generation, from an operational perspective. The currents at the MOX electrode can be easily adjusted in a treatment system within the here given current densities.

Figure 6 shows that charge specific chlorine production rates and current efficiencies are similar for currents of 2, 4, and 6 A with the applied cell. At higher currents, here shown with 8 A, the efficiency drops especially at higher chloride concentrations. This was related to the formation of gas bubbles at the anode surface.

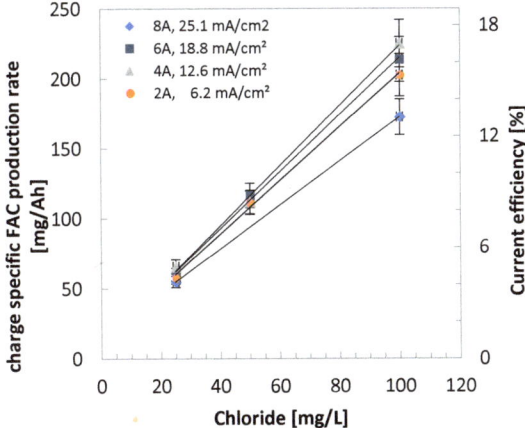

Figure 6. Relation of chloride concentration (with charge specific FAC production rate and current efficiency ($n = 9$) ($Q = 75$ L/h, EC = 400 µS/cm, T = 19 ± 1 °C).

Figure 7 shows the linear relation between the electrolytically produced FAC concentration and the resulting reduction in FAC when irradiated with 1 and 2 UV reactors.

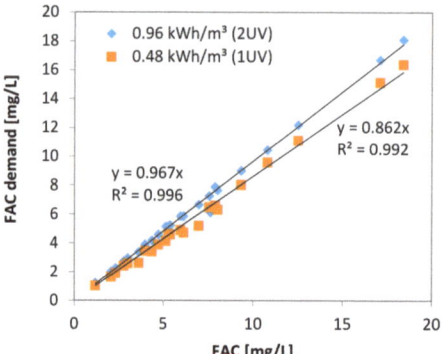

Figure 7. Chlorine consumption by UV treatment in dependence of FAC (Q = 75 L/h, EC = 400 µS/cm, T = 19 ± 1 °C).

At the given FAC concentrations any additional FAC is readily photolyzed independent of its concentration. Already with the energy input of one UV lamp (0.48 kWh/m³) 84% of the FAC is consumed. On average the chlorine concentration was reduced to 1.0 ± 0.5 mg/L. The trials with two UV lamps achieved a FAC reduction of 96% and the chlorine concentration was further reduced on average to 0.2 ± 0.1 mg/L, which would make FAC quenching as suggested by [24] dispensable. Whether the second UV lamp and with that the extra energy demanded is required, depends on the site-specific treatment targets, with regards to TOrC reduction, DBP formation potential, residual chlorine concentration and the design of the used UV-reactor.

Figure 8 shows the relation between FAC degradation and the radical formation by means of chlorine photolysis at the given pH values.

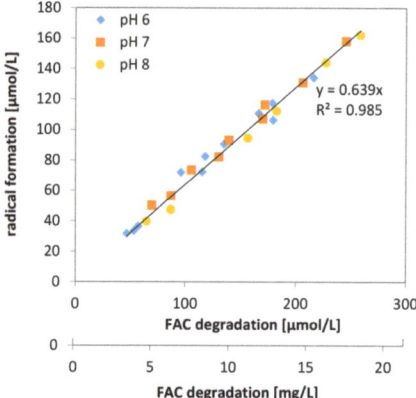

Figure 8. Radical formation in dependence of FAC demand at different pH (Q = 75 L/h, EC = 400 µS/L, T = 19 ± 1 °C) (n = 26).

The pH has no observable effect on the formation of radicals from photolyzed FAC. This relation has already been described by [28]. The average radical yield factor was constant over the here given FAC degradation range and with 64% (n = 26) significantly higher than the reported values of [45].

From an operational perspective the here achieved constant radical yield factor in the examined FAC concentration range offers the possibility to evaluate the degradation of TOrC by monitoring the chlorine degradation.

However, with rising pH the concentration of OCl⁻ is increasing (Equation (6)). As OCl⁻ has a slightly higher adsorption coefficient at 254 nm [28] slightly higher radical concentrations can be expected at higher pH values. On the other hand radical scavenger effects of HOCl and ClO⁻ may result in additional chlorine degradation and the formation of oxy-chlorine radicals. Such scavenger effects are stronger at basic pH [29].

3.2. Field Tests

The total chlorine and FAC concentration measured directly after the ECl$_2$ or the respective number of UV lamps (SP 3) during the treatment of Elbe river water is shown in Figure 9.

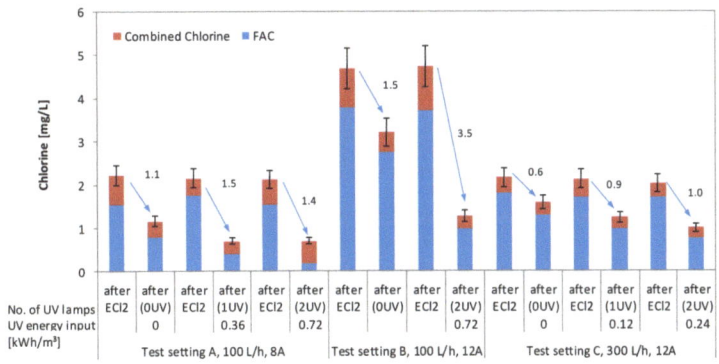

Figure 9. FAC and total chlorine concentrations measured in three different settings during field tests with Elbe river water.

The average concentrations of FAC directly after the ECl$_2$ reached 1.6 (n = 3), 3.8 (n = 2), and 1.7 mg/L (n = 3) for test settings A, B and C, respectively. Average total chlorine concentrations were 2.2 (n = 3), 4.7 (n = 2) and 2.1 mg/L (n = 3). The chlorine degradation was dependent on the initial chlorine concentration and the UV energy input into the water as already shown in Figure 7.

The reduction in chlorine concentration of the tested water during the trials without UV radiation indicated with 1.1, 1.4 and 0.6 mg/L a high DBP formation potential, typical for surface waters. The lower chlorine demand values during test setting C were related to the shorter HRT in the final storage tank.

The degradation of the tested TOrCs after ECl$_2$ only and after the photolysis of chlorine with the respective UV energy input is shown in Figure 10a–d.

The input concentrations for gabapentin and oxipurinol were 0.25 and 0.93 µg/L respectively. These substances were nearly completely degraded at all given test settings alone by the produced chlorine. After passing the first UV lamp in test setting A, eventually still available quantities of gabapentin and oxipurinol were degraded below the LoD. In contrast to that benzotriazole and 5-methylbenzotriazole could be degraded by only by 5% and 4% with ECl$_2$ alone during test setting A. Iomeprol and 4-methylbenzotriazole were degraded by 8% and 11%. Higher chlorine concentrations produced during test setting B did not substantially improve the degradation. However, in combination with UV radiation the degradation could be significantly increased. In test setting A the degradation of benzotriazole was increased to 49% and 73% with an energy input of 0.36 (1 UV) and 0.72 kWh/m³ (2 UV), respectively. In combination with UV radiation the additional chlorine made available during test setting B further increased the degradation of benzotriazole to 89%. Iomeprol degradation could be increased to 83% and 82% during test settings A and B, respectively, with the energy of two UV lamps (0.72 kWh/m³). The behavior of 4-methylbenzotriazole and 5-methylbenzotriazole degradation was similar.

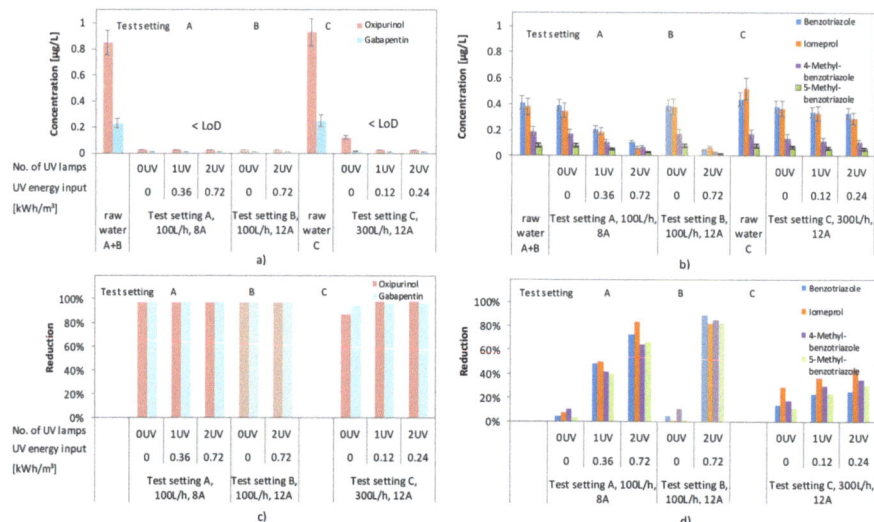

Figure 10. Concentrations of tested TOrCs (**a**,**b**) and degradation percentages (**c**,**d**) concentrations measured in three different settings during field tests with Elbe river water.

Even though nearly equal amounts of chlorine were available, the degradation of all substances susceptible for the UV/ECl$_2$ AOP was largely reduced during test setting C. As also smaller amounts of chlorine were degraded, it can be concluded that at a flow rate of 300 L/h the HRT and by that the UV energy input was not sufficient to produce equal amounts of radicals as in test setting A.

3.3. Behavior of DOC and Formation of Disinfection-by-Products and Metabolites (Toxicity) during Field Tests

The mean DOC of Elbe river water in Dresden, Germany, is about 5.2 mg/L (4.6–6.0 mg/L, $n = 325$) [46]. In test settings A and B the river water showed a slightly elevated DOC of 6.7 mg/L, and 5.6 mg/L in test setting C after ultrafiltration. The comparatively high DOC in test settings A and B was caused by runoff after heavy rainfall during the period of this test, whereas test setting C was performed a rainless period. DOC was reduced by 28% and 23% in test setting A and B and by only 3% in test setting C through ECl$_2$ alone as shown in Figure 11. The degradation of DOC without UV is explained by a direct oxidation of organics during the passage through the electrolytic cell as e.g., described in [47,48]. During test settings A and B the degradation of DOC was substantially higher compared to test setting C. This is related to the higher HRT and the nature of the DOC which is more easily degradable when containing runoff.

Figure 12 shows that DBPs measured as THMs have been formed after ECl$_2$ and for THMs the strict German threshold of 50 µg/L could not be adhered to for the majority of the tested settings. However, it was always possible to adhere to the EU guideline values of < 100 µg/L. The EU guideline values could always be maintained (Table 3). During the trials with UV applied the formation of THM was substantially lower compared to trials without UV radiation. Less THMs were formed because smaller amounts of chlorine were available after photolysis. During test setting A with 2 UV lamps the THM formation made up only 47% compared to the trial without UV radiation, by which the German guideline limits could be maintained. DBP formation is therefore related to residual chlorine levels rather than the AOP itself. This is confirmed by other studies where no significant increase of organic DBP formation during the application of UV/Cl$_2$ was found and most organic DBPs formed were related to the application of chlorine itself [49,50]. Also [24] found no significant quantities of THMs in UV/Cl$_2$ when adding 6 mg/L of chlorine prior to the UV lamp, using simulated wastewater with a DOC of 46 mg/L (COD ~120 mg/L). In this study, residual chlorine was completely quenched after the UV lamp.

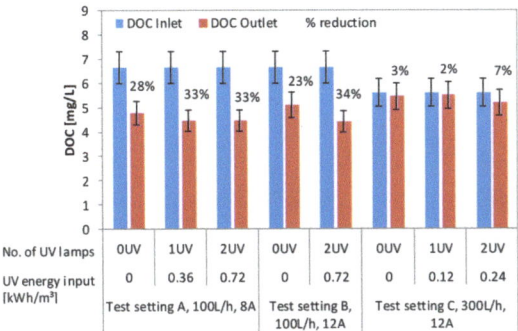

Figure 11. DOC concentrations measured in three different settings during field tests with Elbe river water.

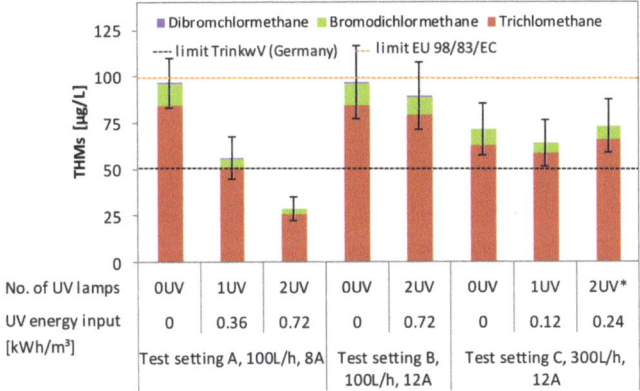

* insufficient quenching of sample caused further generation of THMs and AOX

Figure 12. THM concentrations measured in three different settings during field tests with Elbe river water.

Table 3. Selected guideline values concerning the formation of THMs during drinking water disinfection.

Parameter	Germany [51]	EU [52]	India [53]	WHO [54]
THM (µg/L)	10 [a]/50 [b]	100	-	
Bromoform (µg/L)			100	100
Dibromochloromethane (µg/L)			100	100
Bromodichloromethane (µg/L)			60	60
Chloroform (µg/L)			200	300

[a] At the end of treatment, [b] Point of use.

To prevent excessive THM formation it is therefore suggested to closely monitor the residual chlorine concentration directly after the UV lamps and reduce the concentration whenever required by increasing the UV radiation or the flow rate. The field test data in Figure 9 indicate that residual chlorine concentrations around 0.2 mg/L, as required by many drinking water guidelines, seems to be promising to meet strict German guidelines for THMs.

From an operational perspective such chlorine levels must be monitored by integrating online probes into the algorithm controlling the chlorine production.

3.4. Energy Efficiency

Figure 13a,b and Table A1 (in Appendix A) show the E_{EO} values achieved for the degradation of the analyzed substances under the given conditions differentiating between the energy input into the chlorine production by means of ECl_2 and photolysis of chlorine through UV radiation.

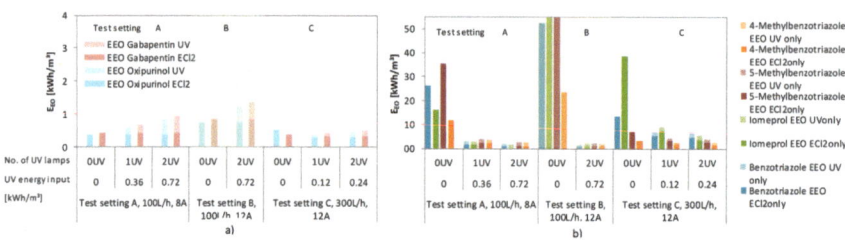

Figure 13. E_{EO}s for gabapentin and oxipurinol (**a**) and 4-methylbenzotriazole, 5-methylbenzotriazole iomeprol and benzotriazole (**b**) calculated from three different settings during field tests with Elbe river water.

For test setting A the E_{EO} for oxipurinol and gabapentin increased from 0.38 to 0.84 and from 0.42 to 0.95 kWh m^{-3} order^{-1} by increasing the energy input for chlorine photolysis from 0 to 0.72 kWh/m^3. This increase is based on the fact that the additional energy input of the UV lamps does not substantially contribute to the degradation, as those substances are readily degraded by ECl_2 alone as shown in Figure 10a. In contrast to that, the chlorine photolysis substantially reduced the E_{EO} for benzotriazole, iomeprol, 5-methylbenzotriazole, and 4-methylbenzotriazole during all test settings. In test setting A the E_{EO} for e.g., 5-methylbenzotriazole was reduced from 35.6 to 2.7 kWh m^{-3} order^{-1} with an energy input for chlorine photolysis caused by the UV lamps of 0.72 kWh/m^3 (2 UV). For benzotriazole the E_{EO} was reduced from 26.5 to 2.3 kWh m^{-3} order^{-1} with the same energy input. During test setting B the UV lamps reduced the E_{EO} for benzotriazole from a rather theoretical value of 52.5 without UV lamps to 1.9 kWh m^{-3} order^{-1} with an UV energy input of 0.72 kWh/m^3 (2 UV). This shows the positive effect of adding a minor quantity of additional energy in the form of UV light into the setting in order to substantially reducing the E_{EO}.

Despite the substantial improvement of TOrC degradation through the photolysis of chlorine the achieved E_{EO} values are higher compared to the literature reported E_{EO} values for e.g., UV/H_2O_2 AOP. [24] reported an E_{EO} for benzotriazole for H_2O_2/UV of 0.52 kWh m^{-3} order^{-1}, neglecting hereby the energy required to produce and supply H_2O_2. However, even after removing the energy required for ECl_2 from the UV/ECl_2 AOP the resulting E_{EO} for e.g., benzotriazole was still higher with 1.0 and 1.2 kWh m^{-3} order^{-1} for test settings A and B. Whether this was related to the different water qualities tested cannot be evaluated here. Due to the limited data available the E_{EO} values calculated and the energy consumed can only be taken as an indication and further studies are required. A reduction of the energy demand by 30–70% by using UV/Cl_2 instead of UV/H_2O_2 as estimated by [24] seems not likely with the settings examined in this work.

3.5. Operational Experience and Optimization

The long term UV/ECl_2 trial was conducted over a period of 10 months during which 1023 m^3 (3.4 m^3/d) of water were treated. The system operated without any interruption caused by technical reasons. The only maintenance required was the manual cleaning of the quartz glass sleeves, from which deposits needed to be removed in a one-month interval, as those have reduced the UV dose. The components, and especially the electrolytic cells, did not show any sign of wear by the time the trial was finished. The concentration of total hardness found in Elbe river water (Table 2) have proven to be harmless with regards the formation of calcareous deposits in the cell surface with the selected polarity inversion intervals of three hours. In future tests these intervals should be extended. No chemicals for the operation and maintenance of the system were required.

Considering the high degradation rates of TOrCs, the comparably low formation of organic DBPs and the low E_{EO} for the setting with a flow rate of 100 L/h, a current at the MOX of 8 A and the UV radiation of 2 × 36 W lamps (0.72 kWh/m^3) was most promising and should constitute the base for further optimization. At this setting a total energy demand for the AOP process including pumping would sum up to 1.35 kWh/m^3. In remote regions this energy can be supplied easily with renewable PV Energy systems as already shown in [55].

4. Conclusions

The photolysis of chlorine (UV/Cl$_2$) was identified as promising AOP for the degradation of TOrCs present in drinking water sources. In the here presented work the chlorine required to run an UV/Cl$_2$ AOP was produced by means of inline electrolysis from the natural chloride content of the water and by that substituted the external supply of chlorine, allowing the operation of an AOP without any additional need of chemicals.

For FAC concentrations between 1 and 18 mg/L an UV energy input of 0.48 kWh/m^3 photolyzed 84% of the chlorine to form OH and Cl radicals. The degradation could be increased to 97% by doubling the UV energy input for the photolysis. The tests have further shown that the molar radical yield factor is about 64% and also very constant over the tested FAC range of 1–18 mg/L. In principle this allows to control the radical formation by adjusting the chlorine concentration simply through variation of the electric current at the MOX electrode.

By combining ECl$_2$ with UV the degradation of benzotriazole and iomeprol can be increased from 5% to 89% and from 8% to 84%, respectively, compared to the application of ECl$_2$ alone.

In most of the test settings the formation of organic DBPs measured as THMs have reached concentrations above the strict German guideline values for drinking water of 50 µg/L. EU guideline values could be adhered to with all of the tested settings. The formation of such DBPs was related to elevated residual chlorine concentrations after the photolysis of chlorine and can be controlled by reducing the residual chlorine concentrations to levels that are adequate for drinking water disinfection. This requires site specific adaption of the treatment process.

The calculated E_{EO} values have found to be higher than literature values of alternative AOPs. The identified optimization potential should be considered to reduce the overall energy requirement of this technology. The main advantage of the here presented approach is the application of an effective AOP to degrade TOrC independent of any external chemical input. By that the UV/ECl$_2$ AOP constitutes an alternative treatment approach especially for decentralized applications.

Further data and research is required to confirm the here presented results and the effect of the AOP on other TOrCs. Optimization potential is given by increasing the HRT in the UV reactors by adapting the reactor design.

Interesting for future application may also be further treatment of treated wastewater prior to its discharge into the environment.

Author Contributions: Conceptualization, methodology, investigation, P.O., F.-M.K., K.M., A.G., F.B., P.M. and T.G. (Tobias Gerhardt); writing—original draft preparation, P.O. and R.W.; writing—review and editing, P.O., T.G. (Thomas Grischek), F.-M.K., K.M. and T.G. (Tobias Gerhardt); supervision, T.G. (Thomas Grischek); funding acquisition, project administration, P.O. and F.-M.K. All authors have read and agreed to the published version of the manuscript.

Funding: This research was funded by the German Ministry of Education and Research (BMBF), program "KMU-Innovativ" grant no. 02WQ1395. The APC was funded by the Open Access Publication Fund of Hochschule für Technik und Wirtschaft Dresden, University of Applied Sciences, and the Deutsche Forschungsgemeinschaft (DFG, German Research Foundation)—432908064.

Conflicts of Interest: There are no conflicts to declare.

Appendix A

Table A1. E_{EO} values achieved during field tests with Elbe river water considering ECl_2 and UV/ECl_2.

	Field test								Literature values	
	$E_{EO\text{-}(ECl_2)}$ [kWh/(order·m³)]		Elimination [%]		$E_{EO\text{-}(ECl_2 + UV)}$ [kWh/(Order·m³)]		Elimination [%]		$E_{EO\text{-}UV/Cl_2}$ [kWh/(Order·m³)]	
	min	max	min	max	min	max	min	max		
Benzotriazole	13.6	52.5	5%	14%	1.9	7.3	25%	89%	0.5	[24]
4-Methyl-benzotriazole	1.3	23.6	11%	18%	0.6	1.5	29%	85%	n.a.	
5-Methyl-benzotriazole	7.2	213.9	1%	11%	0.9	1.6	24%	83%	n.a.	
Iomeprol	16.1	996.2	0%	29%	1.7	9.3	37%	83%	n.a.	
Gabapentin	0.4	≤0.8	94%	≥98%	≤0.4	1.4	≥96%	≥98%	n.a.	
Oxipurinol	≤0.4	≤0.7	87%	≥97%	≤0.3	≤0.7	≥97%	≥97%	n.a.	

References

1. Kümmerer, K. *Pharmaceuticals in the Environment*; Springer Berlin Heidelberg: Berlin/Heidelberg, Germany, 2008; ISBN 978-3-540-74663-8.
2. Ternes, T.; Joss, A. (Eds.) *Human Pharmaceuticals, Hormones and Fragrances. The Challenge of Micropollutants in Urban Water Management*; IWA Publ: London, UK, 2008; ISBN 9781843390930.
3. Houtman, C.J. Emerging contaminants in surface waters and their relevance for the production of drinking water in Europe. *J. Integr. Environ. Sci.* **2010**, *7*, 271–295. [CrossRef]
4. Montes-Grajales, D.; Fennix-Agudelo, M.; Miranda-Castro, W. Occurrence of personal care products as emerging chemicals of concern in water resources: A review. *Sci. Total Environ.* **2017**, *595*, 601–614. [CrossRef] [PubMed]
5. Mohapatra, S.; Huang, C.-H.; Mukherji, S.; Padhye, L.P. Occurrence and fate of pharmaceuticals in WWTPs in India and comparison with a similar study in the United States. *Chemosphere* **2016**, *159*, 526–535. [CrossRef] [PubMed]
6. Tran, N.H.; Reinhard, M.; Gin, K.Y.-H. Occurrence and fate of emerging contaminants in municipal wastewater treatment plants from different geographical regions-a review. *Water Res.* **2018**, *133*, 182–207. [CrossRef]
7. Kibuye, F.A.; Gall, H.E.; Elkin, K.R.; Ayers, B.; Veith, T.L.; Miller, M.; Jacob, S.; Hayden, K.R.; Watson, J.E.; Elliott, H.A. Fate of pharmaceuticals in a spray-irrigation system: From wastewater to groundwater. *Sci. Total Environ.* **2019**, *654*, 197–208. [CrossRef]
8. Balakrishna, K.; Rath, A.; Praveenkumarreddy, Y.; Guruge, K.S.; Subedi, B. A review of the occurrence of pharmaceuticals and personal care products in Indian water bodies. *Ecotoxicol. Environ. Saf.* **2017**, *137*, 113–120. [CrossRef]
9. Hellauer, K.; Mergel, D.; Ruhl, A.; Filter, J.; Hübner, U.; Jekel, M.; Drewes, J. Advancing Sequential Managed Aquifer Recharge Technology (SMART) using different intermediate oxidation processes. *Water* **2017**, *9*, 221. [CrossRef]
10. Hirsch, R.; Ternes, T.; Haberer, K.; Kratz, K.-L. Occurrence of antibiotics in the aquatic environment. *Sci. Total Environ.* **1999**, *225*, 109–118. [CrossRef]
11. Kolpin, D.W.; Furlong, E.T.; Meyer, M.T.; Thurman, E.M.; Zaugg, S.D.; Barber, L.B.; Buxton, H.T. Pharmaceuticals, hormones, and other organic wastewater contaminants in U.S. streams, 1999-2000: A national reconnaissance. *Environ. Sci. Technol.* **2002**, *36*, 1202–1211. [CrossRef]
12. Schwab, B.W.; Hayes, E.P.; Fiori, J.M.; Mastrocco, F.J.; Roden, N.M.; Cragin, D.; Meyerhoff, R.D.; D'Aco, V.J.; Anderson, P.D. Human pharmaceuticals in US surface waters: A human health risk assessment. *Regul. Toxicol. Pharmacol.* **2005**, *42*, 296–312. [CrossRef]
13. Avisar, D.; Levin, G.; Gozlan, I. The processes affecting oxytetracycline contamination of groundwater in a phreatic aquifer underlying industrial fish ponds in Israel. *Environ. Earth Sci.* **2009**, *59*, 939–945. [CrossRef]
14. Ebele, A.J.; Abou-Elwafa Abdallah, M.; Harrad, S. Pharmaceuticals and personal care products (PPCPs) in the freshwater aquatic environment. *Emerg. Contam.* **2017**, *3*, 1–16. [CrossRef]
15. Bolong, N.; Ismail, A.F.; Salim, M.R.; Matsuura, T. A review of the effects of emerging contaminants in wastewater and options for their removal. *Desalination* **2009**, *239*, 229–246. [CrossRef]
16. Dolar, D.; Gros, M.; Rodriguez-Mozaz, S.; Moreno, J.; Comas, J.; Rodriguez-Roda, I.; Barceló, D. Removal of emerging contaminants from municipal wastewater with an integrated membrane system, MBR-RO. *J. Hazard. Mater.* **2012**, *239–240*, 64–69. [CrossRef] [PubMed]
17. Snyder, S.A.; Adham, S.; Redding, A.M.; Cannon, F.S.; DeCarolis, J.; Oppenheimer, J.; Wert, E.C.; Yoon, Y. Role of membranes and activated carbon in the removal of endocrine disruptors and pharmaceuticals. *Desalination* **2007**, *202*, 156–181. [CrossRef]
18. Yang, Y.; Pignatello, J.J.; Ma, J.; Mitch, W.A. Comparison of halide impacts on the efficiency of contaminant degradation by sulfate and hydroxyl radical-based advanced oxidation processes (AOPs). *Environ. Sci. Technol.* **2014**, *48*, 2344–2351. [CrossRef]
19. Rosenfeldt, E.J.; Linden, K.G.; Canonica, S.; Gunten, U. von. Comparison of the efficiency of *OH radical formation during ozonation and the advanced oxidation processes O_3/H_2O_2 and UV/H_2O_2. *Water Res.* **2006**, *40*, 3695–3704. [CrossRef]
20. Fang, J.; Fu, Y.; Shang, C. The roles of reactive species in micropollutant degradation in the UV/free chlorine system. *Environ. Sci. Technol.* **2014**, *48*, 1859–1868. [CrossRef]

21. Wang, W.-L.; Wu, Q.-Y.; Huang, N.; Wang, T.; Hu, H.-Y. Synergistic effect between UV and chlorine (UV/chlorine) on the degradation of carbamazepine: Influence factors and radical species. *Water Res.* **2016**, *98*, 190–198. [CrossRef]
22. Xiang, Y.; Fang, J.; Shang, C. Kinetics and pathways of ibuprofen degradation by the UV/chlorine advanced oxidation process. *Water Res.* **2016**, *90*, 301–308. [CrossRef]
23. Rott, E.; Kuch, B.; Lange, C.; Richter, P.; Kugele, A.; Minke, R. Removal of emerging contaminants and estrogenic activity from wastewater treatment plant effluent with UV/chlorine and UV/H_2O_2 advanced oxidation treatment at pilot scale. *Int. J. Environ. Res. Public Health* **2018**, *15*, 935. [CrossRef] [PubMed]
24. Sichel, C.; Garcia, C.; Andre, K. Feasibility studies: UV/chlorine advanced oxidation treatment for the removal of emerging contaminants. *Water Res.* **2011**, *45*, 6371–6380. [CrossRef] [PubMed]
25. Latimer, W.M. *The Oxidation States of the Elements and Their Potentials in Aqueous Solutions*, 2nd ed.; Prentice-Hall: New York, NY, USA, 1952.
26. Watts, M.J.; Linden, K.G. Chlorine photolysis and subsequent OH radical production during UV treatment of chlorinated water. *Water Res.* **2007**, *41*, 2871–2878. [CrossRef] [PubMed]
27. Baxendale, J.H.; Wilson, J.A. The photolysis of hydrogen peroxide at high light intensities. *Trans. Faraday Soc.* **1957**, *53*, 344. [CrossRef]
28. Feng, Y.; Smith, D.W.; Bolton, J.R. Photolysis of aqueous free chlorine species (HOCl and OCl) with 254 nm ultraviolet light. *J. Environ. Eng. Sci.* **2007**, *6*, 277–284. [CrossRef]
29. Kishimoto, N. State of the art of UV/chlorine Advanced Oxidation Processes: Their mechanism, byproducts formation, process variation, and applications. *J. Wat. Environ. Tech.* **2019**, *17*, 302–335. [CrossRef]
30. Grebel, J.E.; Pignatello, J.J.; Mitch, W.A. Effect of halide ions and carbonates on organic contaminant degradation by hydroxyl radical-based advanced oxidation processes in saline waters. *Environ. Sci. Technol.* **2010**, *44*, 6822–6828. [CrossRef]
31. Haaken, D.; Dittmar, T.; Schmalz, V.; Worch, E. Influence of operating conditions and wastewater-specific parameters on the electrochemical bulk disinfection of biologically treated sewage at boron-doped diamond (BDD) electrodes. *Desalin. Water Treat.* **2012**, *46*, 160–167. [CrossRef]
32. Haaken, D.; Dittmar, T.; Schmalz, V.; Worch, E. Disinfection of biologically treated wastewater and prevention of biofouling by UV/electrolysis hybrid technology: Influence factors and limits for domestic wastewater reuse. *Water Res.* **2014**, *52*, 20–28. [CrossRef]
33. Kraft, A. Electrochemical water disinfection: A short review. *Platinum Metals Review* **2008**, *52*, 177–185. [CrossRef]
34. Kraft, A.; Stadelmann, M.; Blaschke, M.; Kreysig, D.; Sandt, B.; Schröder, F.; Rennau, J. Electrochemical water disinfection: Part I: Hypochlorite production from very dilute chloride solutions. *J. Appl. Electrochem.* **1999**, *29*, 859–866. [CrossRef]
35. Buxton, G.V.; Subhani, M.S. Radiation chemistry and photochemistry of oxychlorine ions. Part 2. Photodecomposition of aqueous solutions of hypochlorite ions. *J. Chem. Soc. Faraday Trans.* **1972**, *68*, 958. [CrossRef]
36. Asmus, K.D.; Moeckel, H.; Henglein, A. Pulse radiolytic study of the site of hydroxyl radical attack on aliphatic alcohols in aqueous solution. *J. Phys. Chem.* **1973**, *77*, 1218–1221. [CrossRef]
37. Abdelraheem, W.H.; Nadagouda, M.N.; Dionysiou, D.D. Solar light-assisted remediation of domestic wastewater by NB-TiO_2 nanoparticles for potable reuse. *Applied Catalysis B Environ.* **2020**, *269*, 118807. [CrossRef]
38. LfULG, online data base iDA, Saxon State Authority for Environment, Agriculture and Geology. 2020. Available online: https://www.umwelt.sachsen.de/umwelt/infosysteme/ida/ (accessed on 5 August 2020).
39. Bolton, J.R.; Bircher, K.G.; Tumas, W.; Tolman, C.A. Figures-of-merit for the technical development and application of advanced oxidation technologies for both electric- and solar-driven systems. *Pure Appl. Chem.* **2001**, *2001*, 627–637. [CrossRef]
40. Bolton, J.R.; Stefan, M.I. Fundamental photochemical approach to the concepts of fluence (UV dose) and electrical energy efficiency in photochemical degradation reactions. *Res. Chem. Intermed.* **2002**, *28*, 857–870. [CrossRef]
41. Monod, A.; Chebbi, A.; Durand-Jolibois, R.; Carlier, P. Oxidation of methanol by hydroxyl radicals in aqueous solution under simulated cloud droplet conditions. *Atmos. Environ.* **2000**, *34*, 5283–5294. [CrossRef]
42. Nash, T. Colorimetric determination of formaldehyde under mild conditions. *Nature* **1952**, *170*, 976. [CrossRef]

43. Payne, W.A.; Brunning, J.; Mitchell, M.B.; Stief, L.J. Kinetics of the reactions of atomic chlorine with methanol and the hydroxymethyl radical with molecular oxygen at 298 K. *Int. J. Chem. Kinet.* **1988**, *20*, 63–74. [CrossRef]
44. Kraft, A.; Blaschke, M.; Kreysig, D.; Sandt, B.; Schröder, F.; Rennau, J. Electrochemical water disinfection. Part II: Hypochlorite production from potable water, chlorine consumption and the problem of calcareous deposits. *J. Appl. Electrochem.* **1999**, *29*, 895–902. [CrossRef]
45. Jin, J.; El-Din, M.G.; Bolton, J.R. Assessment of the UV/chlorine process as an advanced oxidation process. *Water Res.* **2011**, *45*, 1890–1896. [CrossRef] [PubMed]
46. Paufler, S.; Grischek, T.; Benso, M.; Seidel, N.; Fischer, T. The impact of river discharge and water temperature on manganese release from the riverbed during riverbank filtration: A case study from Dresden, Germany. *Water* **2018**, *10*, 1476. [CrossRef]
47. Houk, L.L.; Johnson, S.K.; Feng, J.; Houk, R.S.; Johnson, D.C. Electrochemical incineration of benzoquinone in aqueous media using a quaternary metal oxide electrode in the absence of a soluble supporting electrolyte. *J. Appl. Electrochem.* **1998**, *28*, 1167–1177. [CrossRef]
48. Johnson, S.K.; Houk, L.L.; Feng, J.; Houk, R.S.; Johnson, D.C. Electrochemical incineration of 4-chlorophenol and the identification of products and intermediates by mass spectrometry. *Environ. Sci. Technol.* **1999**, *33*, 2638–2644. [CrossRef]
49. Wang, D.; Bolton, J.R.; Andrews, S.A.; Hofmann, R. Formation of disinfection by-products in the ultraviolet/chlorine advanced oxidation process. *Sci. Total Environ.* **2015**, *518–519*, 49–57. [CrossRef]
50. Yang, X.; Sun, J.; Fu, W.; Shang, C.; Li, Y.; Chen, Y.; Gan, W.; Fang, J. PPCP degradation by UV/chlorine treatment and its impact on DBP formation potential in real waters. *Water Res.* **2016**, *98*, 309–318. [CrossRef]
51. Bundesministerium für Justiz und Verbraucherschutz. Verordnung über die Qualität von Wasser für den menschlichen Gebrauch (Trinkwasserverordnung-TrinkwV). 2001. Gesetze im Internet. Available online: https://www.gesetze-im-internet.de/trinkwv_2001/BJNR095910001.html (accessed on 18 November 2020).
52. European Commission. COUNCIL DIRECTIVE 98/83/EC of 3 November 1998 on the quality of water intended for human consumption. 1998. EUR-lex. Available online: https://eur-lex.europa.eu/legal-content/en/TXT/?uri=CELEX:52017PC0753 (accessed on 18 November 2020).
53. IS 10500. *Drinking Water—Specification*; Indian Standard; Bureau of Indian Standards: New Delhi, India, (Second Revision); 2012; Law-resource; Available online: https://law.resource.org/pub/in/bis/S06/is.10500.2012.pdf (accessed on 18 November 2020)Law-resource.
54. *Guidelines for Drinking-Water Quality*, 4th ed; World Health Organization: Geneva, Switzerland, Incorporating 1st Addendum; 2017; Available online: https://apps.who.int/iris/bitstream/handle/10665/44584/9789241548151_eng.pdf;jsessionid=592F2A53BA42E1A3A26CABA04706353E?sequence=1 ISBN 978-9241549950.
55. Otter, P.; Malakar, P.; Jana, B.; Grischek, T.; Benz, F.; Goldmaier, A.; Feistel, U.; Jana, J.; Lahiri, S.; Alvarez, J. Arsenic removal from groundwater by solar driven inline-electrolytic induced co-precipitation and filtration—A long term field test conducted in West Bengal. *IJERPH* **2017**, *14*, 1167. [CrossRef]

Publisher's Note: MDPI stays neutral with regard to jurisdictional claims in published maps and institutional affiliations.

© 2020 by the authors. Licensee MDPI, Basel, Switzerland. This article is an open access article distributed under the terms and conditions of the Creative Commons Attribution (CC BY) license (http://creativecommons.org/licenses/by/4.0/).

Article

Effect of Zr Impregnation on Clay-Based Materials for H_2O_2-Assisted Photocatalytic Wet Oxidation of Winery Wastewater

Vanessa Guimarães *[ID], Ana R. Teixeira, Marco S. Lucas[ID] and José A. Peres[ID]

Vila Real Chemistry Center (CQVR), University of Trás-os-Montes and Alto Douro (UTAD), Quinta de Prados, 5000-801 Vila Real, Portugal; ritamourateixeira@gmail.com (A.R.T.); mlucas@utad.pt (M.S.L.); jperes@utad.pt (J.A.P.)
* Correspondence: guimavs@gmail.com

Received: 26 October 2020; Accepted: 30 November 2020; Published: 2 December 2020

Abstract: UV-activated Zr-doped composites were successfully produced through the impregnation of Zr on the crystal lattice of different clay materials by a one-step route. Fixing the amount of Zr available for dopage (4%), the influence of different supports, submitted to different chemical treatments, on the photocatalytic activity of the resulting Zr-doped pillared clay materials (PILC) was assessed. Both chemical characterization and structural characterization suggest that the immobilization of Zr on montmorillonite and PILC structures occurred through isomorphic substitution between Si and Zr in the tetrahedral sheet of the clay material. This structural change was demonstrated by significant modifications on Si-OH stretching vibrations (1016 cm^{-1}, 1100 cm^{-1} and 1150 cm^{-1}), and resulted in improved textural properties, with an increase in surface area from 8 m^2/g (natural montmorillonite) to 107 m^2/g after the pillaring process, and to 118 m^2/g after the pillaring and Zr-doping processes ((Zr)Al-Cu-PILC). These materials were tested in the UV-photodegradation of agro-industrial wastewater (AIW), characterized by high concentrations of recalcitrant contaminants. After Zr-dopage on AlCu-PILC heterogeneous catalyst, the total organic carbon (*TOC*) removals of 8.9% and 10.4% were obtained through adsorption and 77% and 86% by photocatalytic oxidation, at pH 4 and 7, respectively. These results suggest a synergetic effect deriving from the combination of Zr and Cu on the photocatalytic degradation process.

Keywords: Zr-doped materials; pillared clays; advanced oxidation processes; photocatalysis; agro-industrial wastewater

---

1. Introduction

Agro-industrial activities are one of the main sources of wastewater pollution and its impact on the environment has received special attention in recent years [1,2]. Winery wastewater (WW) is characterized by high load of recalcitrant organic compounds [1,3], and its unregulated discharge represents a great threat to aquatic ecosystems and human health [4]. In this regard, the development of effective and low cost methods for the treatment of WW is now imperative.

Currently, different techniques have been developed to treat this type of effluent, including adsorption [5], coagulation [6] and biological processes [7]. However, some of the drawbacks include the limited adsorption capacity and the formation of a potential second pollution source, since these processes only transfer contaminants from one phase to another instead of destroying them [8]. Biological degradation is the most common process applied, however, the microbial activity can be inhibited by the recalcitrant character and toxicity of the organic contaminants [9]. To overcome these problems, advanced oxidation processes (AOPs) have been proposed as effective, fast and non-expensive technologies for the degradation of recalcitrant contaminants [10–12]. Different homogeneous AOPs have already been applied in the treatment of agro-industrial wastewaters, particularly ozonation [13], Fenton [14] and photo-Fenton (solar and UV-A LEDs) processes [15]. Nonetheless, despite the interesting results obtained, Fenton processes have important limitations, namely the acidic conditions needed to improve the degradation efficiency, the additional procedure to remove the homogeneous catalyst from treated effluent, and the neutralization of the treated effluent to meet the legal discharge limits (pH 6.0–9.0) [16]. In order to overcome these drawbacks, heterogeneous AOPs have been the main focus of research interest in the last years, due to the substantial reduction in the effective costs associated with the sludge treatment, as well as the easy catalyst recovery and potential reuse [17,18].

The Catalytic Wet Peroxide Oxidation (CWPO) process is one of the most efficient, economical and environmental-friendly advanced oxidation processes for the treatment of non-biodegradable pollutants under milder conditions, and was successfully applied in the treatment of several organic contaminants, using different types of supports [19–22]. Considering the recalcitrant character of some type of effluents and the fairly poor results that have been obtained so far with the conventional Fenton process, the combination of UV light irradiation in the oxidation processes was proposed with a significant improvement in degradation efficiency [22–26].

The application of CWPO process in the treatment of WW is quite limited. Among our previous research studies, where different clay-based supported catalysts were applied for the first time in the heterogeneous UV/H_2O_2-assisted treatment of a real winery wastewater [27,28], only few heterogeneous catalysts including Fe-graphite [29] and natural clay [30] were applied to improve the efficiency of CWPO in the treatment of a winery wastewater. The results obtained by other studies are very interesting, with significant *TOC* removals, 80% and 55%, respectively. However, the authors did not explore the influence of crucial operational conditions, namely the variation of pH conditions, which may affect the efficiency of the photo-catalytic process.

According to our previous studies, the application of AlCu pillared clay (PILC) as heterogeneous catalyst revealed great stability along the treatment process and a high performance at neutral pH conditions, reaching a *TOC* removal of 83% ([H_2O_2]$_0$ = 98 mM; catalyst dosage = 3.00 g/L). This is particularly important, once it allows the catalyst reuse and eliminates the cost of effluent neutralization before its discharge. Thus, in this work, a novel Zr-doped AlCu-pillared clay ((Zr)AlCu-PILC) was prepared, attending to the ZrO_2 excellent electrical, mechanical, chemical and photocatalytic properties. Accordingly, AlCu-PILC was chosen for this purpose owing to their low cost, environmental stability, high surface area and adsorption capacity, as well as great photo-catalytic activity, which combined with Zr may be significantly enhanced [20,21]. Zr-nanocomposites have been prepared by quite a few methods, including the sol–gel process [31,32], combustion [33], the hydrothermal method [34], microwave irradiation [35], etc. However, it continues to be a challenge to find a simple, efficient and low cost methodology to prepare these nanocomposites.

This work intends to develop a one-step route to incorporate Zr onto clay lattice, promoting great stability and improved photocatalytic activity. The resulting photocatalyst will be tested in the photodegradation of a real WW under UV-C irradiation, and the influence of Zr immobilization on the properties and photoactivity of the heterogeneous catalysts will be discussed.

Different models have been employed to describe the kinetics of catalytic processes involving a heterogeneous liquid–solid system [36–38]. Reaction control models, such as pseudo-first-order and pseudo-second-order models, were considered unsuitable to describe the kinetics of heterogeneous photocatalytic processes, because the two separated linear regression analyses obtained did not take into account relevant factors, namely, the transient period between each linear region, the non-linear behavior during the induction period, and the objective determination of each region, which is subjective when applying two separated linear regressions. The Fermi's model provides a single fit to experimental results showing a transition between the induction period (slow degradation) and the subsequent rapid degradation step of an organic compound (inverted S-shaped transient curve) [39–41]. Considering that the degradation process does not have to follow any particular kinetics or reaction order, it is worth noting that Fermis based model was specifically developed to describe the kinetics of complex systems, involving mixtures of unknown pollutants and several reaction intermediates formed during the photocatalytic process. Therefore, it includes lumped analytical parameters, such as TOC, that can be derived in groups of compounds with different reactivity [42].

In a previous work [42], a lumped kinetic model based on Fermi's equation was developed to describe the TOC histories for the degradation of a dye by catalytic wet peroxide oxidation, as shown in Equation (1)

$$\frac{TOC}{TOC_0} = \frac{1 - x_{TOC}}{1 + \exp\left[k_{TOC}(t - t^*_{TOC})\right]} + x_{TOC} \quad (1)$$

where k_{TOC} corresponds to the apparent reaction rate constant; t^*_{TOC} represents the transition time related to the TOC content curve's inflection point, and x_{TOC} corresponds to the fraction of non-oxidazable compounds that are formed during the reaction.

The Lumped kinetic model based on Fermi's equation has successfully described the kinetics of our previous experiments using pillared clays in the H_2O_2-assisted photocatalytic wet oxidation of WW and, therefore, it is intended to apply this method in order to describe the kinetics of the WW degradation process using the new proposed materials as heterogeneous catalysts.

2. Materials and Methods

2.1. Reagents and Winery Wastewater Sampling

$ZrOCl_2.8H_2O$ (99%) was supplied by Alfa-Aesar, $CuCl_2.2H_2O$ (99%) by Panreac, H_2O_2 (30% w/v) by Sigma-Aldrich. NaOH and H_2SO_4 (95%) were both obtained from Analar NORMAPUR. Deionized water was used to prepare the respective solutions. The agro-industrial wastewater (AIW) was collected from a Portuguese winery cellar located in the Douro region (Northeast of Portugal). The main chemical parameters measured are shown in Table 1. Prior to the oxidation process, the wastewater was submitted to a primary treatment, where the suspended solids were removed from the effluent.

Table 1. Agro-industrial wastewater characterization.

Parameter	Value
pH	3.8 ± 0.1
Chemical Oxygen Demand (mg O_2/L)	1420 ± 45
Biochemical Oxygen Demand (mg O_2/L)	610 ± 15
Total Organic Carbon (mg C/L)	500 ± 12
Total Polyphenols (mg gallic acid/L)	105 ± 3
Phosphates (mg P_2O_5/L)	2.7 ± 0.2
Sulphates (mg SO_4^{2-}/L)	17.8 ± 1.0
Total Iron (mg Fe/L)	0.45 ± 0.02
Aluminium (µg Al/L)	17.5 ± 0.9
Cadmium (µg Cd/L)	2.1 ± 0.1
Copper (µg Cu/L)	400 ± 18
Chromium (µg Cr/L)	0.05 ± 0.003
Manganese (µg Mn/L)	29 ± 1.4
Zinc (µg Zn/L)	4200 ± 200

2.2. Clay Mineral

Natural montmorillonite (MT) was purchased from Fluka, Alfa-Aesar. The chemical composition and main surface properties of natural clay mineral are listed in Tables 2 and 3, respectively. The chemical data was determined by energy dispersive X-ray spectroscopy (EDS/EDAX, FEI QUANTA–400). The total iron expressed as Fe_2O_3 content in raw-montmorillonite was found to be 4.28%. The cation exchange capacity (CEC) of the mineral fractions was measured following the ammonium acetate method proposed by Chapman [43].

Table 2. Main chemical compositions of raw montmorillonite and its derived catalysts, obtained by EDS/EDAX (wt.%).

Sample	SiO_2 (%)	Al_2O_3 (%)	Fe_2O_3 (%)	MgO (%)	Na_2O (%)	CaO (%)	K_2O (%)	CuO (%)	ZrO_2 (%)	Al/Si	Zr/Si	CEC (meq/g)
MT	68.80	21.97	1.58	3.13	2.54	0.95	0.31	-	-	0.32	-	0.61
Zr-MT	63.35	20.09	1.50	2.90	2.90	0.85	0.31	-	5.54	0.32	11.44	0.22
AlCu-PILC	64.15	26.36	1.36	2.53	0.97	0.26	0.37	1.32	-	0.41	-	0.23
(Zr)Al-PILC	60.66	25.62	1.69	2.15	1.17	0.56	0.6	-	5.14	0.42	11.80	0.21
(Zr)AlCu-PILC	60.46	25.64	1.03	1.97	1.02	0.24	0.23	1.30	5.41	0.42	11.18	0.22

Table 3. Specific surface areas and pore characteristics of MT and their respective catalysts.

Sample	S_{BET} (m^2/g)	$V_{total\ pore}$ (cm^3/g)
MT	8.5	0.047
Zr-MT	65	0.109
Al-Cu-PILC	107	0.202
(Zr)Al-PILC	81	0.146
(Zr)Al-Cu-PILC	118	0.217

2.3. Analytical Techniques

Several physical-chemical parameters were measured in order to characterize the agro-industrial wastewater, namely the chemical oxygen demand (COD), the biological oxygen demand (BOD_5), the total organic carbon (TOC) and the total polyphenols (mg gallic acid/L) presented in Table 1. The COD and BOD_5 were determined according to Standard Methods (5220D; 5210D; respectively) [44]. COD analysis was carried out in a COD reactor from HACH Co. and a HACH DR 2400 spectrophotometer was used for colorimetric measurement. Biochemical oxygen demand (BOD_5) was determined using a respirometric OxiTop system. pH evolution was followed by means of a pH-meter (HANNA Instruments, Rhode Island, USA). The TOC content (mg C/L) was determined using a Shimadzu TOC-L CSH analyzer (Tokyo, Japan). Total polyphenols were evaluated following the Folin–Ciocalteu method [45].

2.4. Catalysts Preparation

The preparation of the pillared clays was carried out following a conventional procedure described in detail by Molina, et al. [46]. AlCu-PILC was prepared through the intercalation between montmorillonite fractions and poly(hydroxy)aluminium ($Al_3(OH)_4^{5+}$) and copper $Cu_3(OH)_4^{2+}$ species. The pillaring solution was prepared by slow addition of a 0.2 M NaOH solution to a mixture of 0.1 M $AlCl_3$ and 0.1 M $CuCl_2$ (Cu/(Al+Cu) = 0.1), under constant stirring until the molar ratio OH/Al = 2.5 was reached. The resulting solution was adjusted to pH 6 and was further aged for 8 h at 298 K. The intercalation process was initiated by the addition of a suspension of 0.1 wt.% montmorillonite in deionized water to the pillaring solution, applying the stoichiometry of 10 mmol Al/g clay. The cationic exchange process was carried out at room temperature for 12 h under constant stirring. The resulting suspension was washed by centrifugation with deionized water in order to reach ionic conductivity values lower than 10 µS. After air-drying, the resulted material was calcinated for 2 h at 400 °C.

The Al-Cu oligomeric solution was adjusted to pH 6 in order to achieve the higher proportion of oligomeric species: 100% of both $Al_3(OH)_4^{5+}$ and species. The aqueous speciation was calculated by Visual MINTEQ, version 3.0. After the pillaring process, Cu^{2+} oligomeric species were converted to the respective metal oxide clusters by dehydration and dehydroxylation along the calcination process. The Al-PILC was prepared following the same procedure adopted to AlCu-PILC, but only using poly(hydroxy)aluminium ($Al_3(OH)_4^{5+}$) species.

The Zr-doped catalysts (Zr-MT, (Zr)Al-PILC and (Zr)AlCu-PILC)) were prepared by the incipient wetness impregnation method. The precursor solution was prepared with $ZrOCl_2 \cdot 8H_2O$ in order to obtain a zirconium load of 4 wt.%. After impregnation process, the doped catalysts were dried at 100 °C overnight and calcinated for 3 h at 400 °C. The results obtained from the chemical characterization (Table 2) confirm that 4 wt.% of zirconium were successfully immobilized on different heterogeneous catalysts.

2.5. Catalysts Characterization

The FTIR spectra were obtained by mixing 1 mg natural montmorillonite with 200 mg KBr. The powder mixtures were then inserted into molds and pressed at 10 ton/cm^2 to obtain the transparent pellets. The samples were analyzed with a Bruker Tensor 27 spectrometer and the infrared spectra in transmission mode were recorded in the 4000–400 cm^{-1} frequency region. The microstructural characterization was carried out by scanning electron microscopy (SEM/ESEM FEI QUANTA 400) and the chemical composition of the different catalysts was estimated (Table 2) using energy dispersive X-ray spectroscopy (EDS/EDAX).

The textural parameters of samples were obtained from N_2 adsorption–desorption isotherms at 77 K using a Micromeritics ASAP 2020 apparatus (Norcross, Georgia, USA). The samples were degassed at 150 °C up to 10^{-4} Torr before analysis. The specific surface area (S_{BET}) was determined by applying the Gurevitsch's rule at a relative pressure $p/p_0 = 0.30$ and according to the Brunauer, Emmet, Teller (BET) method from the linear part of the nitrogen adsorption isotherms. Different pore volumes were determined by the Barrett, Joyner, Halenda model (BJH model).

2.6. Adsorption Tests

Different adsorption tests were carried out in order to predict the amount of organic carbon removed through adsorption. The adsorption batch experiments were carried out at different pH conditions (pH 4.0 and pH 7.0) by adding 3.00 g/L of each heterogeneous catalyst into 500 mL of WW (500 mg C/L). The temperature was kept constant throughout the experiments. After the adsorption runs, the samples were centrifuged and the *TOC* content of the supernatant solution was measured. The percentage of organic carbon removed through adsorption was calculated according to Equation (2) [47,48]:

$$TOC_{rem.}(\%) = \frac{TOC_0 - TOC_t}{TOC_0} \times 100 \qquad (2)$$

where TOC_0 is the initial *TOC* content (mg C/L) and TOC_t is *TOC* value at instant t (mg C/L).

2.7. Photocatalytic Experiments and Kinetic Modelling

The photocatalytic experiments were performed in a batch cylindrical photoreactor (600 cm^3) equipped with a UV-C low pressure mercury vapour lamp (TNN 15/32)—working power = 15 W (795.8 W/m^2) and $\lambda_{max} = 254$ nm (Heraeus, Germany). The UV absorption spectrum of the AIW reveals a maximum at ca. 275 nm (with and without the catalysts) and a high absorption at the wavelength where the UV-C lamp emits. In a typical run, 3.0 g/L of catalyst was mixed with 500 mL of the AIW (TOC = 500 mg C/L) for 15 min. After this, a specific amount of H_2O_2 (98 mM) was added to the suspension and the UV light was turned on at the same time. The initial pH varied from 4.0 to 7.0, and was adjusted by adding 1 M of H_2SO_4 or 1 M of NaOH. After the reaction has started, 20 mL of solution was withdrawn for *TOC* measurements at different reaction times, completing a total period of 240 min. The samples were centrifuged and the Zr and Cu concentrations were analyzed by atomic absorption spectroscopy (AAS) using a Thermo Scientific iCE 3000 SERIES. All experiments were performed in triplicate and the observed standard deviation was always less than 5% of the reported values.

A kinetic modelling based on a lumped kinetic model traduced by Fermi's equation was carried out in order to describe the WW degradation process. The experiments were conducted at different pH conditions (pH 4 and pH 7), where temperature, effluent volume, contaminant concentration, H_2O_2 concentration and catalyst dosage were kept constant. A nonlinear least squares regression, based on the Levenberg–Marquardt (LM) algorithm, was applied using the OriginPro 8.5 "Sigmoidal Fit Tool". As a result, a unique semi-empirical function is applied to simultaneously describe the initial low *TOC* conversion (induction period) and subsequent rapid degradation step. Therefore, both the initial transition period and pseudo-first order kinetic period can be expressed with the proposed model [49].

3. Results and Discussion

3.1. Catalysts Characterization

The X-ray diffractograms corresponding to natural montmorillonite (MT) and to Zr-doped and undoped PILCs, are shown in Figure 1. The hkl reflections associated with MT diffraction pattern are characteristic of a montmorillonite clay mineral with mixed interlayer composition including different proportions of Na^+ and Ca^{2+} ions (13.08 Å). This assumption is in agreement with the chemical characterization data, which shows proportions of 0.95% and 2.54% of CaO and Na_2O (Table 2), respectively. The MT samples modified with previously synthesized oligomeric species ($Cu_3(OH)_4^{2+}$ and/or $Al_3(OH)_4^{5+}$) show a shift of the basal reflection d001 from 13.08 Å (MT) to 18.02 Å and to 17.01 Å for (Zr)AlCu-PILC and (Zr)Al-PILC, respectively, confirming the insertion of the oligomeric species in the interlayer region of montmorillonite and the successful pillaring process. The higher expansion observed when both Cu- and Al-oligomeric species were intercalated on montmorillonite results from the higher pillars formed, indicating that the number, charge, size and shape of the oligomeric species affect the pillar size. These results were also suggested by the textural properties obtained for these materials (Table 3), since the (Zr)AlCu-PILC has higher surface area (ABET=x) and higher number of total pore volume than (Zr)Al-PILC, suggesting an increase in contact area available for absorption due to the higher pillars formed.

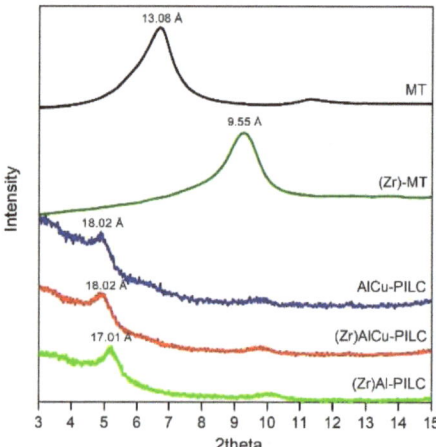

Figure 1. X-ray diffraction results obtained before (MT) and after the pillaring process (AlCu-PILC) and Zr-dopage ((Zr)-Al-PILC, (Zr)AlCu-PILC).

Comparing both (Zr)AlCu-PILC and AlCu-PILC diffraction patterns, it is possible to observe an identical behaviour, confirming that Zr was probably incorporated into the AlCu-PILC lattice without structural modification. The chemical composition of both samples before and after the Zr-doping process is also in agreement with this previous conclusion, given that the increase in Zr amount in doped-clay minerals is accompanied by a decrease in Si proportion, suggesting the isomorphic substitution between Si and Zr in the tetrahedral sheet of the pillared clay. This mechanism is triggered by the similar ionic radii of both cations, where the new one may have identical or lower ionic charge than the replaced one. In this case, both Si and Zr have similar ionic radii and the same ionic charge (+4) and, therefore, no structural charge was developed and no significant structural changes have occurred. This is particularly important, because once the AlCu-PILC has not been structurally affected by the doping process, the adsorption capacity, which is crucial for its catalytic activity, was also not negatively affected. Moreover, enhanced catalyst stability is expected; once Zr is directly incorporated on the crystal lattice of montmorillonite, the risk of metal leaching is significantly lower.

Comparing both (Zr)-MT and MT spectra, no additional conclusions are achieved, since, after the Zr impregnation process, sample (Zr)-MT was submitted to the calcination process, which resulted in the total interlayer collapse to 9.55 Å by dehydration. Therefore, independently of the position of Zr (tetrahedral sheet or interlayer region) on the MT structure, the structural collapse will occur and avoid additional conclusion by means of X-ray diffraction (XRD).

Figure 2 depicts the FTIR spectra obtained before (MT) and after the pillaring process (AlCu-PILC), as well as before (AlCu-PILC) and after the Zr-doping process ((Zr)AlCu-PILC). The results show some structural alterations on montmorillonite after the uptake of metal poly(hydroxy)-complexes and consequent formation of pillars on its internal surface. This is traduced by the decrease in intensity and shift of peaks in the range between 800 and 950 cm^{-1}, after the pillaring process, which are assigned to Al-OH, Fe-OH and Mg-OH vibration modes, at 916 cm^{-1}, 877 cm^{-1} and 849 cm^{-1}, respectively. These structural changes were only observed for PILC samples, which, according to Zhou et al. [50], can be attributed to the interactions between Al or Al/Cu mixed poly(hydroxy) species and the alumina octahedral layers.

Significant modifications on Si-OH stretching vibrations were observed after the Zr-doping process, due to the shift of the main band from 1016 cm^{-1} to 1040 cm^{-1}, and the reduction in intensity of the additional stretching vibrations assigned to the Si-O group, at 1150 cm^{-1} and 1100 cm^{-1}, confirming the incorporation of Zr ions directly in the crystal lattice by isomorphic substitution of Si ions in the tetrahedral sheet of montmorillonite. On the other hand, no additional changes in the vibrations associated with the octahedral sheets of montmorillonite (800–950 cm^{-1}) were observed after this process.

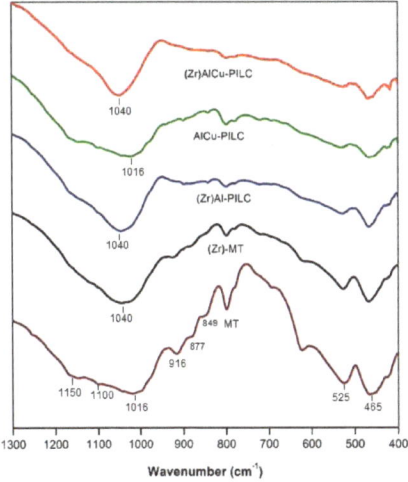

Figure 2. FTIR results obtained before (MT) and after pillaring process (AlCu-PILC) and Zr-dopage ((Zr)-Al-PILC, (Zr)AlCu-PILC), Zr-MT.

The specific surface area and total pore volume of the original montmorillonite and doped and undoped materials are shown in Table 3. These results suggest significant alterations on montmorillonite after the pillaring and Zr-doping processes, resulting in significant and progressive increases in surface area and total pore volume. Accordingly, the surface area increased from 8 m^2/g (MT) to 107 m^2/g after the pillaring process, and to 118 m^2/g after the pillaring and Zr-doping processes, whereas the total pore volume increased from 0.05 cm^3/g (MT) to 0.20 cm^3/g and to 0.22 cm^3/g for AlCu-PILC and (Zr)AlCu-PILC, respectively. The respective isotherms can be classified as type II, where unrestricted monolayer–multilayer adsorption occurs, and the behaviour of the hysteresis loops can be associated with type H3, which usually corresponds to aggregates of plate-like particles forming slit-like pores, which is in agreement with these material structures.

3.2. Adsorption vs. Reaction

Adsorption experiments were carried out using the pillared and Zr-doped catalysts to evaluate the effect of their surface chemistry on the contaminant adsorption and TOC removal. According to previous studies, adsorption plays an important role as the main mechanism involved in the initial induction period, which corresponds to the period necessary for catalyst surface activation [51]. In the present study, part of the mechanism associated with the induction period is probably associated with the adsorption of H_2O_2 and organic compounds onto the catalyst surface, producing surface complexes which promote the activation of the oxidation process through the generation of HO• radicals. Our previous research assessed, for the first time, the application of natural pillared clays (PILCs: Al-Cu-ST and Al-Fe-ST) as heterogeneous photocatalysts for the H_2O_2-assisted treatment of a real AIW [27]. The results indicated that the transition point between the induction period and surface activation and the production of HO• species was directly influenced by the amount of H_2O_2 initially dosed to the process. Accordingly, a decrease in the transition period (t*) TOC from 136 to 96 min was observed, using Al-Cu-ST as the heterogeneous catalyst (3.0 g/L), when H_2O_2 concentration increased from 29 to 98 mM, reducing the period necessary for the surface activation and, therefore, the period required to initiate the degradation process. Considering our previous conclusions, the influence of different catalysts, as well as the effect of Zr-dopage on these supports, were evaluated, taking into account the optimal experimental conditions obtained before, namely $[H_2O_2]_0$ = 98 mM and catalyst dosage = 3.0 g/L.

The evolution of TOC removal through adsorption at different pH conditions and using the different catalysts is shown in Figure 3. As expected, both pH conditions imposed and catalyst textural properties affected the catalyst adsorption capacity. The lowest contaminant adsorption was obtained for Zr-MT and (Zr)Al-PILC, at pH 4 and pH 7, respectively, which correspond to the catalysts with lower surface area (65 m^2/g and 81 m^2/g, respectively). On the contrary, the higher adsorption capacity was obtained for (Zr)AlCu-PILC, at both pH conditions, corresponding to the sample with the highest surface area (118 m^2/g) and total pore volume (0.22 cm^3/g) obtained. This behaviour has particularly impact on the TOC removal efficiency along the oxidation process (Figure 4), as it is observed that the catalyst with enhanced adsorption capacity at both pH conditions has greater activity in the degradation process. Accordingly, the TOC removals obtained using (Zr)AlCu-PILC/UV after 4 h, corresponds to 77% and 86%, at pH 4 and 7, respectively. The incorporation of Zr on montmorillonite lattice (Zr-MT) has contributed to the significant improvement of TOC removals, when compared with raw-montmorillonite, with an increase from 46% to 60% and from 37% to 61% at pH 4 and 7, respectively. The development of AlCu pillared structures had additional advantages considering the improvement of montmorillonite textural properties, resulting in additional stability and catalyst activity. This is traduced by the increase in TOC removals to 69% and 73% at pH 4 and 7, respectively, with a maximum of 33% removal in the non-catalytic UV-C/H_2O_2 experiments performed at both pH conditions. As previously observed, the Zr-dopage on AlCu-PILC has also improved its catalytic activity, promoting an increase in TOC removals to 77% and 86%, at pH 4 and pH 7, respectively, suggesting a synergetic effect of both Zr and Cu on the photocatalytic degradation process. In this case, Zr acts as a semiconductor that is excited by photons with an energy greater than its band gap (5.8–7.1 eV), generating electron–hole pairs, which migrates to the photocatalyst surface yielding radical species that can react with organic molecules upon redox reactions. The electron transfer process is enhanced by the successive redox of Cu^{2+}, a transition metal which is continuously releasing e- species, induced by the presence of a permanent irradiation source (UV-C).

During these processes, a decrease in catalyst stability, caused by the increase in Cu leaching levels from 0.0 to 1.4 mg/L, was also observed from the acidic to the neutral conditions. However, all the Cu leaching concentrations are lower than or very close to the legal discharge limit imposed by EU legislation (1.0 mg Cu/L), and only 5.0% of the Cu immobilized was released at neutral conditions, confirming that Cu immobilization on pillared clay support was successful.

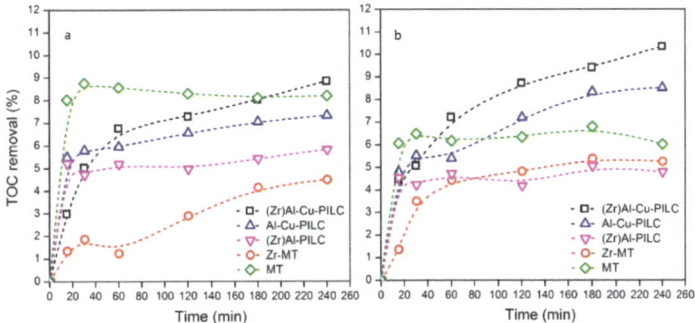

Figure 3. Evolution of *TOC* removal by adsorption: (**a**) pH 4.0 and (**b**) pH 7.0 (catalyst dosage = 3.0 g/L, $[TOC]_0$ = 500 mg C/L).

Comparing both Zr-MT and (Zr)Al-PILC performances at both pH conditions, it is possible to conclude that Zr-MT has shown increased catalytic activity, mainly from 180 min, with *TOC* removals of 60% at both pH conditions for Zr-MT, and 46% at both pH conditions for (Zr)Al-PILC. This behavior could be explained by the competition between Zr and an excess of Al for the Si tetrahedral sites on (Zr)Al-PILC, which may have hampered the Zr incorporation onto the montmorillonite crystal lattice. The lower catalytic activity observed using (Zr)Al-PILC is even more pronounced in acidic conditions, where the *TOC* conversion is very similar to MT along the treatment process. This could be explained by the increased adsorption capacity obtained by MT at pH 4.0 (Figure 3), associated with its higher CEC (Table 2), which despite the lower BET surface area of MT, when compared with (Zr)Al-PILC, has contributed to enhanced adsorption capacity and improved catalytic activity.

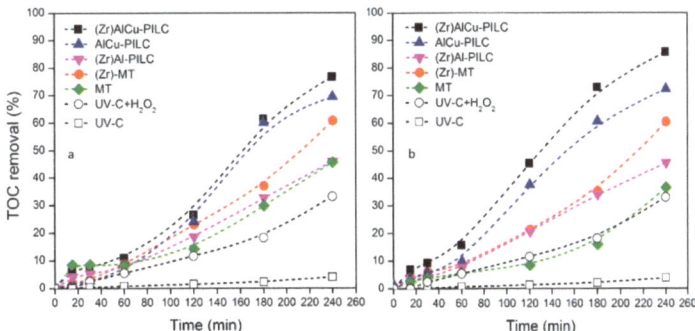

Figure 4. Evolution of *TOC* removal throughout H_2O_2-assisted photocatalytic process: (**a**) pH 4 and (**b**) pH 7 (catalyst dosage = 3.0 g/L, $[H_2O_2]_0$ = 98 mM, $[TOC]_0$ = 500 mg C/L).

3.3. Kinetic Study

In order to better understand the effect of the operational conditions on the induction period, the transition time between the induction period and the fast oxidation reaction ($t * TOC$) was obtained through the fitting of Fermi's equation based on lumped kinetic model to the experimental data. The results obtained for the different parameters are displayed in Table 4, and the fittings obtained for different catalysts are illustrated in Figure 5. The results reveal a good fitting of the kinetic model to the experimental data obtained for different catalysts, with R^2 values higher than 0.983. Considering the different parameters obtained from the modelling, it is assumed that the experimental conditions influenced the kinetic performance of our processes.

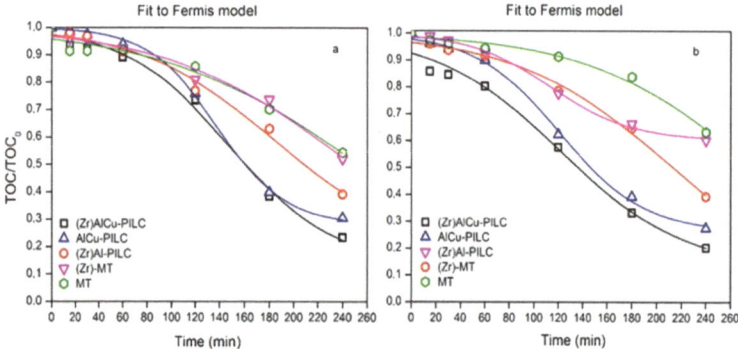

Figure 5. Normalized TOC content (TOC/TOC_0) as a function of time, using different pH conditions: (a) pH 4, (b) pH 7 (catalyst dosage = 3.0 g/L, $[H_2O_2]_0$ = 98 mM, $[TOC]_0$ = 500 mg C/L). Fit of Fermi's equation based on lumped kinetic model to the experimental data.

As observed in Figure 5, the transition point is significantly affected by the pH conditions, as well as the heterogeneous catalyst used. Accordingly, a significant decrease in $t * TOC$ was observed, from 279 to 121 min (pH 7) and from 253 to 137 min (pH 4), using raw-montmorillonite and AlCu-MT, respectively, suggesting a significant reduction in the surface activation period when the heterogeneous catalyst was applied, as well as a quicker production of $HO^•$ species. This tendency was even more pronounced after Zr-dopage at neutral conditions, once the transition time has decreased from 121 to 119 min using (Zr)AlCu-PILC.

In both cases, the evolution of the H_2O_2 concentration along the photocatalytic experiments shows a decrease and subsequent increase in concentration in the first 60 min (Table 5), which suggests a possible adsorption and desorption of H_2O_2 on pillared montmorillonite. Therefore, the results suggest that the first 60 min of reaction were mainly associated with the formation of surface complexes between the H_2O_2 and catalyst surface, whereas the additional period, which completes the total induction period, may be associated with the time required for the surface activation, i.e., the time required for catalytic decomposition of the oxidant in the presence of the active phase (production of $HO^•$).

Concerning the reaction rates of the different photocatalytic processes, higher reaction rates were observed using both (Zr)AlCu-PILC and AlCu-PILC catalysts when compared with the other catalysts applied (Table 4), confirming an improvement of catalytic performance during the oxidation processes. Comparing both catalysts, AlCu-PILC has the higher reaction rates at both pH conditions, 3.54×10^{-2} min^{-1} and 2.87×10^{-2} min^{-1}, at pH 4 and 7, respectively. However, (Zr)AlCu-PILC ($k_{TOC} = 2.58 \times 10^{-2}$ min^{-1}) has contributed to lower fractions of non-oxidazable compounds formed during the reaction ($x_{TOC} = 0.11$, pH 7), when compared with AlCu-PILC ($x_{TOC} = 0.26$, pH 7), which is in agreement with the higher TOC removals obtained.

Table 4. Kinetic parameters obtained by fitting Fermi's model to the experimental data (TOC/TOC_0 as function of time) using different catalysts.

Heterogeneous Catalyst	Variables	Kinetic Parameters			
		k_{TOC} (min^{-1})	t^*_{TOC} (min)	x_{TOC}	r^2
(Zr)AlCu-PILC	pH 4.0	2.35 × 10^{-2}	148	0.14	0.990
	pH 7.0	2.58 × 10^{-2}	119	0.11	0.996
AlCu-PILC	pH 4.0	3.54 × 10^{-2}	137	0.28	0.998
	pH 7.0	2.87 × 10^{-2}	121	0.26	0.997
(Zr)Al-PILC	pH 4.0	1.90 × 10^{-2}	158	0.45	0.992
	pH 7.0	2.16 × 10^{-2}	140	0.50	0.993
Zr-MT	pH 4.0	1.83 × 10^{-2}	183	0.19	0.983
	pH 7.0	1.56 × 10^{-2}	214	0.0	0.990
MT	pH 4.0	1.23 × 10^{-2}	253	0.0	0.960
	pH 7.0	1.50 × 10^{-2}	279	0.0	0.980

Table 5. H_2O_2 removal in the photocatalytic experiments using different catalysts. General conditions: pH 4.0 and pH 7.0, catalyst dosage = 3.00 g/L, $[TOC]_0$ = 500 mg C/L, UV-C irradiation.

	Experiment Time (min)	H_2O_2 Removal (%)		
		Blank (H_2O_2 Only)	(Zr)AlCu-PILC	AlCu-PILC
pH 4.0	0	0.00	0.00	0.00
	15	2.03	14.03	11.18
	30	9.03	10.88	9.73
	60	20.2	54.38	35.63
	120	38.3	47.39	41.89
	180	51.1	79.04	83.58
	240	62.0	96.86	93.13
pH 7.0	0	0.00	0.0	0.00
	15	12.7	16.0	12.06
	30	8.61	12.9	10.05
	60	16.3	8.19	49.30
	120	28.5	45.6	53.29
	180	37.0	69.3	83.07
	240	46.0	88.4	94.69

3.4. Catalyst Regeneration

Considering the best performance of (Zr)AlCu-PILC, the catalyst reuse capacity was evaluated throughout three consecutive cycles of H_2O_2-assisted photocatalytic AIW treatment. The experiments were carried out at pH 7.0, using a catalyst dosage of 3.0 g/L and a H_2O_2 concentration of 98 mM. The results show that the TOC removal obtained using (Zr)AlCu-PILC corresponds to 86%, 66% and 63% after 240 min, for the first, second and third cycles, respectively. In general, a decrease in efficiency was observed from the first to the second cycle (with a loss of 20% in TOC removal). However, no additional loss of activity was observed from the second to the third cycle. All the leaching concentrations along the different cycles were very close to the legal limits imposed (1.0 mg Cu/L) and tended to decrease as the number of cycles increased, from 1.4 to 0.97 mg/L of Cu, from the first to the third cycle, revealing that the catalyst stability is not affected along the cycles.

4. Conclusions

Different catalysts submitted to different chemical treatments and/or the Zr-dopage process, were applied in the H_2O_2-assisted treatment of recalcitrant winery wastewater in order to evaluate the influence of the surface chemical properties of the doped supports on their adsorption and catalytic properties.

FTIR results show that the incorporation of Zr in the crystal lattice of montmorillonite and PILC through isomorphic substitution between Si and Zr was traduced by significant modifications on Si-OH stretching vibrations, due to the shift of the main band from 1016 cm^{-1} to 1040 cm^{-1}, and the decrease in intensity of the additional stretching vibrations assigned to the Si-O group, at, respectively, 1150 cm^{-1} and 1100 cm^{-1}.

In general, the results show that Zr-dopage on AlCu-PILC has improved its adsorption and catalytic activity, promoting an increase in TOC removals to 77% and 86%, with 8.85% and 10.35% of TOC removed through adsorption, at pH 4 and pH 7, respectively. It suggests a synergetic effect caused by the combination of Zr and Cu on the photocatalytic degradation process, once the semiconductor electron transfer process is enhanced by the successive redox of Cu(II), induced by the presence of the UV-C irradiation source.

A significant decrease in $t * TOC$ was observed for AlCu-PILC and (Zr)AlCu-PILC, at both pH conditions, suggesting a significant reduction in the surface activation period when the heterogeneous catalyst was applied, as well as a quicker production of HO$^{\bullet}$ species. As a result, higher reaction rates were obtained using both (Zr)AlCu-PILC (2.58×10^{-2} min^{-1}) and AlCu-PILC ($k_{TOC} = 3.54 \times 10^{-2}$ min^{-1}) catalysts, confirming an improvement in catalytic performance along the oxidation processes. Comparing both catalysts, AlCu-PILC has the higher reaction rates at both pH conditions. However, (Zr)AlCu-PILC has contributed to lower fractions of non-oxidazable compounds formed during the reaction ($x_{TOC} = 0.11$, pH 7.0), making it a more efficient process.

Author Contributions: Conceptualization, V.G. and A.R.T.; methodology, V.G. and A.R.T.; validation, V.G. and A.R.T.; investigation, V.G. and A.R.T.; writing—original draft preparation, V.G. and A.R.T.; writing—review and editing, V.G., A.R.T., M.S.L. and J.A.P.; visualization, V.G., M.S.L. and J.A.P.; supervision, M.S.L. and J.A.P.; project administration, J.A.P. All authors have read and agreed to the published version of the manuscript.

Funding: The authors thank the North Regional Operational Program (NORTE 2020) and the European Regional Development Fund (ERDF), and express their appreciation for the financial support of the Project INNOVINE&WINE (BPD/UTAD/INNOVINE&WINE/WINEMAKING/754/2016), Project AgriFood XXI—NORTE-01-0145-FEDER-000041 and to the Fundação para a Ciência e a Tecnologia (FCT) for the financial support provided to CQVR through UIDB/00616/2020. Marco S. Lucas also thanks the FCT for the financial support provided through the Investigador FCT-IF/00802/2015 project.

Acknowledgments: The authors thank the North Regional Operational Program (NORTE 2020) and the European Regional Development Fund (ERDF), for financial support of the Project INNOVINE&WINE (BPD/UTAD/INNOVINE&WINE/WINEMAKING/754/2016).

Conflicts of Interest: The authors declare no conflict of interest.

References

1. Amor, C.; Marchão, L.; Lucas, M.S.; Peres, J.A. Application of Advanced Oxidation Processes for the Treatment of Recalcitrant Agro-Industrial Wastewater: A Review. *Water* **2019**, *11*, 205. [CrossRef]
2. Ferreira, L.C.; Fernandes, J.R.; Rodríguez-Chueca, J.; Peres, J.A.; Lucas, M.S.; Tavares, P.B. Photocatalytic degradation of an agro-industrial wastewater model compound using a UV LEDs system: Kinetic study. *J. Environ. Manag.* **2020**, *269*, 110740. [CrossRef]
3. Noukeu, N.A.; Gouado, I.; Priso, R.J.; Ndongo, D.; Taffouo, V.D.; Dibong, S.D.; Ekodeck, G.E. Characterization of effluent from food processing industries and stillage treatment trial with *Eichhornia crassipes* (Mart.) and *Panicum maximum* (Jacq.). *Water Resour. Ind.* **2016**, *16*, 1–18. [CrossRef]
4. Sousa, R.M.O.F.; Amaral, C.; Fernandes, J.M.C.; Fraga, I.; Semitela, S.; Braga, F.; Coimbra, A.M.; Dias, A.A.; Bezerra, R.M.; Sampaio, A. Hazardous impact of vinasse from distilled winemaking by-products in terrestrial plants and aquatic organisms. *Ecotoxicol. Environ. Saf.* **2019**, *183*, 109493. [CrossRef]
5. Al Bsoul, A.; Hailat, M.; Abdelhay, A.; Tawalbeh, M.; Jum'h, I.; Bani-Melhem, K. Treatment of olive mill effluent by adsorption on titanium oxide nanoparticles. *Sci. Total Environ.* **2019**, *688*, 1327–1334. [CrossRef]
6. Chen, B.; Jiang, C.; Yu, D.; Wang, Y.; Xu, T. Design of an alternative approach for synergistic removal of multiple contaminants: Water splitting coagulation. *Chem. Eng. J.* **2020**, *380*, 122531. [CrossRef]

7. Candia-Onfray, C.; Espinoza, N.; Sabino da Silva, E.B.; Toledo-Neira, C.; Espinoza, L.C.; Santander, R.; García, V.; Salazar, R. Treatment of winery wastewater by anodic oxidation using BDD electrode. *Chemosphere* **2018**, *206*, 709–717. [CrossRef]
8. Aziz, H.A.; Abu Amr, S.S. *Advanced Oxidation Processes (AOPs) in Water and Wastewater Treatment*; IGI Global: Hershey, PA, USA, 2019. [CrossRef]
9. Trapido, M.; Tenno, T.; Goi, A.; Dulova, N.; Kattel, E.; Klauson, D.; Klein, K.; Tenno, T.; Viisimaa, M. Bio-recalcitrant pollutants removal from wastewater with combination of the Fenton treatment and biological oxidation. *J. Water Process Eng.* **2017**, *16*, 277–282. [CrossRef]
10. Amor, C.; Rodríguez-Chueca, J.; Fernandes, J.L.; Domínguez, J.R.; Lucas, M.S.; Peres, J.A. Winery wastewater treatment by sulphate radical based-advanced oxidation processes (SR-AOP): Thermally vs UV-assisted persulphate activation. *Process Saf. Environ. Prot.* **2019**, *122*, 94–101. [CrossRef]
11. Lucas, M.S.; Peres, J.A. Removal of Emerging Contaminants by Fenton and UV-Driven Advanced Oxidation Processes. *Water Air Soil Pollut.* **2015**, *226*, 273. [CrossRef]
12. Rodríguez-Chueca, J.; Amor, C.; Mota, J.; Lucas, M.S.; Peres, J.A. Oxidation of winery wastewater by sulphate radicals: Catalytic and solar photocatalytic activations. *Environ. Sci. Pollut. Res.* **2017**, *24*, 22414–22426. [CrossRef]
13. Lucas, M.S.; Peres, J.A.; Li Puma, G. Treatment of winery wastewater by ozone-based advanced oxidation processes (O_3, O_3/UV and O_3/UV/H_2O_2) in a pilot-scale bubble column reactor and process economics. *Sep. Purif. Technol.* **2010**, *72*, 235–241. [CrossRef]
14. Amor, C.; Lucas, M.S.; García, J.; Dominguez, J.R.; De Heredia, J.B.; Peres, J.A. Combined treatment of olive mill wastewater by Fenton's reagent and anaerobic biological process. *J. Environ. Sci. Health Part A* **2015**, *50*, 161–168. [CrossRef]
15. Rodríguez-Chueca, J.; Amor, C.; Fernandes, J.R.; Tavares, P.B.; Lucas, M.S.; Peres, J.A. Treatment of crystallized-fruit wastewater by UV-A LED photo-Fenton and coagulation–flocculation. *Chemosphere* **2016**, *145*, 351–359. [CrossRef]
16. Brink, A.; Sheridan, C.; Harding, K. Combined biological and advance oxidation processes for paper and pulp effluent treatment. *South Afr. J. Chem. Eng.* **2018**, *25*, 116–122. [CrossRef]
17. M'Arimi, M.M.; Mecha, C.A.; Kiprop, A.K.; Ramkat, R. Recent trends in applications of advanced oxidation processes (AOPs) in bioenergy production: Review. *Renew. Sustain. Energy Rev.* **2020**, *121*, 109669. [CrossRef]
18. Luo, H.; Zeng, Y.; He, D.; Pan, X. Application of iron-based materials in heterogeneous advanced oxidation processes for wastewater treatment: A review. *Chem. Eng. J.* **2020**. [CrossRef]
19. Rueda Márquez, J.J.; Levchuk, I.; Sillanpää, M. Application of Catalytic Wet Peroxide Oxidation for Industrial and Urban Wastewater Treatment: A Review. *Catalysts* **2018**, *8*, 673. [CrossRef]
20. Gil, A.; Galeano, L.A.; Vicente, M.Á. *Applications of Advanced Oxidation Processes (AOPs) in Drinking Water Treatment*; Springer International Publishing: New York, NY, USA, 2018. [CrossRef]
21. Ameta, S.C.; Ameta, R. *Advanced Oxidation Processes for Wastewater Treatment: Emerging Green Chemical Technology*; Elsevier Science: Amsterdam, The Netherlands, 2018; ISBN 9780128105252.
22. Santos Silva, A.; Seitovna Kalmakhanova, M.; Kabykenovna Massalimova, B.G.; Sgorlon, J.; Jose Luis, D.T.; Gomes, H.T. Wet Peroxide Oxidation of Paracetamol Using Acid Activated and Fe/Co-Pillared Clay Catalysts Prepared from Natural Clays. *Catalysts* **2019**, *9*, 705. [CrossRef]
23. Ormad, M.P.; Mosteo, R.; Ibarz, C.; Ovelleiro, J.L. Multivariate approach to the photo-Fenton process applied to the degradation of winery wastewaters. *Appl. Catal. B Environ.* **2006**, *66*, 58–63. [CrossRef]
24. Khare, P.; Patel, R.K.; Sharan, S.; Shankar, R. 8—Recent trends in advanced oxidation process for treatment of recalcitrant industrial effluents. In *Advanced Oxidation Processes for Effluent Treatment Plants*; Shah, M.P., Ed.; Elsevier: Amsterdam, The Netherlands, 2021; pp. 137–160. [CrossRef]
25. Jaén-Gil, A.; Buttiglieri, G.; Benito, A.; Mir-Tutusaus, J.A.; Gonzalez-Olmos, R.; Caminal, G.; Barceló, D.; Sarrà, M.; Rodriguez-Mozaz, S. Combining biological processes with UV/H_2O_2 for metoprolol and metoprolol acid removal in hospital wastewater. *Chem. Eng. J.* **2021**, *404*, 126482. [CrossRef]
26. Giannakis, S.; Jovic, M.; Gasilova, N.; Pastor Gelabert, M.; Schindelholz, S.; Furbringer, J.-M.; Girault, H.; Pulgarin, C. Iohexol degradation in wastewater and urine by UV-based Advanced Oxidation Processes (AOPs): Process modeling and by-products identification. *J. Environ. Manag.* **2017**, *195*, 174–185. [CrossRef]

27. Guimarães, V.; Teixeira, A.R.; Lucas, M.S.; Silva, A.M.T.; Peres, J.A. Pillared interlayered natural clays as heterogeneous photocatalysts for H_2O_2-assisted treatment of a winery wastewater. *Sep. Purif. Technol.* **2019**, *228*, 115768. [CrossRef]
28. Guimarães, V.; Lucas, M.S.; Peres, J.A. Combination of adsorption and heterogeneous photo-Fenton processes for the treatment of winery wastewater. *Environ. Sci. Pollut. Res.* **2019**, *26*, 31000–31013. [CrossRef]
29. Domínguez, C.M.; Quintanilla, A.; Casas, J.A.; Rodriguez, J.J. Treatment of real winery wastewater by wet oxidation at mild temperature. *Sep. Purif. Technol.* **2014**, *129*, 121–128. [CrossRef]
30. Mosteo, R.; Ormad, P.; Mozas, E.; Sarasa, J.; Ovelleiro, J.L. Factorial experimental design of winery wastewaters treatment by heterogeneous photo-Fenton process. *Water Res.* **2006**, *40*, 1561–1568. [CrossRef]
31. Tyagi, B.; Sidhpuria, K.; Shaik, B.; Jasra, R.V. Synthesis of Nanocrystalline Zirconia Using Sol–Gel and Precipitation Techniques. *Ind. Eng. Chem. Res.* **2006**, *45*, 8643–8650. [CrossRef]
32. Samadi, S.; Yousefi, M.; Khalilian, F.; Tabatabaee, A. Synthesis, characterization, and application of Nd, Zr–TiO_2/SiO_2 nanocomposite thin films as visible light active photocatalyst. *J. Nanostruct. Chem.* **2015**, *5*, 7–15. [CrossRef]
33. Sayagués, M.J.; Avilés, M.A.; Córdoba, J.M.; Gotor, F.J. Self-propagating combustion synthesis via an MSR process: An efficient and simple method to prepare (Ti, Zr, Hf)B_2–Al_2O_3 powder nanocomposites. *Powder Technol.* **2014**, *256*, 244–250. [CrossRef]
34. Teymourian, H.; Salimi, A.; Firoozi, S.; Korani, A.; Soltanian, S. One-pot hydrothermal synthesis of zirconium dioxide nanoparticles decorated reduced graphene oxide composite as high performance electrochemical sensing and biosensing platform. *Electrochim. Acta* **2014**, *143*, 196–206. [CrossRef]
35. Rajabi, M.; Khodai, M.M.; Askari, N. Microwave-assisted sintering of Al–ZrO_2 nano-composites. *J. Mater. Sci. Mater. Electron.* **2014**, *25*, 4577–4584. [CrossRef]
36. Sannino, D.; Vaiano, V.; Ciambelli, P.; Isupova, L.A. Mathematical modelling of the heterogeneous photo-Fenton oxidation of acetic acid on structured catalysts. *Chem. Eng. J.* **2013**, *224*, 53–58. [CrossRef]
37. He, J.; Yang, X.; Men, B.; Wang, D. Interfacial mechanisms of heterogeneous Fenton reactions catalyzed by iron-based materials: A review. *J. Environ. Sci.* **2016**, *39*, 97–109. [CrossRef]
38. Kalmakhanova, M.S.; Diaz de Tuesta, J.L.; Massalimova, B.K.; Gomes, H.T. Pillared clays from natural resources as catalysts for catalytic wet peroxide oxidation: Characterization and kinetic insights. *Environ. Eng. Res.* **2020**, *25*, 186–196. [CrossRef]
39. Rache, M.L.; García, A.R.; Zea, H.R.; Silva, A.M.T.; Madeira, L.M.; Ramírez, J.H. Azo-dye orange II degradation by the heterogeneous Fenton-like process using a zeolite Y-Fe catalyst—Kinetics with a model based on the Fermi's equation. *Appl. Catal. B Environ.* **2014**, *146*, 192–200. [CrossRef]
40. Ghime, D.; Ghosh, P. Decolorization of diazo dye trypan blue by electrochemical oxidation: Kinetics with a model based on the Fermi's equation. *J. Environ. Chem. Eng.* **2020**, *8*, 102792. [CrossRef]
41. Herney-Ramirez, J.; Silva, A.M.T.; Vicente, M.A.; Costa, C.A.; Madeira, L.M. Degradation of Acid Orange 7 using a saponite-based catalyst in wet hydrogen peroxide oxidation: Kinetic study with the Fermi's equation. *Appl. Catal. B Environ.* **2011**, *101*, 197–205. [CrossRef]
42. Silva, A.M.T.; Herney-Ramirez, J.; Söylemez, U.; Madeira, L.M. A lumped kinetic model based on the Fermi's equation applied to the catalytic wet hydrogen peroxide oxidation of Acid Orange 7. *Appl. Catal. B Environ.* **2012**, *121–122*, 10–19. [CrossRef]
43. Chapman, H.D. Cation exchange capacity. *Methods Soil Anal. Chem. Microbiol. Prop.* **1965**, *9*, 891–901.
44. APHA. *Standard Methods for the Examination of Water and Wastewater*, 21st ed.; American Public Health Association: Washington, DC, USA, 2005.
45. Lowry, O.H.; Rosebrough, N.J.; Farr, A.L.; Randall, R.J. Protein measurement with the Folin phenol reagent. *J. Biol. Chem* **1951**, *193*, 265–275.
46. Molina, C.B.; Casas, J.A.; Zazo, J.A.; Rodríguez, J.J. A comparison of Al-Fe and Zr-Fe pillared clays for catalytic wet peroxide oxidation. *Chem. Eng. J.* **2006**, *118*, 29–35. [CrossRef]
47. Guimarães, V.; Rodríguez-Castellón, E.; Algarra, M.; Rocha, F.; Bobos, I. Influence of pH, layer charge location and crystal thickness distribution on U(VI) sorption onto heterogeneous dioctahedral smectite. *J. Hazard. Mater.* **2016**, *317*, 246–258. [CrossRef]
48. Wang, Y.; Wang, W.; Wang, A. Efficient adsorption of methylene blue on an alginate-based nanocomposite hydrogel enhanced by organo-illite/smectite clay. *Chem. Eng. J.* **2013**, *228*, 132–139. [CrossRef]

49. Hou, B.; Han, H.; Jia, S.; Zhuang, H.; Xu, P.; Wang, D. Heterogeneous electro-Fenton oxidation of catechol catalyzed by nano-Fe$_3$O$_4$: Kinetics with the Fermi's equation. *J. Taiwan Inst. Chem. Eng.* **2015**, *56*, 138–147. [CrossRef]
50. Zhou, J.; Wu, P.; Dang, Z.; Zhu, N.; Li, P.; Wu, J.; Wang, X. Polymeric Fe/Zr pillared montmorillonite for the removal of Cr(VI) from aqueous solutions. *Chem. Eng. J.* **2010**, *162*, 1035–1044. [CrossRef]
51. Carriazo, J.G.; Guelou, E.; Barrault, J.; Tatibouët, J.M.; Moreno, S. Catalytic wet peroxide oxidation of phenol over Al–Cu or Al–Fe modified clays. *Appl. Clay Sci.* **2003**, *22*, 303–308. [CrossRef]

Publisher's Note: MDPI stays neutral with regard to jurisdictional claims in published maps and institutional affiliations.

© 2020 by the authors. Licensee MDPI, Basel, Switzerland. This article is an open access article distributed under the terms and conditions of the Creative Commons Attribution (CC BY) license (http://creativecommons.org/licenses/by/4.0/).

Article

Heterogeneous Catalytic Ozonation of Aniline-Contaminated Waters: A Three-Phase Modelling Approach Using TiO$_2$/GAC

Cristian Ferreiro [1,*], Natalia Villota [2], José Ignacio Lombraña [1] and María J. Rivero [3]

1. Department of Chemical Engineering, Faculty of Science and Technology, University of the Basque Country UPV/EHU, Barrio Sarriena s/n, 48940 Leioa, Spain; ji.lombrana@ehu.eus
2. Department of Chemical and Environmental Engineering, Faculty of Engineering Vitoria-Gasteiz, University of the Basque Country UPV/EHU, Nieves Cano 12, 01006 Vitoria-Gasteiz, Spain; natalia.villota@ehu.eus
3. Department of Chemical and Biomolecular Engineering, University of Cantabria, 39005 Santander, Spain; riveromj@unican.es
* Correspondence: cristian.ferreiro@ehu.eus; Tel.: +34-946-012-512

Received: 30 October 2020; Accepted: 3 December 2020; Published: 8 December 2020

Abstract: This work aims to study the sustainable catalytic ozonation of aniline promoted by granular active carbon (GAC) doped with TiO$_2$. Aniline was selected as a model compound for the accelerator manufacturing industries used in the manufacture of rubber due to its environmental impact, low biodegradability, and harmful genotoxic effects on human health. Based on the evolution of total organic carbon (TOC), aniline concentration measured using high performance liquid chromatography (HPLC), pH and ozone concentration in liquid and gas phase, and catalyst loading, a three-phase reaction system has been modelled. The proposed three-phase model related the ozone transfer parameters and the pseudo-first order kinetic constants through three coefficients that involve the adsorption process, oxidation in the liquid, and the solid catalyst. The interpretation of the kinetic constants of the process allowed the predominance of the mechanism of Langmuir–Hinshelwood or modified Eley–Rideal to be elucidated. Seven intermediate aromatic reaction products, representative of the direct action of ozone and the radical pathway, were identified and quantified, as well as precursors of the appearance of turbidity, with which two possible routes of degradation of aniline being proposed.

Keywords: aniline; catalytic ozonation; degradation routes; industrial wastewater; three-phase system; TiO$_2$/GAC

---

1. Introduction

At present, as part of corporate social responsibility, manufacturing industries must compromise in the short-term in order to carry out environmental protection actions. A report from the United Nations has indicated that, in 2015, more than 80% of the wastewaters of worldwide human activities were being discharged into rivers and sea without the removal of polluting substances [1]. Despite the fact that manufacturing companies have been opting for environmentally sustainable processes, the Sustainable Development Goals (SDGs) are unfortunately far from being met in 2030 [2]. In this work, the removal of aniline, as a model pollutant of environmental concern, is studied. Aniline (ANI) is mainly used in the synthesis of methylene diphenyl isocyanate to produce polyurethane foams, antioxidants, activators, and accelerators in the rubber industry, as well as in the synthesis of indigo and other dyes [3]. It is also employed as a raw material in the manufacturing of different types of fungicides in the agricultural and pharmaceutical industries [4,5].

The uncontrolled discharge of industrial effluents containing aniline is harmful to both humans and the environment. In humans, it has been well-documented to induce carcinogenic, teratogenic, and mutagenic effects [6]. Acute exposure to high amounts of aniline (>1 g) for a 75 kg person could lead to coma or even death [7]. In addition, uncontrolled discharges containing aniline could disturb the aquatic environment, causing mortality in aquatic animals and plants [4]. High concentrations of aniline in rivers and groundwater could affect crop safety and, consequently, human health. Studies of aniline genotoxicity in plants have also been carried out, which concluded that it significantly inhibits the growth of wheat crops irrigated with water contaminated with aniline at a concentration of 25 mg L^{-1} [8]. Therefore, aniline has garnered great attention and has been classified as a persistent organic pollutant by the U.S. Environmental Protection Agency [9]. Conventional treatment systems based on biological processes are not suitable for the treatment of wastewater contaminated with aniline, due to micro-organism deactivation. Hence, new treatment technologies are needed, in order to transform aniline into biodegradable substances of lower toxicity before conducting biological treatment [6,8].

The removal of aniline from contaminated water has been a topic of concern for several research groups, as shown in Table 1. According to Table 1, in most studies a high degree of mineralization and degradation of the aniline was not achieved. Additionally, most of the treatment technologies for aniline removal require complex equipment, which are costly and have limited potential for full-scale implementation in wastewater treatment plants. Among such technologies, heterogeneous catalytic ozonation based on the use of TiO_2/granular active carbons (GAC) has emerged as a sustainable and cost-effective (0.64 Euro per kg of total organic carbon (TOC) eliminated) treatment alternative for the removal of aniline from contaminated waters [4]. An increase in the number of scientific reports has indicated the successful development of new activated carbon types modified through the incorporation of metal oxides with enhanced catalytic activity, thus endorsing the application of this catalytic technology at industrial scale [10,11]. The improvement in the removal yields can be attributed to the ozonation mechanisms that take place on catalytic surfaces. These mechanisms have not yet been fully defined [12]. One mechanism postulated is that the catalyst acts as an adsorbent where organic contaminants are first adsorbed on the catalyst surface and then removed [13]. The other mechanism proposed suggests that the titanium oxides favours ozone mass transport and initiates its decomposition. This mechanism assumes that the hydroxyl groups on the catalytic surface play an important role in the generation of the radical species [14].

In order to fulfil the industrial implementation requirements of this new catalytic technology, the prediction of operating behaviour under different working conditions becomes a critical point, which is yet to be solved. Delmas et al. [15] modelled a sequential process based on adsorption onto activated carbon followed by a wet catalytic oxidation with ozone. However, only a few studies have focused on the prediction and estimation of the effects of the main operational variables (e.g., pH, ozone dose, and catalyst load), as well as the contribution of chemical surface properties of TiO_2/GAC composite during the heterogeneous catalytic ozonation of wastewater containing aniline [16–19]. The study presented here addresses such challenges. In particular, we aim to develop a three-phase modelling approach that includes mass transfer parameters and rate constants from both surface and liquid bulk reactions which allow for the establishment of operating conditions that: (i) enhance radical generation due to ozone decomposition promoted by the TiO_2/GAC composite and (ii) avoid both catalyst deactivation and deterioration of the physical–chemical properties of the GAC.

Table 1. Previous studies regarding the treatment of industrial wastewater containing aniline.

Treatment	Catalyst	Operating Conditions	Comments	References
Ozone	—	(Time) = 120 min; C_0 = 103.81 mg L^{-1}; pH = 7.0; T = 20 °C; F_G = 2.5 g h^{-1}; $C_{O_3,G}$ = 22.0 mg L^{-1}; (%) = 93.56%; (% COD) = 31.03%	Studied the effect of operational variables on the biodegradability of aniline oxidation by-products, highlighting among them diacid butane, oxalic acid, and formic acid.	[6]
US/O$_3$	—	(Time) = 30 min; C_0 = 100 mg L^{-1}; pH$_0$ = 7.0; T = 25 °C; F_G = 12 mg min^{-1}; US$_{density}$ = 0.1 W mL^{-1}; (%) = 99%; (% TOC) = 51%	The synergistic effect improved the degradation and mineralization of aniline by 64% and 110% respectively in terms of total organic carbon (TOC) compared to simple ozonation.	[20]
O$_3$/GAC	GAC (Norit® 1240 Plus granular activated carbon (Cabot Norit Americas, Inc., Marshall, TX, USA))	(Time) = 30 min; C_0 = 102.44 mg L^{-1}; pH = 7.0; T = 25 °C; F_G = 150 cm^3 min^{-1}; $C_{O_3,G}$ = 50.0 mg L^{-1}; (%) = 100%; (% TOC) = 56%; M_{CAT} = 500 mg L^{-1}	Studied the catalytic effect of GAC on the ozonisation process. Basic GACs had a higher capacity for decomposition of O$_3$ and organics adsorption.	[21]
O$_3$/TiO$_2$-GAC	TiO$_2$/GAC (Nanocyl® 3100 activated carbon doped with TiO$_2$ by hydration–dehydration method (Nanocyl SA, Sambreville, Belgium))	(Time) = 60 min; C_0 = 93.13 mg L^{-1}; pH = 5.6; T = 25 °C; F_G = 150 cm^3 min^{-1}; $C_{O_3,G}$ = 50.0 mg L^{-1}; (%) = 100%; (% TOC) = 57%; M_{CAT} = 500 mg L^{-1}	A higher mineralization was observed when doping the GAC with TiO$_2$ oxides. The absence of NH$_4^+$ promoted a different oxidation mechanism compared to pristine GAC.	[22]
	TiO$_2$/GAC (Norit® 1240 Plus granular activated carbon doped with TiO$_2$ by precipitation method (Cabot Norit Americas, Inc., Marshall, TX, USA))	(Time) = 45 min; C_0 = 20.0 mg L^{-1}; pH = 7.0; T = 18 °C; F_G = 2.5 g h^{-1}; $C_{O_3,G}$ = 5.4 mg L^{-1}; (%) = 100%; (% TOC) = 80.24%; M_{CAT} = 3.33 g L^{-1}	Through a novel method of synthesis by precipitation, a high yield was obtained in terms of degradation and mineralization.	[4]
TiO$_2$/UV	Hybrid Suspended-Supported TiO$_2$	(Time) = 4.73 h; C_0 = 22 mg L^{-1}; M_{CAT} = 60 mg L^{-1}; (Supp. Cat.) = 2.3 mg cm^{-2}; pH = 12.0; T = 25 °C; (%) = 99%	Under favourable operating conditions, using a hybrid system with suspended TiO$_2$ catalyst, a 23% improvement in the elimination of aniline was observed compared to supported catalyst.	[5]

2. Materials and Methods

2.1. Materials

The granular activated carbon Norit® GAC 1240 Plus was used as a parent material (provided by the Cabot Corporation, Marshall, TX, USA) with an average particle size of 1.4 mm. The GAC was chemically modified and TiO_2 was introduced using a wet precipitation method already described elsewhere [4]; this combination is denoted as TiO_2/GAC. Both activated carbon samples have been previously characterized [4]; Table 2 summarizes their key textural and chemical surface properties. Textural characteristics, such as specific surface area (S_{BET}), micropore (V_{micro}) and mesopore (V_{meso}) volumes, and average pore diameter (D_p), were obtained by observing N_2 adsorption–desorption isotherms at 77 K [4]. The pH of the point of zero charge (pH_{pzc}) was determined using an acidimetric–alkalimetric titration method [5]. The bulk chemical composition of GAC samples was measured using X-ray fluorescence (XRF), as described previously [4]. The XRF results indicate that the parent GAC is mainly composed of SiO_2 (5.61%) and Al_2O_3 (0.45%), followed by Fe_2O_3 (0.28%), CaO (0.12%), S (0.08%), MgO (0.05%), Na_2O (0.04%), TiO_2 (0.03%), and MnO (0.01%). After the applied modification treatment to the parent GAC, the TiO_2/GAC sample was principally comprised of TiO_2 (9.33%) and SiO_2 (7.01%), followed by Al_2O_3 (1.00%), Fe_2O_3 (0.25%), CaO (0.11%), S (0.19%), Cl (2.53%), MgO (0.22%), Na_2O (0.04%), and MnO (0.01%), as determined using XRF.

Table 2. Textural and chemical surface properties of the parent and modified GAC samples [4].

Samples	S_{BET} ($m^2\ g^{-1}$)	V_{micro} ($cm^3\ g^{-1}$)	V_{meso} ($cm^3\ g^{-1}$)	D_p (Å)	pH_{pzc}
Norit® GAC 1240 Plus	967.0	0.32	0.16	36.8	7.4
TiO_2/GAC	985.0	0.29	0.16	33.9	6.4

2.2. Analytical Methods

Aniline concentration was measured using HPLC with a Waters Alliance 2695 liquid chromatograph (Waters, Milford, CT, USA) equipped with an Agilent ZORBAX® Rapid Resolution High Definition (RRHD) Eclipse PAH threaded column (150 mm × 4.6 mm, 5 µm) (Agilent Technologies, Palo Alto, CA, USA) with a guard column and kept at 20 °C. A volume of 20.0 µL of sample was injected. A water:methanol (80:20 v/v) solution was used as a mobile phase with a flow-rate of 1.0 mL min^{-1}. Aniline was detected using a Waters 996 UV-DAD detector (Waters, Milford, CT, USA) at 230 nm.

Aniline oxidation by-products were identified using liquid chromatography coupled to mass spectrometry (MS) in an Agilent 6530 Q-TOF LC/MS (Agilent Technologies, Palo Alto, CA, USA). The separation was carried out using a Kinetex EVO® C18 column (100 mm × 3 mm, 2.6 µm) (Phenomenex, Torrance, CA, USA) kept at 35 °C. A 5.0 µL volume of sample was injected using a flow of 0.3 mL min^{-1}. A mobile phase A consisting of water and a mobile phase B of acetonitrile both containing 0.1% (v/v) HCOOH were used. The elution started with 20% of B and was maintained for 2 min. Then, the concentration of B was increased until it reached 100% at 22 min. The concentration was kept stable for 4 min and a new separation started after 2 min. MS detection was carried out in the positive voltages, following the optimization of electrospray ionization (ESI) parameters. A nitrogen flow of 10 L min^{-1}, a capillary voltage of 3500 V, a nebulizer pressure of 20 psi, and a source temperature of 350 °C were applied. Calibration curves were formed using aqueous solutions of external standards of known composition.

The degree of mineralization was quantified following the total organic carbon (TOC) concentration using a Shimadzu TOC-VCSH analyzer (Izasa Scientific, Alcobendas, Spain). Turbidity was determined using a turbidimeter EUTECH TN-100 from Thermo Scientific (Thermo Scientific, Singapore).

Chemical surface functionalities of pristine and spent GAC samples were identified using Fourier-transform infrared spectroscopy (FTIR). GAC samples were ground in an agate mortar and the resulting powders were mixed with anhydrous KBr to yield a mix with 0.5 % w/w of GAC. A pressed disc of the mixed sample was placed in a disc holder in a JASCO 4200 spectrometer (JASCO Corporation, Tokyo, Japan) equipped with a DLaTGS detector. Spectra were acquired in transmittance mode in the range 4000–400 cm^{-1} with an average of 64 scans and at a resolution of 4 cm^{-1}, using Spectra Manager software V 2.14.02 (JASCO Corporation, Tokyo, Japan). A pressed disc of pure KBr was used as a background for each measurement.

2.3. Experimental Set Up of The Catalytic Ozonation System

Catalytic ozonation experiments were conducted in a 2 L semi-batch jacketed slurry reactor equipped with a magnetic stirrer and several ozone gas diffusers (see Figure 1). The experiments were carried out using a fixed volumetric flow of ozone (Q_G = 4 L min^{-1}) at different pH conditions (3.0, 5.0, 7.0, and 9.0), with varying ozone doses in the gas phase (3.7, 5.4, 11.3, and 20.1 mg L^{-1}) and catalyst load (1.6, 3.3, 6.6, and 13.3 g L^{-1}).

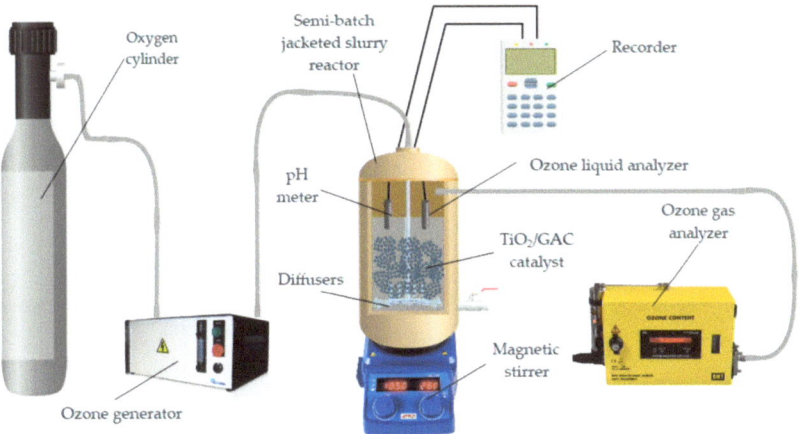

Figure 1. Experimental set-up used to carry out catalytic ozonation tests.

In this study, a concentration of 20.0 mg L^{-1} of aniline—a typical value found in industrial effluent discharges—was used [10,11]. In a typical experiment, the reactor was charged with a 1.5 L solution of aniline at a selected pH value and catalyst load. All experiments were carried out at an authorised discharged temperature of 18 °C and 60 rpm. Ozone was generated in situ from ultra-pure oxygen using a TRIOGEN LAB2B ozone generator (BIO UV, Lunel, France). Ozone concentration in the gas phase was monitored using a BMT 964C ozone analyzer (BMT MESSTECHNIK GMBH, Stahnsdorf, Germany). Dissolved ozone concentration and temperature were measured using a Rosemount 499AOZ-54 dissolved sensor (Emerson, Alcobendas, Spain). The pH value was registered with a Rosemount Analytical model 399 sensor integrated into a Rosemount Analytical Solu Comp II recorder (Emerson, Alcobendas, Spain).

Mass-transfer characterization of the reactor was performed using deionized water in the presence of TiO$_2$/GAC catalyst, following a procedure previously described by Rodríguez et al. [23]. Operating conditions were kept similar to those used in the presence of aniline. Ozone concentrations in the gas and liquid phases were monitored as described above. All experiments were conducted in duplicate with a maximum standard deviation in concentration measurements not exceeding 0.1 mg L^{-1}.

3. Results and Discussion

3.1. Mathematical Modelling Approach Using TiO$_2$/GAC Catalysts

3.1.1. Ozone Kinetic Mass Transfer Modelling

In catalytic ozonation processes, mass transfer is considered to be one of the most important stages in the elimination of organic compounds from industrial wastewater. The transfer of ozone from the gas phase to the aqueous phase is often a controlling step in the process [24]. Therefore, effective ozonation is necessary to improve the oxidation of those compounds that are not highly biodegradable and refractory. Thus, the effect of operational variables involved in the transfer of ozone to the liquid (N_{O_3}, mg L^{-1} min^{-1}) through the overall mass transfer coefficient of ozone gas to water ($K_G a$) or $K_L a$ is expressed as follows:

$$N_{O_3} = K_G a \times (P_{O_3} - P^*_{O_3}) = K_L a \times (C^*_{O_3,L} - C_{O_3,L}) = \frac{Q_G}{V_{reac}} \times (C_{O_3,in} - C_{O_3,out}) \tag{1}$$

For calculation of the mass transfer coefficient ($K_L a$) through the second equality, the contactor was considered to be a perfectly mixed semi-continuous reactor. Equation (1) describes the transfer of ozone from the gas to the aqueous phase during the isothermal catalytic ozonation process, where $K_L a$ is the volumetric ozone mass transfer coefficient (min^{-1}), $C^*_{O_3,L}$ is the concentration of dissolved ozone in the liquid phase at saturation conditions (mg L^{-1}), V_{reac} is the volume of reaction solution (L), Q_G is the flow rate of ozone gas at the inlet (L min^{-1}), and $C_{O_3,in}$ and $C_{O_3,out}$ are the concentrations of ozone in the gas phase at the inlet and outlet, respectively (mg L^{-1}).

According to Rodriguez et al. and Schulz and Prendiville [23,25], in this study, the mass transfer resistance in the gas phase was considered negligible compared to that of the liquid. Consequently, the mass transfer coefficient $K_L a$ can be influenced by the volumetric gas flow, the pH of the solution, the bubble size, and mixing regime, among others. A correctly designed ozone contactor should have a good ozone transfer, avoiding mass transfer control in order to achieve a high mineralization

For determination of the $K_L a$ coefficient, the ozone concentration in the liquid at equilibrium was determined through Henry's law, according to Equation (2):

$$C^*_{O_3,L} = P_{O_3}/He \tag{2}$$

where Henry's solubility constant (atm mole fraction^{-1}) was estimated through Roth and Sullivan's correlation [26]:

$$He = 3.84 \times 10^7 \times C^{0.035}_{OH^-} \times \exp\left(-\frac{2428}{T}\right), \tag{3}$$

where C_{OH^-} is the concentration of hydroxyl ions (mol L^{-1}) and T is the system temperature (K). Through the execution of a calculation program with Scilab® software version 6.1.0 (Scilab Enterprises, Rungis, France), the $K_L a$ values (with a determination coefficient of $R^2 \cong 0.99$) were determined for different pH values and ozone doses at the reactor input. Figure 2 shows the estimated values of the mass transfer constant as a function of pH maintaining a constant ozone dose (11.3 mg L^{-1}) and as a function of ozone dose maintaining a constant neutral pH of 7.0 for an ozonation system with TiO$_2$/GAC catalyst in the absence of a pollutant.

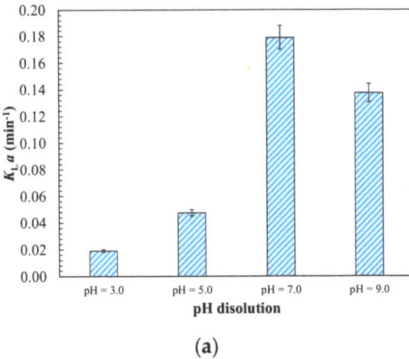
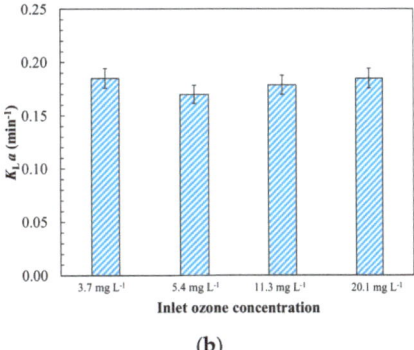

Figure 2. Comparison of volumetric mass transfer coefficient: (**a**) influence of pH on K_La [1] and (**b**) effect of inlet ozone concentration on K_La [2]. Experimental conditions: Q_G = 4 L min^{-1}; M_{CAT} = 3.3 g L^{-1}; T = 18.0 °C; P = 1 atm; V_{reac} = 1.5 L; (Agitation) = 60 rpm. [1] $C_{O_3,in}$ = 11.3 mg L^{-1}; [2] pH = 7.0.

According to Figure 2a, an increase in the mass transfer coefficient was obtained by increasing the pH of the solution. This is due to the fact that the solubility of ozone in water decreases in acid (pH ≅ 2–3) and alkaline (pH ≅ 8–12) solutions [27]. Under acidic conditions, the decrease in solubility is due to the formation of protonated ozone. This is because the protonation of ozone by the hydronium ion (H_3O^+) is thermodynamically unfavourable and the reverse reaction leads to a lower activation energy, which consequently leads to a decrease in solubility [28,29]. On the other hand, under alkaline conditions (pH > 8), the decrease in ozone solubility in water is associated with the self-decomposition of ozone due to the catalytic action of hydroxyl radicals.

The mass transfer coefficients obtained were of the same order of magnitude as those found by other authors [23,30,31], emphasizing the significant increase at pH = 7.0, where Rodríguez et al. [23] obtained a coefficient of 0.073 min^{-1}.

In Figure 2b, the mass transfer coefficients obtained at different ozone doses compared while maintaining a pH of 7.0 are shown. It was found that the dose of ozone had no significant effect on the obtained K_La values. As can be seen, the mass transfer coefficient was independent of the input gas concentration, which was 0.182 min^{-1}. The same K_La obtained at different inlet ozone concentrations is due to the fact that an increase in the ozone concentration in the gas simultaneously produces an increase in the amount adsorbed and in the driving gradient ($C^*_{O_3,L} - C_{O_3,L}$). This confirms that ozone transfer is controlled by the liquid phase, as reported by Berry et al. [32] in a study using membranes and different doses of ozone injected into the reaction system.

According to the results obtained, it is suggested that these transfer coefficients depend mainly on other external factors, such as the system through which the gas is introduced into the contactor or the fluid dynamic conditions of the agitation of the ozonized solution. For better comprehension, Figure S1 shows the evolution of the utilization coefficient as a function of the pH and dose of ozone. The utilization coefficient (%) is defined as follows:

$$U_{O_3} = \frac{C_{O_3,in} - C_{O_3,out}}{C_{O_3,in}} \times 100 \qquad (4)$$

In both cases, two consecutive and distinct steps were observed. In the first step, up to approximately 5 min, it was observed that ozone transferred from the gas phase to the liquid phase, in which ozone becomes a molecular compound in water that previously did not exist. After that, a transitory step occurred, in which the ozone transferred from the gas to the liquid phase is greater than that consumed through self-decomposition and reaction with TiO$_2$/GAC catalyst. Finally, a stationary state was reached, in which ozone accumulated in the system reached saturation and a constant value, resulting in an equilibrium between the transferred and consumed ozone.

In Figure S1a, it can be observed that, after reaching saturation, a utilization coefficient of 32% was obtained at pH = 7.0, compared to 19.5% obtained at pH = 3.0. This difference was due to the different solubility that ozone has in water at different pH values, as explained above. The utilization value of 24.8% obtained at pH = 9.0 may be due to the fact that the transferred ozone instead contributes to the accumulation of ozone in the liquid phase; in this case, it would undergo self-decomposition [12]. In the study of the ozone dose effect (Figure S1b), it can be observed that an increase in the ozone dose reduces the efficiency of ozone use. This can be justified largely by the study carried out by Rodriguez et al. [23]. In their study, the degradation of a dye such as rhodamine 6G was evaluated at different pH values and doses of ozone in the presence of activated carbon. The ozone transfer and subsequent consumption was conditioned by parameters, such as pH, which indirectly affect the reactivity of the compound to be oxidized but not the concentration of ozone dose applied.

Von Sonntag and Von Gunten [33] studied the reactivity of aniline oxidation in an ozonation process at different pH values. It was found that, at pH = 6.5, an oxidation kinetic constant of 1.4×10^7 M^{-1} s^{-1} was obtained versus pH = 1.5, where a constant of 5.9×10^4 $M^{-1}s^{-1}$ was obtained. All of this indicates that, at pH = 7.0, the best operating conditions will be obtained (with regards to the chemical reaction of oxidation and transfer of ozone from the gas phase to the liquid).

3.1.2. Aniline Degradation Kinetic Modelling

Modelling of the catalytic ozonation process was carried out to describe the combined action of ozone and TiO_2/GAC composite and to predict and estimate the overall parameters affecting the process of degradation (and, in particular, the mineralization) of aniline. In this section, we propose applying the adaptation of the three-phase mathematical model described by Ferreiro et al. [19], which considers G-L ozone mass transfer, the adsorption process as well as the parallel oxidation that takes place in the liquid phase and the oxidation at the surface of the catalytic material (TiO_2/GAC). The model proposed allows the evolution of the monitored pollutant or total organic carbon (TOC) during the simultaneous oxidation reaction and adsorption process to be calculated. For application of this simultaneous adsorption–oxidation (Ad/Ox) model, it was assumed that:

- The rate of the global ozonation process, or G-L mass transfer rate, coincides with the consumption of ozone in the parallel (liquid and solid) reactive process.
- The oxidation kinetics of aniline in the liquid and in the solid, in terms of TOC, is considered as pseudo-first order.
- For the TiO_2/GAC composite, it was considered a sufficiently porous material for ozone and aniline diffusion mechanisms to take place also in the internal surface of particle.
- The kinetic constant of the aniline oxidation on the solid, also in terms of TOC, includes desorption of degradation compounds.
- The adsorption process is simultaneous to the reaction process on the solid so that its kinetics are strongly affected by the ozonation conditions.

Based on these considerations, Ferreiro et al. [19], for an Ad/Ox process, deduced Equation (5) to express the decrease in contaminant concentration in the liquid in terms of total organic carbon (C_p). The degradation of the pollutant, r_p, should be explained at two levels: in the liquid (r_p^I) and on the activated carbon (r_p^{II}), as a consequence of the oxidative action of the ozone transferred and consumed in the liquid ($N_{O_3}^I$) and on the solid ($N_{O_3}^{II}$), respectively. In addition to the degradation of the pollutant, the elimination due to adsorption is important in the first stages of the process.

$$-\frac{dC_p}{dt} = r_p^I + r_p^{II} + \left(\frac{dC_p}{dt}\right)_{ads} \qquad (5)$$

The kinetic parameters of oxidation and adsorption can be expressed as a function of the respective kinetic constants ($k_{c,L}$, $k_{c,S}$, and k_{ads}), ozone concentration, and pollutant, according to the following:

$$r_P^I + r_P^{II} = \left(z^I N_{O_3}^I + z^{II} N_{O_3}^{II}\right) = k_{c,L} \times C_{O_3,L} \times C_P + k_{c,S} \times \frac{C_{O_3,L}}{m} \times M_{CAT} \times Z_P \quad (6)$$

$$\left(\frac{dC_P}{dt}\right)_{ads} = k_{ads} \times \left(Z_{p,\infty} - Z_p\right)^2 \times M_{CAT} \quad (7)$$

In Equation (6), the stoichiometric coefficients z^I and z^{II} express the amount of pollutant (mg TOC per mg of ozone) consumed in the liquid and on the catalyst, respectively. The above equation assumes that the concentration of ozone on the catalyst, $C^*_{O_3,S}$ is in equilibrium with the liquid ozone concentration, $C_{O_3,L}$, and can be expressed as a function of this, according to $C_{O_3,L} = m \times C^*_{O_3,S}$, where m (mg L^{-1})/(mg g^{-1} catalyst) is the slope of the corresponding L-S balance. On the other hand, Z_p is the concentration of total organic carbon adsorbed in the catalyst (mg g^{-1}), M_{CAT} is the concentration of catalyst (g L^{-1}), and $Z_{p,\infty}$ is the amount of total organic carbon adsorbed in the catalytic material at equilibrium (mg g^{-1}), according to the Freundlich equation:

$$Z_{p,\infty} = K_F \times C_p^{1/n} \quad (8)$$

where K_F is a Freundlich constant that indicates the adsorption capacity of the adsorbent (mg g^{-1}) (L mg^{-1})$^{1/n}$ and n describes the adsorption intensity. Values of n between 2 and 10 indicate good adsorption intensity [34].

In a semi-continuous process in which an ozone flow is continuously injected, the concentration of ozone in the liquid, C_{O_3}, is assumed to be constant during the reaction [23]. This assumption can be used to define the apparent first-order kinetic constants—k_{oxL} in the liquid and k_{oxS} over the catalyst—to describe the degradation of the aniline, in terms of TOC (min^{-1}). Consequently, the overall pollutant concentration variation (Equation (5)) can be written as follows:

$$-\frac{dC_P}{dt} = k_{oxL} \times C_P + k_{oxS} \times M_{CAT} \times Z_p + k_{ads} \times \left(Z_{p,\infty} - Z_p\right)^2 \times M_{CAT}, \quad (9)$$

where C_p is the concentration of the pollutant (TOC) at a given time t (mg L^{-1}), Z_p is the concentration of total organic carbon adsorbed onto the catalyst (mg g^{-1}), M_{CAT} is the concentration of catalyst (g L^{-1}), $Z_{p,\infty}$ is the amount of total organic carbon adsorbed in the composite at equilibrium (mg g^{-1}), k_{oxL} is the first-order kinetic constant due to the oxidation of aniline in the liquid (min^{-1}), k_{oxS} is the pseudo-first-order kinetic constant due to oxidation on the catalytic material (min^{-1}), and k_{ads} is the aniline adsorption constant (g mg^{-1} min^{-1}). The combination of Equations (8) and (9) provides a description of the variation in the pollutant concentration in the system from the determination of the kinetic parameters (k_{oxL}, k_{oxS}, and k_{ads}) and the equilibrium parameters (K_F and n).

3.1.3. Evaluation of Operating Conditions for Model Validation

Operational parameters such as pH, ozone concentration, or the dose of catalyst used have a considerable effect on the efficiency of the removal of aniline using catalytic ozonation processes with TiO$_2$/GAC catalyst. Using the Ad/Ox model described above, it is possible to explain the effects of the different parameters, such as the behaviour of the ozone or the catalyst dosage. In Figure 3, the evolution of aniline degradation is shown, as well as the mineralization under various pH values (3.0, 5.0, 7.0, and 9.0), ozone concentrations at the inlet (3.7, 5.4, 11.3, and 20.1 mg L^{-1}), and catalyst doses (1.6, 3.3, 6.6, and 13.3 g L^{-1}).

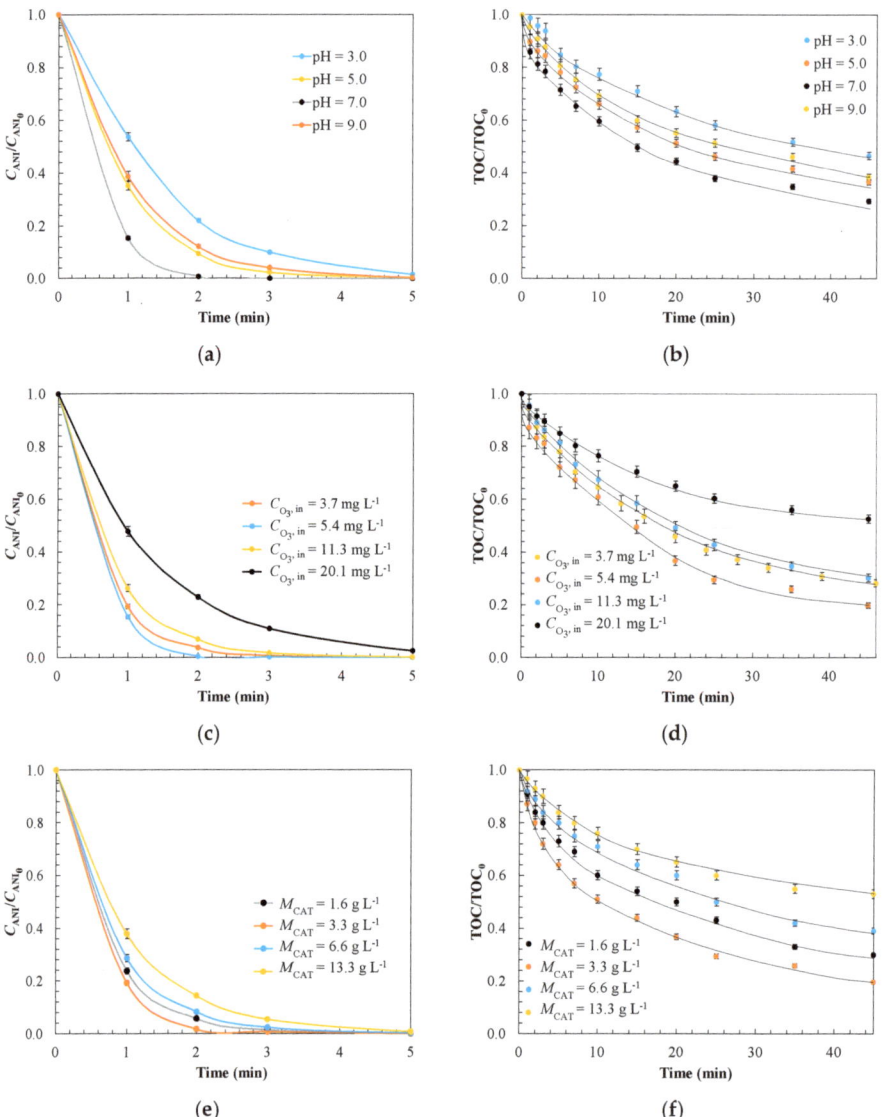

Figure 3. Effect of operating conditions on catalytic ozonation with TiO$_2$/GAC: (**a**) pH [1], (**c**) inlet ozone concentration [2], and (**e**) mass of TiO$_2$/GAC catalyst in the aniline removal [3] and (**b**) [1], (**d**) [2] and (**f**) in TOC (mineralization), showing experimental and modelled profiles fitted to the Ad/Ox process [3]. Experimental conditions: $Q_G = 4$ L min^{-1}; $T = 18.0$ °C; $P = 1$ atm; $V_{reac} = 1.5$ L; (Agitation) = 60 rpm. [1] $C_{O_3,in} = 5.4$ mg L^{-1}, $M_{CAT} = 3.3$ g L^{-1}; [2] pH = 7.0, $M_{CAT} = 3.3$ g L^{-1}; [3] pH = 7.0, $C_{O_3,in} = 5.4$ mg L^{-1}.

The pH has a significant effect on the catalyst, as it directly affects the pathway through which the ozone acts on the organic compound. Figure 3a,b shows that the best conditions for the effective removal of aniline are present at a neutral pH of 7.0. The change in basicity associated with pollutant removal at that pH, or specific basicity loss, is optimal [35]. In addition, it was observed that those experiments conducted at slightly alkaline pH (pH = 9.0) obtained a higher oxidation rate because, under these conditions, the indirect pathway for hydroxyl radical generation is promoted by the interaction between ozone and the TiO$_2$/GAC catalyst. On the other hand, at acidic pH (pH = 3.0), a more limited oxidation and mineralization rate was observed because, at this pH value, the attack through the molecular pathway between ozone and aniline is mostly promoted [36,37]. Although this pathway is more selective than the radical one, the oxidation power of molecular ozone (2.07 V) is lower than that of the hydroxyl radicals (2.80 V) [28].

On the other hand, the dose of ozone in the catalytic ozonation reaction system is a critical parameter for the degradation and mineralization of aniline. In Figure 3c,d, it is shown that, with an ozone dose of 5.4 mg L^{-1}, the highest aniline mineralization was achieved, with a value of 80.2%. However, upon increasing the ozone dose considerably (to 20.1 mg L^{-1}), a negative effect was obtained as opposed to the generation of a larger amount of hydroxyl radicals. This was due to the fact that, on the surface of the TiO$_2$/GAC catalyst, instead of generating hydroxyl radicals for the degradation of the organic compound, other species were generated, such as hydroperoxyl radicals—which have a lower oxidation potential (1.70 V) than hydroxyl radicals (2.80 V) or ozone (2.07 V) [38]. Moreover, a very high ozone dose (20.1 mg L^{-1}) could transform the basic sites into acidic sites, as the oxidation of the carbonaceous support can generate acidic groups such as lactones or carboxylic acid, which could negatively influence the adsorptive properties of the TiO$_2$/GAC composite itself [39]. This last aspect will be analysed in detail later on.

In Figure 3e,f, the effect of the catalyst dose on the aniline removal rate is shown, providing information on the optimal use of the TiO$_2$/GAC material. An increase in catalyst dosage led to increased degradation and mineralization of the aniline, with a favourable dosage of 3.3 g L^{-1}, where mineralization was close to 80% and complete degradation of the aniline was achieved in less than 10 min. In general, a higher catalyst dose increases the number of active sites on the surface of the composite, thus facilitating the further decomposition of ozone into hydroxyl radicals [40]. However, above a certain critical value, a decrease in TOC removal was observed, which may be due to the adsorption of a higher proportion of pollutant on the catalyst, thus reducing the catalytic effect [38] in favour of the adsorption. This indicates that a small amount of catalyst is sufficient to induce a radical chain reaction that appears to be more effective on the dissolved than adsorbed pollutant. Moussavi and Khavanin [41] studied the effect of the dose of activated carbon in a catalytic ozonation process for the removal of phenol. It was observed that, from the optimal dose of catalyst, no further improvement in mineralization was observed, as with the TiO$_2$/GAC composite. Consequently, it was concluded that there is an optimal dose of catalyst, which changes depending on the type of catalyst, the organic compound to be eliminated, the operational conditions of the reaction, and the desired cost-effectiveness.

The TOC removal kinetics was fitted to the model proposed in Section 3.1.2 by solving Equations (8) and (9). Before analysing the results obtained from the Ad/Ox model, it was verified that under the operating conditions studied the assumption of control of the chemical reaction during catalytic ozonation was satisfied.

The control regime was verified through the U_{O_3} profiles shown in Figure S2. It can be seen at the lowest ozone doses ($C_{O_3,in}$ = 3.7 mg L^{-1}), in the initial instants, the ozone concentration in the gaseous phase is practically zero (U_{O_3} = 100%). This indicates that the process is controlled by the mass transfer. However, after 4 min it changes to be controlled by the chemical reaction, once $C_{O_3,L}$ starts rising. For resolution of the TOC removal kinetics, the calculation tool Scilab® was used, with which the corresponding adjustment to the experimental data was simulated to obtain the adsorption and ozonation kinetics constants during the removal of aniline with the previous determination of equilibrium adsorption constants. As for the initial conditions of the dependent variables involved in the differential equation, at the initial time, an aniline concentration of 20.0 mg L^{-1} was considered, corresponding to an initial TOC concentration of $C_{p,0}$ = 12.50 mg L^{-1} and the TiO$_2$/GAC free of aniline or other organics ($Z_{p,0}$ = 0.0 mg g^{-1}). C_p^* vs. t modelled profiles were fitted to the experimental ones (of N values), in order to minimize the weighted standard deviation, σ, given by Equation (10). Thus, the estimated values of k_{ads}, k_{oxL}, and k_{oxS} are listed in Table 3 [42]:

$$\sigma = \sqrt{\frac{\sum_{i=1}^{N}\left(\frac{(C_p - C_p^*)}{C_p}\right)^2}{N-1}} \quad (10)$$

Table 3. Summary of the adsorption and oxidation kinetic constants for the mineralization of wastewater containing aniline using Norit® GAC 1240 Plus and TiO$_2$/GAC composite by catalytic ozonation.

	Catalyst Comparison	
Kinetic Parameter	Norit® GAC 1240 Plus [1,2]	TiO$_2$/GAC [1,2]
$k_{ads} \times 10^{-4}$, g mg^{-1} min^{-1}	2.4	3.5
$k_{oxL} \times 10^{1}$, min^{-1}	8.1	5.9
$k_{oxS} \times 10^{1}$, min^{-1}	0.0030	2.3
σ	0.050	0.046

	TiO$_2$/GAC Composite Analysis			
	Effect of pH			
Kinetic Parameter	3.0	5.0	7.0	9.0
$k_{ads} \times 10^{-4}$, g mg^{-1} min^{-1}	3.6	3.1	2.9	2.5
$k_{oxL} \times 10^{1}$, min^{-1}	3.4	6.5	8.3	8.6
$k_{oxS} \times 10^{1}$, min^{-1}	0.20	1.1	1.2	0.90
σ	0.061	0.053	0.059	0.048

	Effect of Ozone Inlet Concentration, mg L^{-1}			
Kinetic Parameter	3.7	5.4	11.3	20.1
$k_{ads} \times 10^{-4}$, g mg^{-1} min^{-1}	3.2	3.5	2.9	1.8
$k_{oxL} \times 10^{1}$, min^{-1}	4.2	5.9	8.3	11.5
$k_{oxS} \times 10^{1}$, min^{-1}	1.5	2.3	1.2	0.50
σ	0.052	0.046	0.057	0.045

	Effect of Catalyst Dose, g L^{-1}			
Kinetic Parameter	1.6	3.3	6.6	13.3
$k_{ads} \times 10^{-4}$, g mg^{-1} min^{-1}	1.8	3.5	4.4	8.4
$k_{oxL} \times 10^{1}$, min^{-1}	3.8	5.9	2.2	1.0
$k_{oxS} \times 10^{1}$, min^{-1}	1.9	2.3	1.5	0.90
σ	0.054	0.046	0.055	0.052

[1] Experimental conditions: $Q_G = 4$ L min^{-1}; $C_{O_3,in} = 5.4$ mg L^{-1}; pH$_0 = 7.0$; $M_{CAT} = 3.3$ g L^{-1}; $T = 18.0$ °C; $P = 1$ atm; $V_{reac} = 1.5$ L; (Agitation) = 60 rpm. [2] Freundlich equilibrium parameters: $K_F = 33.07$ (mg g^{-1})(L mg^{-1})$^{1/n}$, $n = 2.39$ for Norit® GAC 1240 Plus and $K_F = 44.02$ (mg g^{-1})(L mg^{-1})$^{1/n}$, $n = 3.97$ for TiO$_2$/GAC composite.

The Freundlich parameters of the catalytic materials tested are given in Table 3, which were obtained experimentally in the previous adsorption tests. The C_p values had an acceptable fit to the Ad/Ox model with a weighted standard deviation of $\sigma \cong 0.05$ for the mineralization kinetics.

In Table 3 it was observed that the k_{oxS} constant was lower than k_{oxL}. This is because in the solid involved an oxidation reaction in conjunction with the adsorption phenomena in parallel. From the comparison of the catalytic ozonation with Norit® GAC 1240 Plus activated carbon and TiO$_2$/GAC composite, it can be seen that the TiO$_2$/GAC catalysts kinetically favour adsorption, with $k_{ads} = 3.5 \times 10^{-4}$ g mg^{-1} min^{-1}, compared to commercial carbon ($k_{ads} = 2.4 \times 10^{-4}$ g mg^{-1} min^{-1}). Its higher adsorption capacity, rather than the specific surface ($S_{EXT} = 298.9$ m^2 g^{-1})—which is very similar to that of commercial activated carbon ($S_{EXT} = 224.4$ m^2 g^{-1})—explains this behaviour, despite the fact that TiO$_2$/GAC has a 25% larger external surface, as seen in the Material and Methods section [43]. The adsorption properties of metal oxides, such as TiO_2, deposited onto GAC, appear to also enhance the catalytic activity. Other authors [37,44] have highlighted that metal oxides have a high adsorption capacity, which is due to ligand exchange reactions that provide strong bonds between the ionized species and the active site of the metal oxide surface. The increase in the oxidation kinetic constant in the solid, from a value of $k_{oxS} = 0.003 \times 10^1$ to 2.3×10^1 min^{-1}, in the TiO_2/GAC composite is remarkable. This increase is due to the contribution of TiO_2 in the activated carbon, improving its reactivity and the number of active sites on the catalyst. The deposited TiO_2 is responsible for efficiently decomposing the ozone, producing HO$^\bullet$ radicals. Other properties, such as pore volume, porosity, pore size distribution, and, particularly, the presence of active sites on the surface (e.g., Lewis acid sites, which are responsible for the catalytic reactions), may be responsible for this increase in k_{oxS} [45]. Acidity or basicity, for example, is key to surface properties. Furthermore, the hydroxyl groups, which are present on all surfaces of the metal oxides, are dependent on the deposited metal oxide [39]. Valdés and Vega [45] studied the effect of the chemical structure of various active carbons on the catalytic activity for the generation of hydroxyl radicals, where they suggested that the presence of iron metal ions played an important role in the decomposition of ozone and hydrogen peroxide towards the generation of radicals. Those carbons with more basic surface functionalities led to a higher radical generation and, consequently, to a higher catalytic activity.

Authors such as Orge et al. [46] and Kasprzyk-Hordern et al. [38] have suggested that the variability of surface properties and interactions between the catalyst and ozone with organic pollutants results in different reaction mechanisms, derived from two main types: Langmuir–Hinshelwood (LH) or modified Eley–Rideal (ER).

The LH stage consists of adsorption, surface reaction, and subsequent desorption [28]:

Adsorption of ozone onto every site, S, of GAC surface:

$$O_3 + S \underset{}{\overset{K_{LH1}}{\rightleftarrows}} O=O-O-S \tag{11}$$

$$O=O-O-S \overset{k_{LH2}}{\rightarrow} O-S + O_2 \tag{12}$$

Adsorption of aniline:

$$C_6H_5NH_2 + S \underset{}{\overset{K_{LH3}}{\rightleftarrows}} C_6H_5NH_2 - S \tag{13}$$

Surface reaction and desorption of oxidation products:

$$\frac{21}{2}O-S + C_6H_5NH_2 - S \overset{k_{LH4}}{\rightarrow} NO_3^- + 2CO_2 + \frac{7}{2}H_2O + 2S \tag{14}$$

The other mechanism (modified ER), proposed by Beltran et al. [44], is specific for metal oxides supported on activated carbon materials and consists of an adsorption stage on the GAC, together with the assumptions of the modified ER mechanism. In this case, these stages are: (i) the adsorption of aniline onto the TiO_2/GAC composite, (ii) the reaction stage between the adsorbed aniline and the ozone, and (iii) the irreversible ozonation reaction of adsorbed pollutant.

Adsorption of aniline on the GAC active sites:

$$C_6H_5NH_2 + S \underset{}{\overset{K_{ER1}}{\rightleftarrows}} C_6H_5NH_2 - S \tag{15}$$

Adsorption of aniline on the TiO_2 active sites:

$$C_6H_5NH_2 + S' \underset{}{\overset{K_{ER2}}{\rightleftarrows}} C_6H_5NH_2 - S' \tag{16}$$

where S' represents any active site on the TiO_2.

Reaction between ozone and adsorbed aniline on TiO_2 active sites:

$$\frac{41}{2}O_3 + C_6H_5NH_2 - S' \xrightarrow{k_{ER3}} 6CO_2 + \frac{7}{2}H_2O + O_2 + NO_3^- + S' \tag{17}$$

In this last mechanism, it was assumed that the CO_2 generated corresponds to that resulting from complete mineralization. Overall, it can be concluded that the removal of aniline is due to an adsorption process on the GAC and to the catalytic ozonation itself, which takes place in the active centres of the TiO_2 metal oxide. Based on the values of the k_{oxS} and k_{oxL} constants obtained from the three-phase model, it is possible to qualitatively determine the predominant mechanism (LH or modified ER). An intensification of the k_{oxS} constant due to the presence of TiO_2 metal oxide against the k_{oxL} constant indicates the predominance of the modified ER mechanism. In contrast, negligible values of the k_{oxS} constant compared to the k_{oxL} constant indicate the predominance of the LH mechanism. Taking into account the mechanical aspects of the reaction system, the effect of operational variables, such as pH, ozone dosage, and TiO_2/GAC catalyst loading were analysed.

Regarding the effect of pH, due to the surface properties of the composite, it was observed that the adsorption constant increased under more acidic pH values, while under alkaline pH values, the quantity of aniline removed decreased. This is due to the speciation of the aniline, with $pK_a = 4.61$ [47], and the character of the surface of the composite through the zero loading point ($pH_{pzc} = 6.4$). The TiO_2/GAC composite, at a pH below 6.4, develops a negative charge on its surface; below pH = 4.61, it will mostly be in ionic form, favouring adsorption. On the other hand, under alkaline pH, the affinity between the aniline and the surface of the material is weak and adsorption is limited. Similarly, Shahamat et al. [48], in a catalytic ozonation process in which a carbon nanocomposite was used for the removal of phenol, observed the same behaviour at a pH between pH_{pzc} and pK_a. The oxidation constant in the liquid, k_{oxL}, increased from a value of 3.4×10^1 to 8.6×10^1 min^{-1}, due to the fact that, under alkaline conditions, the radical pathway in which ozone directly attacks the OH^- generating radicals $HO^\bullet$ is favoured, according to [49]:

$$O_3 + OH^- \rightarrow O_3^{\bullet -} + HO^\bullet, \tag{18}$$

$$O_3^{\bullet -} \rightarrow O^\bullet + O_2, \tag{19}$$

$$O^{\bullet -} + H^+ \rightarrow HO^\bullet \tag{20}$$

As for k_{oxS}, the maximum value observed was 1.2×10^1 min^{-1} at pH = 7.0. This increase was due to the contribution that TiO_2 metal oxide provides to the GAC, thus improving the capacity of transforming ozone into hydroxyl radicals. According to Roshani et al. [50], the surface charge and the capacity of the TiO_2/GAC composite to transfer electrons to the ozone are factors that affect the elimination of TOC. Operating at a pH of 7.0, favourable conditions allow for a positive interaction between the ozone and TiO_2 metal oxide, which allows for the decomposition of ozone and, thus, the generation of a greater number of $HO^\bullet$ radicals in sufficient quantity to oxidize the organic compounds adsorbed on the surface of the GAC. According to Nawrocki and Kasprzyk-Hordernb [51], considering the type of radicals formed on TiO_2 nanoparticles in the presence of ozone, it was concluded that catalytic ozonation was more effective at a pH close to pH_{pzc}. Under these conditions, the presence of neutral hydroxyls are responsible for the formation of the hydroxyl radical.

For the effect of the ozone dose, it was observed that the use of a low ozone dose ($C_{O_3,in} = 3.7$ mg L^{-1}) led to a mineralization yield of 69%, as the amount of hydroxyl radicals generated was low. Taking into account the distribution of the kinetic constants obtained (Table 3), when the dose is insufficient, the adsorbent effect is enhanced but not the oxidation of the organic pollutant. On the other hand, with a high dose ($C_{O_3,in} = 20.1$ mg L^{-1}), ozone accumulates in the system, favouring the generation of the less reactive perhydroxyl radical ($HO_2^\bullet$), according to [49]:

$$O_3 + HO^\bullet \rightarrow O_2 + HO_2^\bullet \tag{21}$$

It seems evident that moderate concentrations favour the adsorption kinetics and result in a sufficient generation of hydroxyl radicals on the surface of the catalyst. A dose of 5.4 mg L^{-1} ensures that the aniline is adsorbed rapidly over the GAC (k_{ads} = 3.5 × 10^{-4} g mg^{-1} min^{-1}) and, at the same time, maximizes oxidation at the solid level (k_{oxS} = 2.3 × 10^{1} min^{-1}), where TiO$_2$ plays an important role. The action of ozone on the adsorbed aniline, according to the modified ER mechanism (see Equation (17)), appears to be the dominant mechanism in this case and responsible for the high k_{oxS} value.

Concerning the effect of catalyst loading, it was observed that, by increasing the dose from 1.6 to 13.3 g, the adsorption constant increased from 1.8 × 10^{-4} to 8.4 × 10^{-4} g mg^{-1} min^{-1}. A higher quantity of adsorbed organics and ozone is to be expected when increasing the amount of catalyst but does not result in further destruction of the aniline. It can be observed that high doses of catalyst lead to a predominance of surface oxidation reactions, as a result of increased TiO$_2$/GAC active sites and an increased amount of adsorbed contaminant [36]. However, the most favourable values were produced with an intermediate amount of catalyst (k_{oxS} = 2.3 × 10^{1} min^{-1}). The negative effect became increasingly evident as the amount of catalyst increased, which was related to a change in the mechanism of the oxidation processes at the solid level. According to the assumption made in Section 3.1.3, a transition from the modified ER to the LH mechanism should take place. In effect, with small amounts of catalyst, a moderate amount of contaminants was adsorbed—preferably on TiO$_2$—through non-associated hydroxyl groups [52] or S' sites (Equation (17)) with favourable incidence in the increase of k_{oxS} and consequently in the degradation of the contaminant. An increase in the amount of catalyst led to a greater proportion of the contaminant being adsorbed, resulting in the occupation of S sites and participation of slow rate reactions (such as that shown in Equation (14)), with a subsequent decrease in k_{oxS}. A decrease in the degradation of the aniline at the liquid phase as k_{oxL} = 1.0 × 10^{1} min^{-1} decreased and the lower proportion of contaminant in the liquid were other negative effects associated with an increase in the amount of catalyst [38]. Moderate amounts of catalyst (3.3 g L^{-1}) favoured oxidation at the solid and liquid level, obtaining the highest mineralization (80% TOC removal) in 45 min.

3.2. Physicochemical Surface Characterization of Spent-Granular Activated Carbon

In order to verify the lower generation of hydroxyl radicals when a high catalyst dose was used and that the aniline would be oxidized under very poor oxidation conditions, an analysis of the physicochemical properties of the TiO$_2$/GAC composite used was carried out, as shown in Table 4.

Table 4 shows a decrease of the specific surface of the TiO$_2$/GAC composite of the experiment performed with a catalyst load of 13.3 g L^{-1} from 985.0 to 901.2 m^2 g^{-1}. The pore volume was also reduced, but the physical properties were practically unchanged with a catalyst dose of 3.3 g L^{-1}. Considering the importance of the adsorption stages contemplated in the Ad/Ox model, the necessary equilibrium of reactions at the liquid and solid level should not be broken. This balance is broken when the catalyst dose is increased, with negative effects on the degradation of the contaminant being key in the optimization of the process.

Concerning the chemical properties (e.g., the point of zero charge), ozone appeared to affect them by a small amount. The carbonaceous support lost active basic sites while the concentration of the acidic functional groups increased, leading to a reduction in adsorption capacity over long periods [53].

Table 4. Physicochemical properties of pristine and spent TiO$_2$/GAC in aniline ozonation with different catalyst doses. Experimental conditions: Q_G = 4 L min^{-1}; $C_{O_3,in}$ = 5.4 mg L^{-1}; pH = 7.0; T = 18.0 °C; P = 1 atm; V_{reac} = 1.5 L; (Agitation) = 60 rpm.

Property	Pristine TiO$_2$/GAC	Spent TiO$_2$/GAC	
		M_{CAT} = 3.3 g L^{-1}	M_{CAT} = 13.3 g L^{-1}
S_{BET}, m^2 g^{-1}	985.0	980.4	901.2
S_{ext}, m^2 g^{-1}	298.9	289.1	267.5
V_T, cm^3 g^{-1}	0.45	0.39	0.32
V_μ, cm^3 g^{-1}	0.29	0.25	0.20
V_M, cm^3 g^{-1}	0.16	0.14	0.12
V_M/V_T, %	35.2	35.9	37.5
V_μ/V_T, %	64.8	64.1	62.5
D_P, Å	33.9	33.0	29.3
pH$_{pzc}$	6.4	6.2	6.3

Moreover, Figure 4 shows FTIR spectra of the non-ozonized and ozonized TiO$_2$/GAC catalyst using a 13.3 g L^{-1} load, in order to confirm the presence of compounds that verify the previous hypothesis. According to Figure 4, it was observed that most of the spectral bands corresponded to organic compounds, highlighting the band from 3300 to 3500 cm^{-1} (–OH stretching), which is due to the presence of water in the sample during preparation. The 500 cm^{-1} spectral band (Ti–O stretching) corresponded to the presence of TiO$_2$ deposited on the GAC [54]. The main spectral modifications, when comparing the sample of the non-ozonized composite and the ozonized one, were detected in the emergence of the 790 cm^{-1} band. The accumulation of intermediate species, such as oxamic acid, and subsequent sorption in the solid explain this FTIR sorption band [55]. The peak near to 2800 cm^{-1} would correspond to aldehyde groups, such as formaldehyde or acetaldehyde, and strongly depended on the ozone dose, adsorbed on the catalyst due to the opening of the aromatic ring [56], while another band at 1102 cm^{-1} was due to the superoxide radical [57].

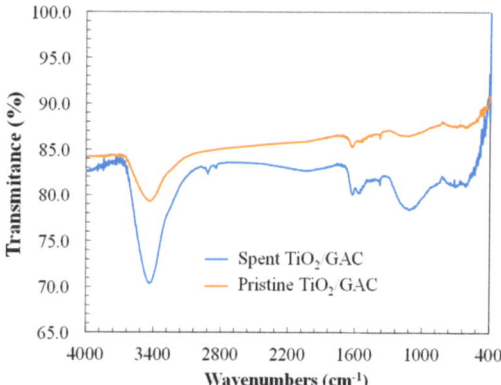

Figure 4. FTIR spectra of pristine and spent TiO$_2$/GAC composite in the aniline catalytic ozonation. Experimental conditions: Q_G = 4 L min^{-1}; $C_{O_3,in}$ = 5.4 mg L^{-1}; pH = 7.0; M_{CAT} = 13.3 g L^{-1}; T = 18.0 °C; P = 1 atm; V_{reac} = 1.5 L; (Agitation) = 60 rpm.

3.3. Degradation Pathway Approach

In order to provide more detail on the types of intermediates formed when increasing the dose of catalyst, other physical–chemical parameters, such as turbidity, were analysed. In Figure 5, the effect of the catalyst dose on turbidity is shown. Solid particles or high molecular weight insoluble degradation products usually cause turbidity [58]. The experiment with the highest amount of catalyst

(13.3 g L^{-1} TiO$_2$/GAC catalyst), with an ozone dose of 5.4 mg L^{-1} and a neutral pH, led to the highest turbidity, with a turbidity of 17 NTU after a reaction time of 45 min. However, with a catalyst dose of 3.3 g L^{-1}, the lowest turbidity was produced, corresponding to the highest mineralization observed. Consequently, the higher turbidity could be associated with the formation of more recalcitrant intermediate products.

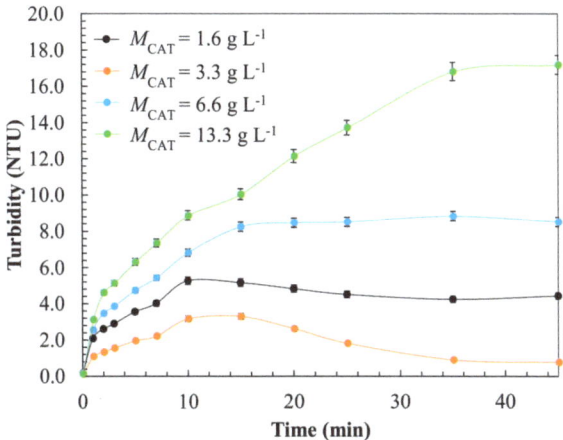

Figure 5. Effect of TiO$_2$/GAC catalyst dosage on turbidity during catalytic ozonation of aniline containing wastewater. Experimental conditions: Q_G = 4 L min^{-1}; $C_{O_3,in}$ = 5.4 mg L^{-1}; pH = 7.0; T = 18.0 °C; P = 1 atm; V_{reac} = 1.5 L; (Agitation) = 60 rpm.

Orge et al. [22] have highlighted oxalic and oxamic acid among the reaction products of aniline with the highest resistance to degradation. Figure 6 shows the results obtained from the identification of aniline, oxamic acid, and oxalic acid at the initial time, after decomposition of the aniline (≈5 min), and after a sufficiently long reaction time. Figure 6d shows the chromatogram obtained at zero time (peak 1) with a retention time of 8.21 min, which was assigned to aniline. After the catalytic ozonation reaction had progressed for 5 min, the peak of the aniline (1) decreased but others appeared, which persisted until sufficiently long reaction times (30–40 min). The mass spectra of the identified peaks are shown in Figure 6a–c. These peaks were assigned to oxalic (2, 1.80 min) and oxamic acid (3, 2.65 min), corresponding to two degradation intermediates formed during the ozonation of the ozone aniline [59].

Oxalic acid under conditions of low hydroxyl radical generation as well as its conjugate base are stable degradation intermediates for a wide variety of organic contaminants, such as pesticides. The accumulation of this refractory compound in the reaction system is due to its very low oxidation constant (k < 0.04 M^{-1} s^{-1} at pH values above 5.0), compared to that of other aniline oxidation products such as hydroquinone (k = 2.3 × 10^6 M^{-1} s^{-1} at pH = 7.0) [33,60]. On the other hand, oxamic acid is another compound present in the degradation of aniline that, under poor oxidation conditions, shows high refractoriness to ozonation [61].

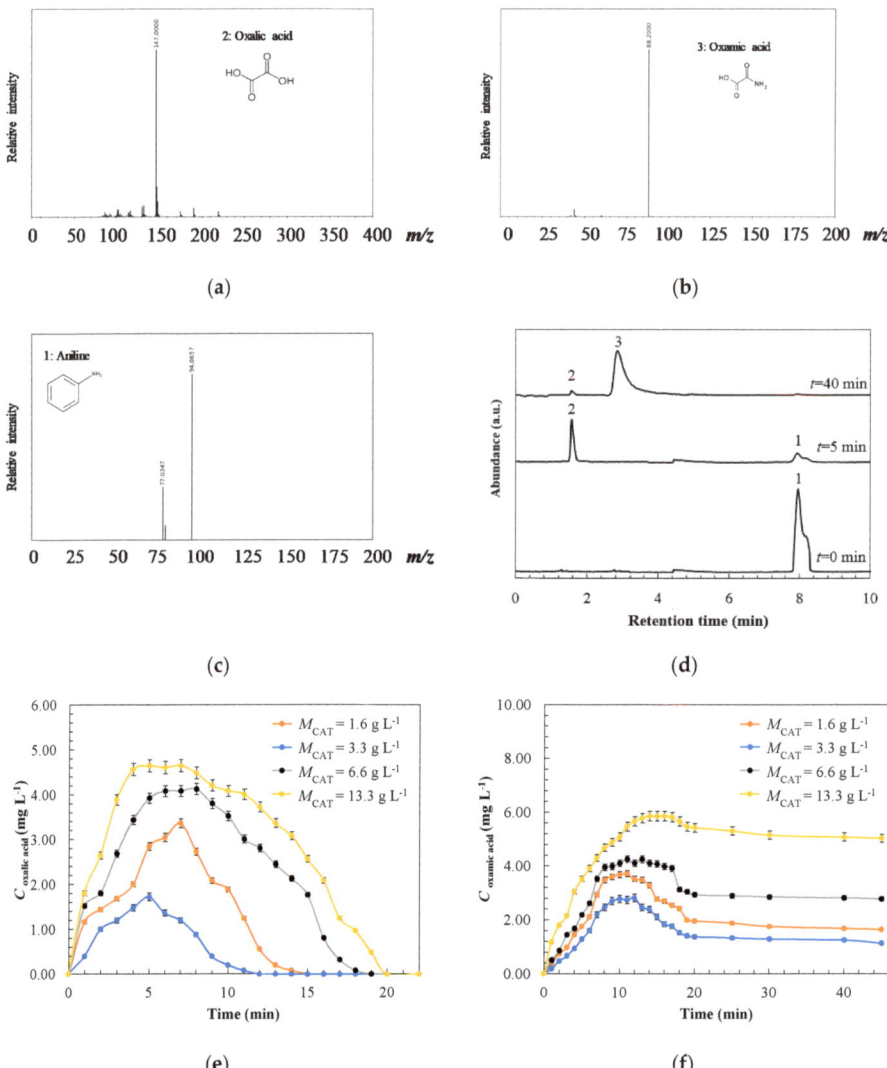

Figure 6. Analysis of some by-products formed during the TiO$_2$/GAC ozonation of aniline: (**a**–**c**) mass spectrometry (MS) of the aniline at different ozonation times, (**d**) liquid chromatography (LC) of the aniline and representative ozonation by-products, and (**e**,**f**) effect of catalyst dosage on the evolution of oxalic and oxamic acid during ozonation. Experimental conditions: $Q_G = 4$ L min^{-1}; $C_{O_3,in} = 5.4$ mg L^{-1}; pH = 7.0; $T = 18.0$ °C; $P = 1$ atm; $V_{reac} = 1.5$ L; (Agitation) = 60 rpm.

Figure 6e,f show the evolution of the concentration of oxamic and oxalic acid for different doses of catalyst. For both acids and for all doses of catalyst studied, a continuous increase corresponding to the accumulation phenomenon was observed during the first 5 min. Then, coinciding with the primary degradation of the aniline (Figure 3e), a maximum was reached, which was higher with an increased catalyst dose. The increase in turbidity observed in Figure 5 at any catalyst dose studied is coincident with that of both acids. The relationship between oxalic acid and turbidity is evident, as well as the amount associated with the mechanism change from modified ER to LH. A Langmuir–Hinshelwood

(LH) type oxidative mechanism was dominant, with high amounts of TiO$_2$/GAC (6.6 or 13.3 g L^{-1}), promoting oxidation products such as oxalic acid [22]. Concerning the evolution of oxamic acid, the initial accumulation was slower, reaching its maximum 7 min later than that observed for oxalic acid. Unlike oxalic acid, oxamic acid could not be removed, explaining the flat tailing observed in the TOC profiles in Figure 3f. In these cases, the hypothesis of a dominant reactive mechanism of type LH at the surface with very low reaction rate (k_{oxS}) seems evident. This situation also led to the high occupation of active centres with a decrease in the radical concentration in the liquid and subsequent decrease in k_{oxL}. Authors such as Faria et al. [61] have reported a similar result during the removal of oxalic and oxamic acid via catalytic ozonation using active carbon. Furthermore, it has been reported that, at neutral pH, oxamic acid is mainly present as a zwitterion ($^-$OOC—CONH^{3+}), which is highly hydrophilic and stable in water. The C–H bonds explain its low reactivity towards hydroxyl radicals, and although it is possible to mineralize it completely (according to Legube and Leitner [14]), it requires a hydroxyl radical concentration approximately 100 times higher than that needed for other organic compounds with the same functional group. Thus, the persistence of oxamic acid in the liquid phase can explain the observed turbidity increase.

Due to the catalytic ozonation process in which the aniline was degraded, some degradation intermediates were formed. Using liquid chromatography (LC), higher concentration degradation products, such as nitrobenzene, phenol, catechol, o-benzoquinone, 1,2,3-benzenetriol, p-benzoquinone, and muconic acid, were detected excluding oxamic and oxalic acid. Other organic compounds, which were detected at lower concentrations, could not be identified. In this section, only the first intermediates (C$_6$) are included, in an attempt to determine the beginning of the first degradation routes. Many of these degradation products showed up in the solution within the first 5 min, through the change from a non-coloured solution to another with reddish, brown, and yellow colouring [6,62,63]. In Figure 7, three cases with the same mass of TiO$_2$/GAC (3.3 g L^{-1}) were selected, representative of the different experimental conditions studied. Among them, an adverse situation was selected with a low generation of hydroxyl radicals at pH = 3.0, as well as another with an excess of oxidant of 20.1 mg L^{-1}—which was ineffective due to the low mineralization achieved—and, finally, with the favourable conditions indicated in the previous section. For additional information, Figures S3 and S4 show the concentrations of each intermediate in terms of TOC. In Figure S4c, it can be observed that the amount of unknown TOC after 5 min of reaction was 5.9 mg L^{-1}.

According to Figure 7, the two possible degradation routes can be differentiated, according to the involvement of the direct and radical ozone pathways. In addition, it was observed that the rupture of the aniline molecule occurred in the bond between the benzene ring and the amino group (—NH$_2$). Other authors, such as Villota et al. [64] and Von Sonntag and Von Gunten [33], have also observed this same rupture.

In Figure 7a, it can be observed that the direct attack of ozone could be responsible for the high selectivity towards the formation of dihydroxy aromatic rings, such as catechol or 1,2,4-benzenetriol, through ortho-, meta-, and para-substituted oxidation pathways and electrophilic substitutions [65]. However, the carboxylic acids or muconic acid detected later (after about 10 min of reaction) were more refractory to direct ozone attack, which explains the 40% mineralization observed after a reaction time of 20 min [66].

However, the results in Figure 7b show a different behaviour, in which the aniline was degraded by another mechanism—attributed to the radical pathway. In this case, we note the presence of nitrobenzene. According to Brillas et al. and Tolosana-Moranchel et al. [67,68], the presence of nitrobenzene could be due to the high selectivity of the hydroperoxyl radical, compared with the hydroxyl radical, to the amino group, which strongly favours its conversion to a nitro group. Despite the last point, considering the radical species generated from direct interaction with TiO$_2$/GAC, Sanchez et al. [69] suggested that the hydroxyl radical is responsible for abstracting hydrogen from the amino group and then substituting it with an iminium radical, thus generating nitrobenzene. The generation of nitrobenzene could explain the low reactivity and inefficiency observed when using a high dose of ozone. With respect to the presence of phenol or p-benzoquinone, Commninellis and Pulgarin [70] attributed it to hydroxylation reactions of these benzene structures with hydroxyl groups.

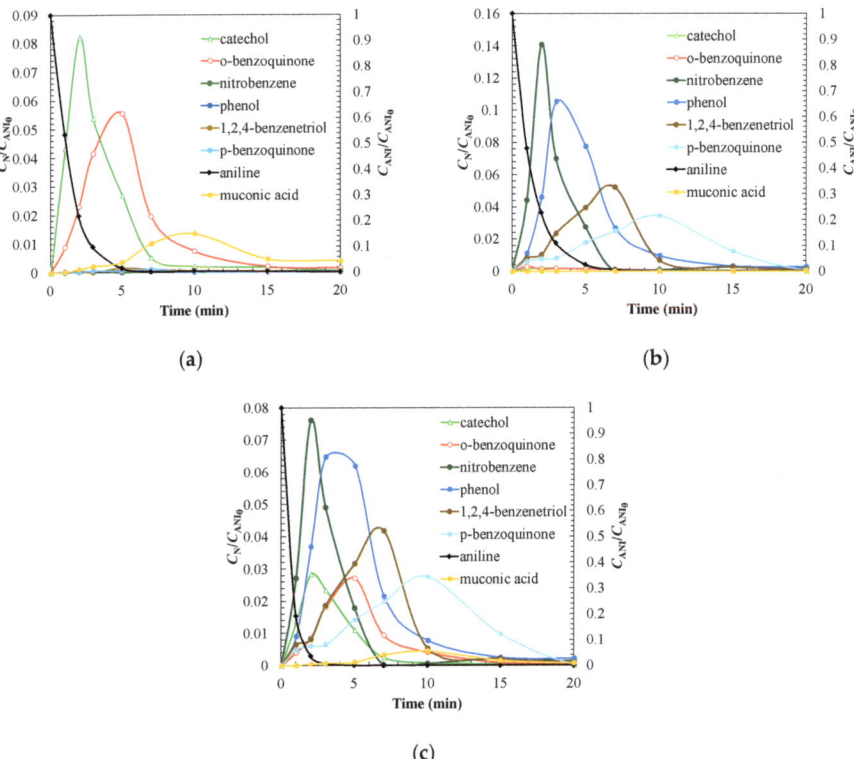

Figure 7. Analysis of the main intermediates in the aniline catalytic ozonation with TiO$_2$/GAC excluding oxalic and oxamic acid in different cases: (**a**) molecular attack of ozone at pH = 3.0, (**b**) ozone in excess and dominance of radicalary attack (pH = 7.0 and ozone dose of 20.1 mg L^{-1}), and (**c**) most favourable conditions (pH = 7.0, 5.4 mg L^{-1} ozone concentration) or combined molecular radicalary attack. Experimental conditions: Q_G = 4 L min^{-1}; M_{CAT} = 3.3 g L^{-1}; T = 18.0 °C; P = 1 atm; V_{reac} = 1.5 L; (Agitation) = 60 rpm.

In Figure 7c, an overlap of both radical and molecular pathways can be observed, which is consistent with the kinetic parameters obtained from the Ad/Ox model, where an equilibrium situation was observed between oxidation in the liquid and on the surface of the solid through adsorption. The results obtained in the analyses carried out allow us to propose the mechanism of aniline degradation shown in Figure 8.

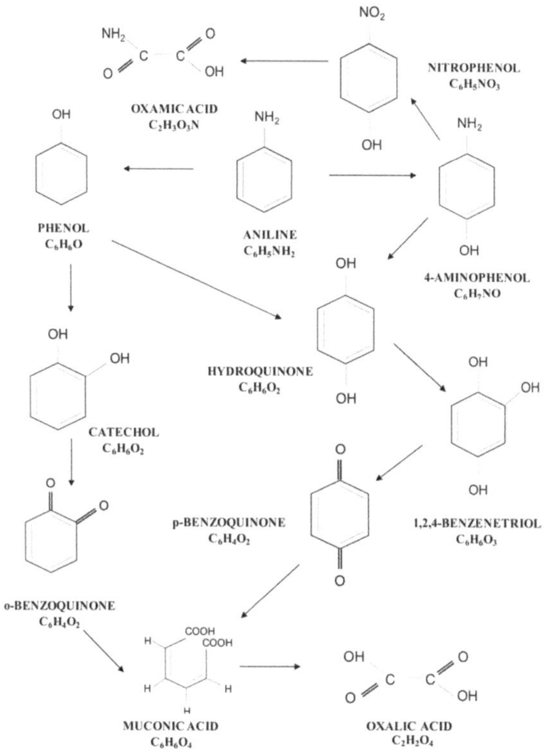

Figure 8. Oxidation pathway proposed for aniline oxidation via catalytic ozonation with TiO$_2$/GAC composites.

4. Conclusions

A three-phase mathematical model reaction (Ad/Ox) was proposed that describes the stages of G–L transfer, adsorption, and oxidation in the liquid and on the surface of the catalyst, in order to study catalytic ozonation using TiO$_2$/GAC composites. The model was verified using experimental aniline ozonation results. Despite the wide variety of conditions studied, the model provided a good fit to the experimental data, obtaining a weighted standard deviation lower than 0.05 in all cases. The model has been applied to evaluate the effect of the main operational variables on the G-L mass transfer. The analysis resulted in K_La values of 0.18 min^{-1} at neutral pH, with no significance of other process conditions and no effect on the overall process rate (chemical reaction control).

Catalytic oxidation using commercial activated carbons, such as Norit® GAC 1240 Plus, was proved to occur through a Langmuir–Hinshelwood mechanism with preferential oxidation in the liquid phase. With the TiO$_2$/GAC composite, the estimated oxidation constants—in the liquid and in the solid—suggest a modified Eley–Rideal type mechanism, obtaining 80.2% mineralization in the most favourable conditions. Estimation of the oxidation constants allowed us to deduce that ozone acts mainly in the liquid at acidic pH, whereas under basic pH values, oxidation happens either on the solid or in the liquid. The use of high doses of ozone limits the kinetics and adsorption capacity of aniline and its degradation oxidation products, with oxidation in the liquid being the principal route of degradation. On the other hand, at moderate ozone doses, a greater role of the adsorption and oxidation mechanisms of the TiO$_2$ deposited in the GAC was observed. An excess in the catalyst load was ineffective and led to an increase in turbidity by inducing degradation pathways that ended with oxidation products such as oxalic acid and, especially, oxamic acid. Finally, the model allowed us to analyse the significance of the different stages involved in the catalytic ozonation. Additionally, the most favourable operating conditions for the potentiation of the TiO_2 deposited on the GAC were found (pH = 7.0, 5.4 mg L^{-1} ozone dose, and 3.3 g L^{-1} catalyst load). TiO_2 contributed to a greater capacity of the material to adsorb the pollutant and subsequent predisposition to be attacked by the ozone through the hydroxyl radicals generated on its surface. From the identification and analysis of the degradation intermediates, two possible routes—which occur simultaneously under the most favourable conditions mentioned above—were proposed. This model could be applied at an industrial level with new catalysts for the prediction of operating behaviour under different working conditions.

Supplementary Materials: The following are available online at http://www.mdpi.com/2073-4441/12/12/3448/s1, Figure S1: Utilization efficiencies of ozone in a TiO_2/GAC catalytic system at different pHs, Figure S2: Effect of the ozone dose on the control stage during catalytic ozonation of aniline with the TiO_2/GAC catalyst, Figure S3: Analysis of some by-products formed during the TiO_2/GAC ozonation of aniline in terms of TOC, Figure S4: Analysis of the main intermediates in the aniline catalytic ozonation with TiO_2/GAC in terms of TOC, excluding oxalic and oxamic acid in different cases.

Author Contributions: J.I.L. and C.F. performed the conceptualization; M.J.R. and J.I.L. carried out the design of the methodology and analyses; C.F. and N.V. contributed to the model validation; C.F. carried out the formal analysis; C.F. performed the investigation; C.F. and J.I.L. prepared the original draft; M.J.R., C.F. and N.V. reviewed and edited the manuscript; J.I.L. and M.J.R. supervised the experimentation; M.J.R., N.V., and J.I.L. acquired the funding. All authors have read and agreed to the published version of the manuscript.

Funding: The authors are grateful to the University of the Basque Country for their financial support of this study through the PPGA19/63 project and C. Ferreiro's predoctoral PIF grant (PIF16/367).

Acknowledgments: The authors are thankful for the technical and human support provided by The Singular Coupled Multispectroscopy Laboratory (Raman-LASPEA) and Central Analysis Service (Araba Unit), part of the General Research Services (SGIker) of the UPV/EHU, and Cabot Corporation for supplying the sample of activated carbon used in this work.

Conflicts of Interest: The authors declare no conflict of interest.

Nomenclature

σ	Weighted standard deviation
$C^*_{O_3,L}$	Concentration of ozone in the equilibrium with the ozone adsorbed on the activated carbon, mg L^{-1}
$C^*_{O_3,S}$	Concentration of ozone on the catalyst in equilibrium with the liquid ozone concentration, mg L^{-1}
C^*_p	Calculate pollutant concentration in the liquid in terms of total organic carbon, mg L^{-1}
C_N	Concentration degradation products, mg L^{-1}
$C_{O_3,in}$	Concentrations of ozone in the gas phase at the inlet, mg L^{-1}
$C_{O_3,L}$	Ozone concentration in liquid, mg L^{-1}
$C_{O_3,out}$	Concentration of ozone in the gas phase at the outlet, mg L^{-1}
C_{OH^-}	Concentration of hydroxyl ions, mol L^{-1}
C_p	Pollutant concentration in the liquid in terms of total organic carbon, mg L^{-1}
He	Henry's constant, bar L mg^{-1}
k_{ads}	Kinetic constant of aniline adsorption, g mg^{-1} min^{-1}
$k_{c,L}$	Elemental kinetic constant for the ozonation in the liquid, L mg^{-1} min^{-1}
$k_{c,S}$	Elemental kinetic constant for the ozonation in the solid, L mg^{-1} min^{-1}
K_F	Freundlich constant, (mg g^{-1}) (L mg^{-1})$^{1/nF}$
$K_G a$	Overall mass transfer coefficient of ozone gas to water, min^{-1}
$K_L a$	Volumetric ozone mass transfer coefficient, min^{-1}
k_{oxL}	Apparent first-order kinetic constant in the liquid in terms of TOC, min^{-1}
k_{oxS}	Apparent first-order kinetic constant over the catalyst in terms of TOC, min^{-1}
m	Slope of the equilibrium line between the liquid and solid phase
M_{CAT}	Concentration of catalyst, g L^{-1}
n	Heterogeneity factor, dimensionless
N	Number of experimental values
$N^{II}_{O_3}$	Ozone consumption in the solid, mg L^{-1} min^{-1}
$N^{I}_{O_3}$	Ozone consumption in the liquid, mg L^{-1} min^{-1}

N_{O_3}	Whole ozone consumption, mg L^{-1} min^{-1}
$P^*_{O_3}$	Partial pressure of the ozone in equilibrium with the adsorbed ozone on the solid, bar
P_{O_3}	Partial pressure of ozone in the gas phase, bar
Q_G	Ozone gas flow, L min^{-1}
r^{II}_p	Degradation of the pollutant on the activated carbon, mg L^{-1} min^{-1}
r^{I}_p	Degradation of the pollutant in the liquid, mg L^{-1} min^{-1}
T	Temperature, K
t	Time, min
U_{O_3}	Utilization coefficient, %
V_{reac}	Volume of dissolution, L
z^I	Amount of pollutant consumed in the liquid, mg TOC mg^{-1} O$_3$
z^{II}	Amount of pollutant consumed on the catalyst, mg TOC mg^{-1} O$_3$
Z_p	Concentration of pollutant in the solid, mg g^{-1}
$Z_{p,\infty}$	Amount of pollutant adsorbed in the solid in equilibrium, mg g^{-1}

References

1. United Nations. About the Sustainable Development Goals. Available online: https://www.un.org/sustainabledevelopment/sustainable-development-goals/ (accessed on 27 November 2020).
2. Zouboulis, A.I.; Katsoyiannis, I.A. Recent Advances in Water and Wastewater Treatment with Emphasis in Membrane Treatment Operations. *Water* **2019**, *11*, 45. [CrossRef]
3. Anotai, J.; Jevprasesphant, A.; Lin, Y.-M.; Lu, M.-C. Oxidation of aniline by titanium dioxide activated with visible light. *Sep. Purif. Technol.* **2012**, *84*, 132–137. [CrossRef]
4. Ferreiro, C.; Villota, N.; Lombraña, J.I.; Rivero, M.J. An efficient catalytic process for the treatment of genotoxic aniline wastewater using a new granular activated carbon-supported titanium dioxide composite. *J. Clean. Prod.* **2019**, *228*, 1282–1295. [CrossRef]
5. Ferreiro, C.; Villota, N.; Lombraña, J.I.; Rivero, M.J.; Zúñiga, V.; Rituerto, J.M. Analysis of a Hybrid Suspended-Supported Photocatalytic Reactor for the Treatment of Wastewater Containing Benzothiazole and Aniline. *Water* **2019**, *11*, 337. [CrossRef]
6. Jing, Z.; Cao, S.; Yu, T.; Hu, J. Degradation Characteristics of Aniline with Ozonation and Subsequent Treatment Analysis. *J. Chem.* **2015**, 905921. [CrossRef]
7. European Commission. *European Union Risk Assessment Report aniline*; Office for Official Publications of the European Communities: Luxembourg, 2004; Volume 50.
8. Tao, N.; Liu, G.; Bai, L.; Tang, L.; Guo, C. Genotoxicity and growth inhibition effects of aniline on wheat. *Chemosphere* **2017**, *169*, 467–473. [CrossRef]
9. United State Environmental Protection Agency. Contaminant Candidate List 4-CCL 4. Available online: https://www.epa.gov/ccl/contaminant-candidate-list-4-ccl-4-0 (accessed on 9 October 2020).
10. Álvarez, P.M.; Beltrán, F.J.; Masa, F.J.; Pocostales, J.P. A comparison between catalytic ozonation and activated carbon adsorption/ozone-regeneration processes for wastewater treatment. *Appl. Catal. B Environ.* **2009**, *92*, 393–400. [CrossRef]
11. Cheng, W.; Quan, X.; Li, R.; Wu, J.; Zhao, Q. Ozonation of Phenol-Containing Wastewater Using O$_3$/Ca(OH)$_2$ System in a Micro Bubble Gas-Liquid Reactor. *Ozone-Sci. Eng.* **2018**, *40*, 173–182. [CrossRef]
12. Rodríguez, A.; Rosal, R.; Perdigón-Melón, J.A.; Mezcua, M.; Agüera, A.; Hernando, M.D.; Letón, P.; Fernández-Alba, A.R.; García-Calvo, E. Ozone-Based Technologies in Water and Wastewater Treatment. In *Emerging Contaminants from Industrial and Municipal Waste. Removal Technologies*; Barceló, D., Petrovic, M., Eds.; Springer-Verlag: Heidelberg, Germany, 2008; Volume 5S/2, pp. 127–175. ISBN 978-3-540-79209-3.
13. Guo, Y.; Yang, L.; Wang, X. The Application and Reaction Mechanism of Catalytic Ozonation in Water Treatment. *J. Environ. Anal. Toxicol.* **2012**, *2*. [CrossRef]
14. Legube, B.; Karpel Vel Leitner, N. Catalytic ozonation: A promising advanced oxidation technology for water treatment. *Catal. Today* **1999**, *53*, 61–72. [CrossRef]
15. Delmas, H.; Creanga, C.; Julcour-Lebigue, C.; Wilhelm, A.-M. AD-OX: A sequential oxidative process for water treatment-Adsorption and batch CWAO regeneration of activated carbon. *Chem. Eng. J.* **2009**, *152*, 189–194. [CrossRef]

16. Erol, F.; Ozbelge, T.A.; Ozbelge, H.O. Modeling of catalytic ozonation process in a three-phase reactor. *J. Environ. Sci. Health A Tox. Hazard Subst. Environ. Eng.* **2009**, *44*, 295–306. [CrossRef] [PubMed]
17. Cheng, J.; Yang, Z.R.; Chen, H.Q.; Kuo, C.H.; Zappi, E.M. Modeling of organic pollutant destruction in a stirred-tank reactor by ozonation. *J. Environ. Sci. (China)* **2001**, *13*, 449–452.
18. Guelli Ulson de Souza, S.M.d.A.; de Souza, F.B.; Ulson de Souza, A.A. Application of Individual and Simultaneous Ozonation and Adsorption Processes in Batch and Fixed-Bed Reactors for Phenol Removal. *Ozone-Sci. Eng.* **2012**, *34*, 259–268. [CrossRef]
19. Ferreiro, C.; Villota, N.; de Luis, A.; Lombrana, J.I. Analysis of the effect of the operational variants in a combined adsorption-ozonation process with granular activated carbon for the treatment of phenol wastewater. *React. Chem. Eng.* **2020**, *5*, 760–778. [CrossRef]
20. Song, S.; He, Z.; Chen, J. US/O_3 combination degradation of aniline in aqueous solution. *Ultrason. Sonochem.* **2007**, *14*, 84–88. [CrossRef]
21. Faria, P.C.C.; Orfao, J.J.M.; Pereira, M.F.R. Ozonation of aniline promoted by activated carbon. *Chemosphere* **2007**, *67*, 809–815. [CrossRef]
22. Orge, C.A.; Faria, J.L.; Pereira, M.F.R. Photocatalytic ozonation of aniline with TiO_2-carbon composite materials. *J. Environ. Manag.* **2017**, *195*, 208–215. [CrossRef]
23. Rodriguez, C.; Ignacio Lombrana, J.; de Luis, A.; Sanz, J. Oxidizing efficiency analysis of an ozonation process to degrade the dye rhodamine 6G. *J. Chem. Technol. Biotechnol.* **2017**, *92*, 656–665. [CrossRef]
24. Byun, S.; Cho, S.H.; Yoon, J.; Geissen, S.U.; Vogelpohl, A.; Kim, S.M. Influence of mass transfer on the ozonation of wastewater from the glass fiber industry. *Water Sci. Technol.* **2004**, *49*, 31–36. [CrossRef]
25. Christopher, R.; Schulz, P.E.; Paul, W. Prendiville P.E. Designing High Concentration Ozone Contactors for Drinking Water Treatment Plants. *Ozone Sci. Eng.* **1993**, *15*, 245–266. [CrossRef]
26. Roth, J.; Sullivan, D. Solubility of Ozone in Water. *Ind. Eng. Chem. Fundam.* **1981**, *20*, 137–140. [CrossRef]
27. Egorova, G.V.; Voblikova, V.A.; Sabitova, L.V.; Tkachenko, I.S.; Tkachenko, S.N.; Lunin, V.V. Ozone Solubility in Water. *Moscow Univ. Chem. Bull.* **2015**, *70*, 207–210. [CrossRef]
28. Beltran, F.J. *Ozone Reaction Kinetics for Water and Wastewater Systems*; CRC Press: Boca Ratón, FL, USA, 2003; ISBN 978-0-203-50917-3.
29. Cacace, F.; Speranza, M. Protonated Ozone: Experimental Detection of O_3H^+ and Evaluation of the Proton Affinity of Ozone. *Science* **1994**, *265*, 208–209. [CrossRef] [PubMed]
30. Flores-Payán, V.; Herrera-López, E.J.; Navarro-Laboulais, J.; López-López, A. Parametric sensitivity analysis and ozone mass transfer modeling in a gas–liquid reactor for advanced water treatment. *J. Ind. Eng. Chem.* **2015**, *21*, 1270–1276. [CrossRef]
31. Ratnawati, R.; Kusumaningtyas, D.; Suseno, P.; Prasetyaningrum, A. Mass Transfer Coefficient of Ozone in a Bubble Column. *MATEC Web Conf.* **2018**, *156*, 02015. [CrossRef]
32. Berry, M.J.; Taylor, C.M.; King, W.; Chew, Y.M.J.; Wenk, J. Modelling of Ozone Mass-Transfer through Non-Porous Membranes for Water Treatment. *Water* **2017**, *9*, 452. [CrossRef]
33. VonSonntag, C.; VonGunten, U. *Chemistry of Ozone in Water and Wastewater Treatment: From Basic Principles to Applications*; Iwa Publishing: London, UK, 2012; ISBN 978-1-84339-313-9.
34. Masel, R.I. *Principles of Adsorption and Reaction on Solid Surfaces*, 1st ed.; Wiley: Oak Brook, IL, USA, 1996.
35. De Luis, A.; Lombraña, J.I. pH-Based Strategies for an Efficient Addition of H_2O_2 During Ozonation to Improve the Mineralisation of Two Contaminants with Different Degradation Resistances. *Water Air Soil Pollut.* **2018**, *229*, 372. [CrossRef]
36. Wang, B.; Zhang, H.; Wang, F.; Xiong, X.; Tian, K.; Sun, Y.; Yu, T. Application of Heterogeneous Catalytic Ozonation for Refractory Organics in Wastewater. *Catalysts* **2019**, *9*, 241. [CrossRef]
37. Zheng, X.; Yu, N.; Wang, X.; Wang, Y.; Wang, L.; Li, X.; Hu, X. Adsorption Properties of Granular Activated Carbon-Supported Titanium Dioxide Particles for Dyes and Copper Ions. *Sci. Rep.* **2018**, *8*, 6463. [CrossRef]
38. Kasprzyk-Hordern, B.; Ziółek, M.; Nawrocki, J. Catalytic ozonation and methods of enhancing molecular ozone reactions in water treatment. *Appl. Catal. B Environ.* **2003**, *46*, 639–669. [CrossRef]
39. Valdes, H.; Sanchez-Polo, M.; Rivera-Utrilla, J.; Zaror, C.A. Effect of ozone treatment on surface properties of activated carbon. *Langmuir* **2002**, *18*, 2111–2116. [CrossRef]
40. Ghuge, S.P.; Saroha, A.K. Catalytic ozonation for the treatment of synthetic and industrial effluents—Application of mesoporous materials: A review. *J. Environ. Manag.* **2018**, *211*, 83–102. [CrossRef] [PubMed]

41. Moussavi, G.; Khavanin, A.; Alizadeh, R. The investigation of catalytic ozonation and integrated catalytic ozonation/biological processes for the removal of phenol from saline wastewaters. *J. Hazard. Mater.* **2009**, *171*, 175–181. [CrossRef] [PubMed]
42. Sanchez, M.; Rivero, M.J.; Ortiz, I. Kinetics of dodecylbenzenesulphonate mineralisation by TiO_2 photocatalysis. *Appl. Catal. B Environ.* **2011**, *101*, 515–521. [CrossRef]
43. MiarAlipour, S.; Friedmann, D.; Scott, J.; Amal, R. TiO_2/porous adsorbents: Recent advances and novel applications. *J. Hazard. Mater.* **2018**, *341*, 404–423. [CrossRef] [PubMed]
44. Beltran, F.J.; Rivas, F.J.; Montero-de-Espinosa, R. A TiO_2/Al_2O_3 catalyst to improve the ozonation of oxalic acid in water. *Appl. Catal. B-Environ.* **2004**, *47*, 101–109. [CrossRef]
45. Vega, E.; Valdes, H. New evidence of the effect of the chemical structure of activated carbon on the activity to promote radical generation in an advanced oxidation process using hydrogen peroxide. *Microporous Mesoporous Mat.* **2018**, *259*, 1–8. [CrossRef]
46. Orge, C.A.; Sousa, J.P.S.; Gonçalves, F.; Freire, C.; Órfão, J.J.M.; Pereira, M.F.R. Development of Novel Mesoporous Carbon Materials for the Catalytic Ozonation of Organic Pollutants. *Catal. Lett.* **2009**, *132*, 1–9. [CrossRef]
47. Turhan, K.; Uzman, S. The degradation products of aniline in the solutions with ozone and kinetic investigations. *Ann. Chim.* **2007**, *97*, 1129–1138. [CrossRef]
48. Shahamat, Y.D.; Farzadkia, M.; Nasseri, S.; Mahvi, A.H.; Gholami, M.; Esrafili, A. Magnetic heterogeneous catalytic ozonation: A new removal method for phenol in industrial wastewater. *J. Environ. Health Sci. Eng.* **2014**, *12*, 50. [CrossRef] [PubMed]
49. Boczkaj, G.; Fernandes, A. Wastewater treatment by means of advanced oxidation processes at basic pH conditions: A review. *Chem. Eng. J.* **2017**, *320*, 608–633. [CrossRef]
50. Roshani, B.; McMaster, I.; Rezaei, E.; Soltan, J. Catalytic ozonation of benzotriazole over alumina supported transition metal oxide catalysts in water. *Sep. Purif. Technol.* **2014**, *135*, 158–164. [CrossRef]
51. Nawrocki, J.; Kasprzyk-Hordern, B. The efficiency and mechanisms of catalytic ozonation. *Appl. Catalysis B Environ.* **2010**, *99*, 27–42. [CrossRef]
52. Tamura, H.; Mita, K.; Tanaka, A.; Ito, M. Mechanism of Hydroxylation of Metal Oxide Surfaces. *J. Colloid Interface Sci.* **2001**, *243*, 202–207. [CrossRef]
53. Valdés, H.; Sánchez-Polo, M.; Zaror, C.A. Role of oxygen-containing functional surface groups of activated carbons on the elimination of 2-hydroxybenzothiazole from waters in A hybrid heterogeneous ozonation system. *J. Adv. Oxid. Technol.* **2017**, *20*. [CrossRef]
54. Orha, C.; Pode, R.; Manea, F.; Lazau, C.; Bandas, C. Titanium dioxide-modified activated carbon for advanced drinking water treatment. *Process Saf. Environ. Protect.* **2017**, *108*, 26–33. [CrossRef]
55. Beltrán, F.J.; Rivas, J.; Álvarez, P.; Montero-de-Espinosa, R. Kinetics of Heterogeneous Catalytic Ozone Decomposition in Water on an Activated Carbon. *Ozone Sci. Eng.* **2002**, *24*, 227–237. [CrossRef]
56. Leyva, E.; Moctezuma, E.; Noriega, S. Photocatalytic degradation of omeprazole. Intermediates and total reaction mechanism. *J. Chem. Technol. Biotechnol.* **2017**, *92*, 1511–1520. [CrossRef]
57. Bulanin, K.M.; Lavalley, J.C.; Tsyganenko, A.A. Infrared Study of Ozone Adsorption on TiO_2 (Anatase). *J. Phys. Chem.* **1995**, *99*, 10294–10298. [CrossRef]
58. Bodzek, M.; Rajca, M. Photocatalysis in the treatment and disinfection of water. Part I. Theoretical backgrounds. *Ecol. Chem. Eng. S* **2012**, *19*, 489–512. [CrossRef]
59. Orge, C.A.; Faria, J.L.; Pereira, M.F.R. Removal of oxalic acid, oxamic acid and aniline by a combined photolysis and ozonation process. *Environ. Technol.* **2015**, *36*, 1075–1083. [CrossRef] [PubMed]
60. Zazo, J.A.; Casas, J.A.; Mohedano, A.F.; Gilarranz, M.A.; Rodríguez, J.J. Chemical Pathway and Kinetics of Phenol Oxidation by Fenton's Reagent. *Environ. Sci. Technol.* **2005**, *39*, 9295–9302. [CrossRef] [PubMed]
61. Faria, P.C.C.; Órfão, J.J.M.; Pereira, M.F.R. Activated carbon catalytic ozonation of oxamic and oxalic acids. *Appl. Catal. B Environ.* **2008**, *79*, 237–243. [CrossRef]
62. Villota, N.; Lombraña, J.I.; Cruz-Alcalde, A.; Marcé, M.; Esplugas, S. Kinetic study of colored species formation during paracetamol removal from water in a semicontinuous ozonation contactor. *Sci. Total Environ.* **2019**, *649*, 1434–1442. [CrossRef] [PubMed]
63. Mijangos, F.; Varona, F.; Villota, N. Changes in Solution Color During Phenol Oxidation by Fenton Reagent. *Environ. Sci. Technol.* **2006**, *40*, 5538–5543. [CrossRef] [PubMed]

64. Villota, N.; Lomas, J.M.; Camarero, L.M. Study of the paracetamol degradation pathway that generates color and turbidity in oxidized wastewaters by photo-Fenton technology. *J. Photochem. Photobiol. A Chem.* **2016**, *329*, 113–119. [CrossRef]
65. Rasalingam, S.; Kibombo, H.S.; Wu, C.-M.; Peng, R.; Baltrusaitis, J.; Koodali, R.T. Competitive role of structural properties of titania–silica mixed oxides and a mechanistic study of the photocatalytic degradation of phenol. *Appl. Catal. B Environ.* **2014**, *148–149*, 394–405. [CrossRef]
66. Anpo, M.; Kamat, P.V. (Eds.) *Environmentally Benign Photocatalysts: Applications of Titanium Oxide-based Materials*; Nanostructure Science and Technology; Springer: New York, NY, USA, 2010; ISBN 978-0-387-48441-9.
67. Brillas, E.; Bastida, R.M.; Llosa, E.; Casado, J. Electrochemical Destruction of Aniline and 4-Chloroaniline for Wastewater Treatment Using a Carbon-PTFE O_2—Fed Cathode. *J. Electrochem. Soc.* **1995**, *142*, 1733. [CrossRef]
68. Tolosana-Moranchel, A.; Montejano, A.; Casas, J.A.; Bahamonde, A. Elucidation of the photocatalytic-mechanism of phenolic compounds. *J. Environ. Chem. Eng.* **2018**, *6*, 5712–5719. [CrossRef]
69. Sánchez, L.; Peral, J.; Domènech, X. Photocatalyzed destruction of aniline in UV-illuminated aqueous TiO_2 suspensions. *Electrochim. Acta* **1997**, *42*, 1877–1882. [CrossRef]
70. Comninellis, C.; Pulgarin, C. Anodic oxidation of phenol for waste water treatment. *J. Appl. Electrochem.* **1991**, *21*, 703–708. [CrossRef]

Publisher's Note: MDPI stays neutral with regard to jurisdictional claims in published maps and institutional affiliations.

© 2020 by the authors. Licensee MDPI, Basel, Switzerland. This article is an open access article distributed under the terms and conditions of the Creative Commons Attribution (CC BY) license (http://creativecommons.org/licenses/by/4.0/).

Article

Erythromycin Abatement from Water by Electro-Fenton and Peroxyelectrocoagulation Treatments

Anna Serra-Clusellas [1,*], Luca Sbardella [1], Pol Herrero [2], Antoni Delpino-Rius [2], Marc Riu [2], María de Lourdes Correa [1], Anna Casadellà [1], Núria Canela [2] and Xavier Martínez-Lladó [1]

1. Eurecat, Centre Tecnològic de Catalunya, Water, Air and Soil Unit, Plaça de la Ciència 2, 08242 Manresa, Spain; luca.sbardella@eurecat.org (L.S.); lulecorrea10@hotmail.com (M.d.L.C.); anna.casadella@irta.cat (A.C.); xavier.martinez@eurecat.org (X.M.-L.)
2. Eurecat, Centre Tecnològic de Catalunya, Centre for Omic Sciences (Joint Unit Eurecat-Universitat Rovira i Virgili), Unique Scientific and Technical Infrastructure (ICTS), 43204 Reus, Spain; pol.herrero@agilent.com (P.H.); antoni.delpino@eurecat.org (A.D.-R.); marc.riu@eurecat.org (M.R.); nuria.canela@eurecat.org (N.C.)
* Correspondence: anna.serra@eurecat.org

Abstract: Electro-Fenton (EF) and peroxyelectrocoagulation (PEC) processes were investigated to mineralize 10 mg L^{-1} erythromycin from ultrapure water, evaluating the influence of the anode material (BDD and Fe), current density (j_{anode}) (5 mA cm^{-2} and 10 mA cm^{-2}), oxygen flowrate injected to the cathode (0.8 L min^{-1} O$_2$ and 2.0 L min^{-1} O$_2$) and pH (2.8, 5.0 and 7.0) on the process efficiency and the electricity costs. 70% mineralization was reached after applying 0.32 A h L^{-1} under the best operational conditions: PEC treatment at 5 mA cm^{-2}, 2.0 L min^{-1} O$_2$ and pH 2.8. The electricity consumption of the electrochemical cell under these conditions was approximately 0.3 kWh m^{-3}. Early-stage intermediates produced from erythromycin degradation were identified and quantified throughout the treatment and a potential erythromycin degradation pathway was proposed. The most appropriate operational conditions tested with synthetic solutions were applied to treat a real effluent from the tertiary treatment of an urban wastewater treatment plant. All emerging compounds listed in the EU Decision 2018/840 (Watch List 2018) were determined before and after the PEC treatment. All listed pollutants were degraded below their quantification limit, except estrone and 17-α-ethinylestradiol which were 99% removed from water. Electricity consumption of the electrochemical cell was 0.4 kWh m^{-3}. Whilst awaiting future results that demonstrate the innocuity of the generated byproducts, the results of this investigation (high removal yields for emerging pollutants together with the low electricity consumption of the cell) indicate the promising high potential of PEC treatment as a water treatment/remediation/regeneration technology.

Keywords: erythromycin; peroxyelectrocoagulation; electro-Fenton; Watch List 2018; emerging contaminants; antibiotics

1. Introduction

Over the last two decades there has been increasing concern regarding emerging contaminants due to the possible threats they pose to both the human population and the aquatic environment [1]. Wastewater reuse practices, together with more advanced analytical methods to detect substances at levels ranging from ng L^{-1} to µg L^{-1} [2], have resulted in more advanced and stringent regulations aiming at tackling this environmental problem [3]. At the European Union (EU) level, the first watch list of substances for tracking was published in 2015 [4] establishing 10 substances or groups of substances to be monitored. This watch list aimed to gather data for the future prioritisation of emerging compound regulation in the field of water policy. It was updated in 2018 by the EU Decision 2018/840 [5] (called in this paper, Watch List 2018) and just the last year (EU Decision 2020/1161) it was again modified since "according to Article 8b(2) of Directive

2008/105/EC, the duration of a continuous watch list monitoring period for any individual substance shall not exceed four years" [6].

Three macrolide antibiotics (erythromycin, clarithromycin, azithromycin) were included in the EU Decision 2018/840 [5] although their monitoring ceased in 2019 [6]. Antibiotics are a class of pharmaceutical active compounds with high usage and consumption worldwide both for humans and animals [7]. They can enter to the environment via several pathways and can, therefore, contaminate different environmental compartments [8]. Their presence has been detected in surface water, groundwater and drinking water [9]. This fact has further increased concern due to their negative effect, including the spread of antibiotic resistance [10].

Among the macrolide antibiotics listed in the Watch List 2018, erythromycin was detected in diverse water matrices such as surface water (85,200 and 7100 ng L^{-1} [11]), hospital effluents (127–575 ng L^{-1} [12], maximum of 140 ng L^{-1} [13], 1000–27,000 ng L^{-1} [14]), livestock wastewater (139 ng L^{-1} [12]), and urban wastewater (179–185 [15]), among others.

The removal of this pollutant from water by various technologies has been studied. More than 88% and 93% erythromycin removals from a hospital effluent were obtained by powdered activated carbon and ozone, respectively, at pH 8.5–8.8 [13]. However, less than 35% erythromycin degradation was achieved by applying UV-C doses equal to or lower than 7200 J m^{-2} [13], indicating that it is not a photoactive compound at 254 nm. Its total degradation was achieved both by the EreB esterase enzymatic treatment [16] and solar-driven heterogeneous photocatalysis with immobilised TiO$_2$ [15]. However, its degradation percentage diminished by 45% or 22% when solar light was simulated during immobilised [16] or suspended [17] TiO$_2$ photocatalysis, respectively. The solar photo-Fenton process also resulted in 100% erythromycin removal [17].

The results of previous investigations demonstrate that advanced oxidation processes (AOPs) are highly efficient treatments to reach gradual oxidation of erythromycin [15–17] as well as other emerging contaminants [18–21], through the generation of highly reactive oxygen species (ROS) (mostly hydroxyl radical •OH) with high oxidation potentials (e.g., E^0_{NHE} = 2.80 V in the case of •OH) and low selectivity. Among the mechanisms to generate hydroxyl radical, the Fenton process [22] (caused by the decomposition of H$_2$O$_2$ mediated by catalytic quantities of Fe(II) under acidic and dark conditions (Equation (1)) has been widely investigated for wastewater treatment [23,24]:

$$H_2O_{2(aq)} + Fe^{2+}_{(aq)} \rightarrow Fe^{3+}_{(aq)} + OH^-_{(aq)} + {}^\bullet OH_{(aq)} \quad k = 41.7\text{–}79 \text{ M}^{-1} \text{ s}^{-1} \quad (1)$$

However, the major drawback of this process is that the regeneration of Fe(II) takes place by Fenton-like reaction (Equation (2)), which presents a reaction rate much lower than the Fenton reaction:

$$Fe^{3+}_{(aq)} + H_2O_{2(aq)} \rightarrow Fe^{2+}_{(aq)} + HO_2^\bullet{}_{(aq)} \quad k = 9.1 \times 10^{-7} \text{ M}^{-1} \text{ s}^{-1} \quad (2)$$

It is for this reason that solar light is usually applied during the Fenton process so as to accelerate the Fe(II) regeneration and to improve process efficiency [16]. Alternatively, high quantities of iron should be added to the contaminated water, which implies an increase in the consumption of the iron reagent and the generation of an iron oxyhydroxide sludge at the end of the treatment that needs to be managed.

The upgraded version of the Fenton process is the electrochemical generation of H$_2$O$_2$ and Fe(II), where both reagents are in-situ generated in an electrochemical cell. Moreover, Fe(II) can be regenerated by Fe(III) reduction on the cathode (Equation (4)), ensuring the continuity of the Fenton reaction (Equation (1)). This process is known as peroxyelectrocoagulation (PEC) [25–27]. The main reactions occurring at the oxygen/air diffusion cathode (Equations (3) and (4)) and at the Fe anode (Equation (5)) are detailed below:

$$O_{2(g)} + 2H^+_{(aq)} + 2e^- \rightarrow H_2O_{2(aq)} \quad (3)$$

$$Fe^{3+}_{(aq)} + e^- \rightarrow Fe^{2+}_{(aq)} \qquad (4)$$

$$Fe_{(s)} \rightarrow Fe^{2+}_{(aq)} + 2e^- \qquad (5)$$

When an inert material is used as an anode instead of iron, Fe(II) must be added to the system. In this case, other oxidation reactions occur at the anode, depending on its materials. Partial water oxidation to hydroxyl radical (Equation (6)) takes places at non-active boron-doped diamond (BDD) anodes [28]. This process is called electro-Fenton (EF) [29].

$$H_2O_{(l)} \rightarrow {}^\bullet OH_{(aq)} + H^+_{(aq)} + e^- \qquad (6)$$

Although a dark electrochemical process to degrade erythromycin from water was studied, such as a microbial electro-Fenton system with an average removal efficiency of 89% in 48 h [30], oxidation of this compound using the EF and PEC processes has not yet been investigated. Again, solar light was chosen to increase the efficiency of the electrochemical treatment [31]. Approximately 69% of 165 mg L^{-1} erythromycin was mineralized by solar photo electro-Fenton process with Pt/Ti anode, graphite-felt cathode, and applying the following conditions: 27.7 mg L^{-1} Fe(II), pH 3, a current density on cathode ($j_{cathode}$) of 0.16 mA cm^{-2} and a current density on anode (j_{anode}) of 32.8 mA cm^{-2} [31]. This study [31] also examined certain erythromycin transformation byproducts generated at intermediate (30 min) or elevated (120 min) treatment times of the solar photo electro-Fenton process.

Considering all this information, studying the removal of this emerging compound from water in Fenton-based processes in dark conditions was highly recommendable, given that they offer the opportunity of competitive industrial technologies with lower surface occupation and lower investment costs when compared to solar driven treatments [32].

The intention of this article is to describe how to answer this necessity, investigating the mineralization of erythromycin from ultrapure water by PEC and EF processes, comparing the influence of current density, oxygen flowrate injected into the cathode and pH on the process efficiency. The erythromycin concentration (10 mg L^{-1}) used in the current work is higher than those potentially found in real effluents [11–15] in order to identify non-reported erythromycin degradation byproducts generated during the early stage of the treatment and to propose its degradation pathway. Finally, the most appropriate operational conditions detected in synthetic water were applied to treat a real effluent from the tertiary treatment of a municipal wastewater treatment plant (WWTP). The emerging compounds listed in the Watch List 2018 were quantified for this effluent and their degradation percentage, total mineralization, and the economic efficiency of the electrochemical process determined.

2. Materials and Methods

2.1. Chemicals

Erythromycin (purity > 99%) was of reagent grade and supplied by Sigma Aldrich (St. Louis, MO, USA). Anhydrous Na$_2$SO$_4$ (used as background electrolyte) and FeSO$_4$·7H$_2$O (used as catalyst during EF experiments) were of analytical grade and both were purchased from Scharlab S.L. (Sentmenat, Spain). Reagent grade 30% (w/w) H$_2$O$_2$ was also supplied by Scharlab S.L. Analytical grade 96% H$_2$SO$_4$ and NaOH pellets from Scharlab S.L. were used for the preparation of diluted solutions used for the pH adjustment. 1000 mg L^{-1} Ti standard solution (TiSO$_5$ × H$_2$SO$_4$) from Sigma Aldrich and analytical grade 96% H$_2$SO$_4$ were used to determine the H$_2$O$_2$ concentration.

Solutions were prepared with high-purity water from a Milli-Q Academic (Merck Life Science S.L, Madrid, Spain) system (resistivity > 18 MΩ cm, at 25 °C).

Acetonitrile (ACN), methanol (MeOH) and water, all LC-MS grade, were purchased from Scharlab S.L. Formic acid (CH$_2$O$_2$) LC-MS grade were purchased from Fisher Scientific (Loughborough, Leicestershire, UK), and ammonium fluoride (NH$_4$F), dimethylsulfoxide (DMSO) LC-MS grade, phosphate buffered saline (PBS) and ammonium formate (NH$_4$HCO$_2$) were supplied by Sigma-Aldrich.

The standards to determine the compounds described in the EU Decision 2018/840 [5] were supplied by various companies. 17-α-ethinylestradiol (EE2), 17-β-estradiol (E2) and metaflumizone were supplied by Dr. Ehrenstorfer GmbH (Augsburg, Germany); estrone (E1) and erythromycin were supplied by Sigma-Aldrich; clarithromycin, azithromycin, methiocarb, imidacloprid, thiacloprid, thiametoxam, chlothianidin, acetamiprid and ciprofloxacin-HCl were supplied by Neochema GmbH (Bodengeim, Germany) at a concentration of 10 mg L^{-1} with ACN; amoxicillin trihydrate was supplied by Toronto Research Chemicals (Toronto, ON, Canada).

Isotopically labelled compounds were used as internal standards (IS). 17-α-ethinylestradiol-2,4,16,16-d4, 17-β-estradiol-2,4,16,16,17-d5, erythromycin-d3, estrone-2,4-16,16-d4, clarithromycin-d3, imidacloprid-d4, methiocarb-d3 were supplied by Neochema GmbH at a concentration of 10 mg L^{-1} in ACN; thiamethoxam-d4 (oxadiazine-d4) and chlothianidin-d3 (N′-methyl-d3) were supplied by Dr. Ehrenstorfer GmbH; acetamiprid-d3 (N-methyl-d3) (mixture of isomers) was supplied by CDN isotopes (Pointe-Claire, QC, Canada); and amoxicillin–d4 (major) was supplied by Toronto Research Chemical. Standard solutions were prepared in MeOH LC-MS grade, with the exception of amoxicillin trihydrate that it was prepared with DMSO. Calibration mixtures were prepared in water:MeOH (1:1, v/v).

2.2. Instruments and Analytical Procedures

Dissolved organic carbon (DOC), total inorganic carbon (TIC) and total nitrogen (TN) was monitored by an Multi N/C 3100 analyser (Analytik Jena, Jena, Germany). Reproducible values, within ± 2% accuracy, were obtained by injecting 50 µL aliquots previously filtered by 0.45 µm polyvinylidene fluoride (PVDF) membranes. Total and dissolved iron was measured by the spectrophotometric method based on Standard Method 3500-Fe-B [33]. A UV/Vis UV-2450 spectrophotometer (Shimadzu, Kyoto, Japan) was used to determine H_2O_2 concentrations from absorption of the Ti-H_2O_2 coloured complex at λ = 410 nm [34], filtering previously 1 mL of samples by 0.45 µm PVDF filters. A sensiION^{TM} MM374 pH meter and electrical conductivity meter (Hach Lange Spain, S.L.U., Hospitalet de Llobregat, Barcelona, Spain) was used to follow the values of pH and the conductivity.

For the determination of erythromycin and its transformation products, 100 µL of water sample was mixed with 50 µL of internal standard solution (erythromycin-d3) at a concentration of 1 mg mL^{-1} prepared in ultrapure water. The samples were diluted with 850 µL of PBS to adjust the pH of the solution at 7.8 since erythromycin degradation is promoted at acidic pH. Then, samples were vortexed, centrifuged for 5 min at 4 °C and 15,000 rpm and transferred to a chromatographic vial for their analysis.

The chromatographic separation was performed on an UHPLC 1290 Infinity II Series coupled to a QTOF/MS 6550 Series, both from Agilent Technologies (Santa Clara, CA, USA) The chromatographic column was a Kinetex EVO C18 (100 × 2.1 mm) from Phenomenex (Torrance, CA, USA) and the mobile phase was 100% water with 0.1% formic acid (A) and 100% ACN (B). The gradient elution was as follows: 0–0.5 min (0% B isocratic); 0.5–1.0 (0–15% B); 1.0–10.0 (15–60% B); 10.0–11.0 (60–100% B); 11.0–13.0 (100% B isocratic). The flowrate was set at 0.6 mL min^{-1}, the column temperature at 25 °C and the injection volume was 2 µL.

The mass spectrometer operated in positive electrospray ionization (ESI) mode and the source parameters were as follows: gas temperature (260 °C), gas flow (11 L min^{-1}), sheath gas temperature (350 °C), sheath gas flow (12 L min^{-1}), nebulizer (20 psi), capillary voltage (5100 V) and nozzle voltage (500 V). Data acquisition was carried out in full-scan over a mass-range of 100 to 1700 m/z at 2.5 spectra/s and MS/MS spectra were performed at 15, 30 and 45 eV.

The assignment of erythromycin was performed by direct comparison with the commercial standard, whereas the tentative identification of erythromycin degradation products was according to bibliography [16,35–39], exact mass, retention time and tandem mass spectra. These degradation products were semi-quantified using the erythromycin internal standard calibration curve. Additionally, erythromycin E and erythromycin F dehydrogenation TP-2 which have the same exact mass and retention time were semi-quantified by using their product ion spectra with the chromatographic peak of 748.4478→574.3605 and 748.4478→590.3554 transitions, respectively.

A solid phase extraction (SPE) procedure based on a previous publication [40] to analyze the compounds included in the Watch List 2018 was carried out. Oasis HLB SPE cartridges (3 mL, 400 mg) were supplied by Waters (Milford, MA, USA) and the extraction was performed on a Supelco Visiprep SPE Vacuum Manifold (Sigma Aldrich, Saint Louis, MI, USA). The cartridges were previously conditioned with successive volumes of 2 mL of MeOH and 2 mL of Milli-Q water. Then, 100 mL of wastewater samples, previously filtered through a 0.45 µm nylon filter from Millipore (Burlington, MA, USA), were spiked with 0.1 mL of internal standard mixture at a concentration of 1 µg mL^{-1} and loaded to the SPE cartridge. After this, the cartridges were washed with 2 mL of Milli-Q water and dried under vacuum system. The compounds were eluted with 1.5 mL of MeOH. The eluates were evaporated under a nitrogen flow and reconstituted with 200 µL of MeOH:water (80:20, v/v) and transferred to glass vials for LC-MS/MS analysis.

The chromatographic analysis of the compounds of the Watch List 2018 was performed on and UHPLC 1290 Series coupled to a triple quadrupole mass spectrometer (QqQ) 6490 Series, both from Agilent Technologies.

The compounds included in this study were determined in two different chromatographic methods depending on their ESI mode (positive or negative). In both cases, the chromatographic separation was performed using a Acquity UPLC BEH C18, 1.7 µm (100 × 2.1 mm) from Waters. The group of compounds determined on negative ESI mode (ESI-), corresponding to estrogens (detailed in Table 1), were separated using as a mobile phase of 100% water with 5 mmol L^{-1} of ammonium fluoride (A) and 100% MeOH (B). The gradient elution was as follows: 0–1 min (5% B isocratic); 1–2 min (5–50% B), 2–9 min (50–65% B), 9–10 min (65–98% B), 10–12 min (98% B isocratic). The flowrate was set at 0.3 mL min^{-1}, the column temperature at 40 °C and the injection volume was 2 µL. The source parameters were gas temperature (250 °C), gas flow (16 L min^{-1}), sheath gas temperature (375 °C), sheath gas flow (12 L min^{-1}), nebulizer (40 psi), capillary voltage (3000 V) and nozzle voltage (1000 V). Quantification of compounds was performed by internal standard calibration using multiple reaction monitoring acquisition, as detailed in Table 1.

The other compounds described in the Watch List 2018 (detailed Table 1) were determined by positive ESI (ESI+). The compounds were separated using a mobile phase consisting on water:MeOH (98:2, v/v) with 0.1% of formic acid (A) and MeOH:water (99:1) with 0.1% of formic acid and 10 mmol L^{-1} of ammonium formate (B). The gradient elution was as follows: 0–1 min (0% B isocratic); 1–6 min (0–50% B), 6–8.5 min (50–55% B), 8.5–11.5 min (55–75% B), 11.5–13.5 min (75–85% B), 13.5–19.0 min (85% B isocratic), 19.0–19.3 min (85–100% B), 19.3–21.3 min (100% B isocratic), 21.3–21.4 min (100–0% B), 21.4–23.4 min (0% B isocratic). The flowrate was set at 0.5 mL min^{-1}, the column temperature at 40 °C and the injection volume was 2 µL. The source parameters were gas temperature (180 °C), gas flow (20 L min^{-1}), sheath gas temperature (225 °C), sheath gas flow (11 L min^{-1}), nebulizer (40 psi), capillary voltage (4500 V) and nozzle voltage (0 V). Quantification of compounds was performed by internal standard calibration using multiple reaction monitoring acquisition, as detailed in Table 1.

Table 1. Detailed multiple reaction monitoring transitions and retention time for each analyte listed in the EU Decision 2018/840.

Compound	Retention Time (min)	ESI	Quantitative Transition (m/z)	Qualitative Transition (m/z)	Collision Energy (V)	r^2	Quantification Limit (MLOQ) (ng L^{-1})	Detection Limit (MLOD) (ng L^{-1})
17-β-Estradiol-d5	8.70	ESI (−)	276.2→147.1	276.2→187.2	49/41			
17-β-Estradiol (E2)	8.75	ESI (−)	271.2→145.1	271.2→183.1	45/49	0.99	1	0.3
Estrone-d4	8.76	ESI (−)	273.1→147.1	273.1→145.0	49/60			
Estrone (E1)	8.79	ESI (−)	269.1→145.1	269.1→143.1	41/60	0.99	1	0.3
17-α-Ethinylestradiol-d5	8.90	ESI (−)	299.2→147.1	299.2→269.1	33/53			
17-α-Ethinylestradiol (EE2)	8.85	ESI (−)	295.2→145.2	295.17→142.9	49/60	0.99	1	0.03
Amoxicillin-d4	2.71	ESI (+)	370.1→114.0	370.1→353.3	21/9			
Amoxicillin	2.73	ESI (+)	366.1→114.0	366.1→349.2	18/4	0.95	1	0.3
Thiamethoxam-d4	4.78	ESI (+)	296.1→215.0	296.1→183.0	5/21			
Thiamethoxam	4.82	ESI (+)	292.0→210.9	292.0→180.8	9/21	0.99	1	0.3
Ciprofloxacin	5.39	ESI (+)	332.1→231.0	332.0→314.2	45/17	0.99	1	0.3
Clothianidin-d3	4.47	ESI (+)	253.0→172.0	253.0→131.8	9/21			
Clothianidin	4.48	ESI (+)	250.0→131.9	250.0→169.1	13/9	0.99	1	0.3
Imidacloprid-d4	5.49	ESI (+)	260.1→179.2	260.1→213.1	13/9			
Imidacloprid	5.52	ESI (+)	256.1→175.0	256.1→209.1	1721	0.99	10	3
Acetamiprid-d3	5.94	ESI (+)	226.1→126.1	226.1→59.2	21/13			
Acetamiprid	5.95	ESI (+)	223.1→126.1	223.1→56.0	17/13	0.99	1	0.3
Thiacloprid	6.42	ESI (+)	253.0→125.9	253.0→98.9	21/53	0.99	1	0.3
Azithromycin	7.25	ESI (+)	749.5→82.9	749.5→591.3	60/33	0.99	1	0.3
Erythromycin-d3	9.12	ESI (+)	737.5→161.0	737.5→82.9	37/60			
Erythromycin	9.12	ESI (+)	734.4→158.1	734.4→83.1	37/60	0.99	6	1.8
Methiocarb-d3	10.36	ESI (+)	229.1→169.1	229.1→121.0	9/17			
Methiocarb	10.38	ESI (+)	226.1→169.0	226.1→121.1	17/9	0.99	20	6
Clarithromycin-d3	10.84	ESI (+)	751.5→161.1	751.5→83.1	25/57			
Clarithromycin	10.86	ESI (+)	748.5→158.2	748.5→83.1	37/60	0.99	1	0.3
Metaflumizone	14.60	ESI (+)	507.1→178.2	507.1→87.0	33/25	0.99	6	1.8

2.3. Experimental Setup and Operational Conditions

1 L of 10 mg L^{-1} erythromycin (Milli-Q water) and 0.05 M Na$_2$SO$_4$ (background electrolyte) was treated by EF and PEC processes by continuously recirculating the solution from a reservoir (cylindrical Pyrex glass reactor of 2 L and 100 mm of internal diameter) to a mono-compartment filter-press cell at 20 L h^{-1}. Figure 1 shows a scheme of the experimental system.

The electrochemical cell was integrated by an anode and an oxygen-diffusion cathode, both of 100 mm of diameter, separated 5 mm. Two anode materials were tested: Fe and boron doped diamond coated on silicon (BDD/Si), the later with a resistivity of 100 mΩ·cm, and with a 2.5 μm BDD coating ([B] = 700 ppm) (NeoCoat®, Eplatures-Grise, La Chaux-de-Fonds, Switzerland). The oxygen-diffusion cathode was composed by a titanium diffuser, used as electrode collector, and an activated carbon cloth with polytetrafluoroethylene (PTFE) (Fuel Cells Etc, College Station, TX, USA) that promotes the reduction of oxygen to H$_2$O$_2$. Oxygen was fed to the cathode at an overpressure of 4.5 bar, testing two different flowrates: 0.8 L min^{-1} and 2.0 L min^{-1}. The carbon cloth of the oxygen-diffusion cathode was previously activated by electrolyzing 1 L of 0.05 M Na$_2$SO$_4$ at 150 mA cm^{-2} and pH 2.8 during 1 h.

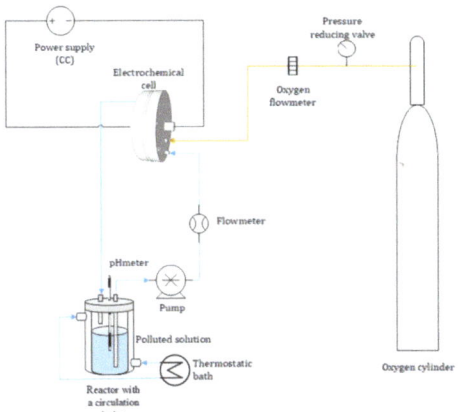

Figure 1. Scheme of the experimental system for the EF and PEC processes.

Experiments were carried out under galvanostatic conditions, using a power supply of Elektro-Automatik GmbH & Co (Viersen, Germany), and directly measuring the potential difference, assessing the influence of two current densities on anode (j_{anode}): 5 mA cm^{-2} and 10 mA cm^{-2}. These current densities were chosen considering that the most suitable j_{anode} = 16.7 mA cm^{-2} for the EF treatment of a complex pharmaceutical mixture [41], therefore smaller current densities were selected to study the treatment of a synthetic water and an urban wastewater.

Solutions were maintained at 25 °C with a water circulation jacket connected to an external thermostat. The pH was also kept constant throughout the trials. Three different pHs were evaluated: 2.8 (optimal pH for the Fenton process [42,43]), near neutral pH (5.0) and neutral pH (7.0), the two last were studied to reduce the use of acid reagent to adjust the pH. During EF experiments, the initial concentration of the catalyst, added externally to the water, was 10 mg L^{-1} Fe(II). This value was chosen considering that the maximum iron concentration of wastewaters to be discharged in the sewer is 10 mg L^{-1} in Catalonia [44]. Therefore, no additional posttreatments to precipitate iron would be needed for treating a wastewater by the EF process at 10 mg L^{-1} Fe(II).

Finally, the electrochemical treatment was applied to a real effluent from a sand filter of the tertiary treatment of an urban WWTP using the optimal operational conditions identified for treating synthetic polluted water. All the experimental conditions are summarized in Table 2.

Table 2. Summary of the experimental conditions.

Experiment Conditions	Type of Water	Anode Type	j_{anode} (mA/cm^2)	Oxygen Flowrate (L/min)	pH
EF-1		BDD	5	2.0	2.8
EF-2		BDD	10	2.0	2.8
PEC-1	Ultrapure water. 10 mg L^{-1} erythromycin and 0.05 M Na$_2$SO$_4$	Fe	5	2.0	2.8
PEC-2		Fe	10	2.0	2.8
PEC-3		Fe	5	0.8	2.8
PEC-4		Fe	10	0.8	2.8
PEC-5		Fe	5	2.0	5.0
PEC-6		Fe	5	2.0	7.0
PEC-7	Effluent from a tertiary treatment of an urban WWTP	Fe	5	2.0	2.8

3. Results and Discussion

3.1. Effect of the Anode Material and the Current Density

Electro-Fenton (using a BDD anode) and PEC (using a Fe anode) were compared applying two different current densities (5 and 10 mA cm^{-2}), at pH 2.8 and injecting 2.0 L min^{-1} O$_2$ into the cathode. Figure 2 plots the evolution of the DOC and the [H$_2$O$_2$] as a function of the applied charge.

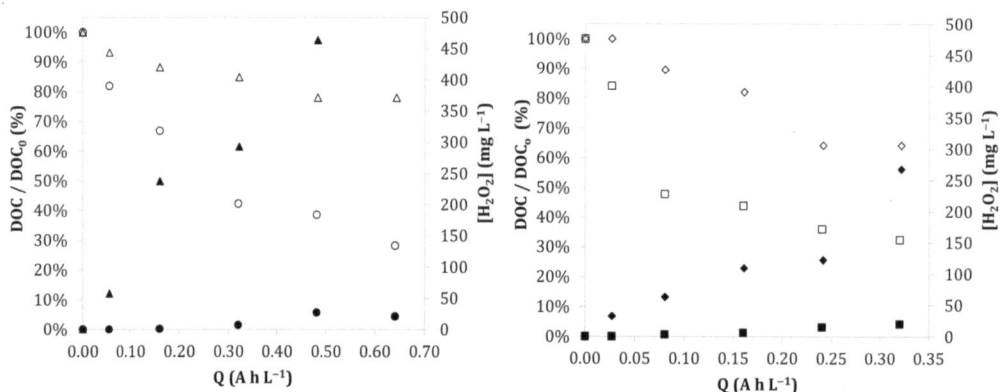

Figure 2. Evolution of dissolved organic carbon (DOC) (△, ○, ◇, □) and [H$_2$O$_2$] remained in water (▲, ●, ◆, ■) through the applied charge in 1 L of 10 mg L^{-1} erythromycin and 0.05M Na$_2$SO$_4$ (ultrapure water) treated by the EF and PEC processes at 25 °C, 2 L min^{-1} O$_2$ and pH = 2.8 under the following conditions: (△, ▲) BDD anode, 10 mA cm^{-2} (left graphic); (○, ●) Fe anode, 10 mA cm^{-2} (left graphic); (◇, ◆) BDD anode, 5 mA cm^{-2} (right graphic); (□, ■) Fe anode, 5 mA cm^{-2} (right graphic).

The mineralization of the organic matter was defined as the complete oxidation of the organic molecules into CO$_2$, inorganic ions, and water, and it was determined by the analysis of the total dissolved organic carbon (DOC) throughout the treatment time.

The mineralization rates of the EF (k_{EF}) and the PEC (k_{PEC}) processes under studied conditions followed a pseudo-first order kinetics (Equation (7)):

$$\ln(C/C_0)/t = k \quad (7)$$

where C and C$_0$ represent the final and initial erythromycin concentration, respectively, and t the treatment time.

Mineralization constant rates obtained by the EF process under the studied conditions were k_{EF1} (5 mA cm^{-2}) = 0.008 s^{-1} and k_{EF2} (10 mA cm^{-2}) = 0.004 s^{-1}. These values were significantly lower than those achieved when iron is used as anode (PEC process), where constant rate increased to 0.020 s^{-1} for both current densities. The mineralization reached in EF process was approximately 35%, whereas it was increased at 70% during PEC treatment, after applying 0.32 A h L^{-1}.

As depicted in the Figure 2, the remained H$_2$O$_2$ concentrations after 1 h of the EF treatment were around 260 mg L^{-1} and 460 mg L^{-1} at 5 and 10 mA cm^{-2} when 0.32 and 0.64 A h L^{-1} were applied, respectively.

These experimental H$_2$O$_2$ concentrations remained throughout the EF were similar than the theoretical ones generated. This fact indicates that only small quantities of H$_2$O$_2$ reacted with the 10 mg L^{-1} Fe(II). Moreover, it should be considered that an excessive H$_2$O$_2$ concentration acts as •OH scavenger (Equation (8)) reducing the efficiency for organic matter oxidation:

$$\text{•OH}_{(aq)} + \text{H}_2\text{O}_{2(aq)} \rightarrow \text{HO}_2\text{•}_{(aq)} + \text{H}_2\text{O}_{(l)} \quad k = 1.7\text{–}4.5 \times 10^7 \text{ M}^{-1}\text{ s}^{-1} \quad (8)$$

Although $^\bullet$OH reacts with organic molecules with rate constants of 10^9–10^{10} M^{-1} s^{-1} [45], it reacts with H$_2$O$_2$ at a 100–1000 times lower rate constant (k = 1.7–4.5 × 10^7 M^{-1} s^{-1} [24]) as the reaction rate depends on the initial concentration of organic matter and H$_2$O$_2$. Considering the initial erythromycin concentration (10 mg L^{-1}) and the remained H$_2$O$_2$ concentrations of this work, the molar H$_2$O$_2$/erythromycin ratio was higher than 100 after the 15 min and 5 min of the EF treatment at 5 and 10 mA cm^{-2}, respectively. Therefore, after these treatment times, hydroxyl radical is expected to react preferentially with H$_2$O$_2$ instead of organic matter. This fact can explain why the k_{EF2} (10 mA cm^{-2}) = 0.004 s^{-1} was lower than k_{EF1} (5 mA cm^{-2}) = 0.008 s^{-1}.

In comparison, most H$_2$O$_2$ was consumed during PEC experiments, remaining around 20 mg L^{-1} at the end of the trials. This is a consequence of the higher Fe concentrations generated during PEC (theoretical concentrations of around 220 mg L^{-1} and 440 mg L^{-1} Fe(III) at 0.32 and 0.64 A h L^{-1}, respectively) compared to that used during EF trials (10 mg L^{-1} Fe(II)). Indeed, in the PEC experimental conditions, the molar H$_2$O$_2$/Fe(III) ratios were around 1.7–2.0, indicating that higher iron concentration resulted in higher reactivity with the hydrogen peroxide. Consequently, higher theoretical concentration of hydroxyl radical (Equation (1)) and other reactive oxygen species that can attack organic compounds can be obtained. Nonetheless, elevated concentrations of Fe(II) or Fe(III) can also act as scavenger of $^\bullet$OH, following Equations (9) and (10) [24]:

$$Fe^{2+}_{(aq)} + {}^\bullet OH_{(aq)} \rightarrow Fe^{3+}_{(aq)} + OH^-_{(aq)} \qquad k = 2.5\text{–}5.1 \times 10^8 \text{ M}^{-1} \text{ s}^{-1} \qquad (9)$$

$$Fe^{3+}_{(aq)} + {}^\bullet OH_{(aq)} \rightarrow Fe^{4+}_{(aq)} + OH^-_{(aq)} \qquad k = 7.9 \times 10^7 \text{ M}^{-1} \text{ s}^{-1} \qquad (10)$$

After 15 min and 8 min of PEC treatment at 5 and 10 mA cm^{-2}, respectively, the molar ratio Fe(III)/erythromycin was higher than 100, therefore the $^\bullet$OH starts to preferentially react with Fe(II) or Fe(III) instead of organic matter.

Moreover, it should be also considered that at elevated concentrations of Fe(III) (higher than 0.6 mg L^{-1}) ferric oxyhydroxides start to precipitate at pH = 2.8 [46], whereas Fe(II) remains in solution at this pH. Related to this, the percentage of dissolved *versus* total iron was around 100–85% throughout the PEC treatment at 5 mA cm^{-2} whereas this value decreased to 75–40% at 10 mA cm^{-2}. These results indicate that most iron was dissolved at 5 mA cm^{-2} and pH 2.8. The quantity of dissolved iron at both current densities were similar, thereby contributing in a similar way to the Fenton (Equation (1)) and Fenton-like (Equation (2)) reactions. Dissolved iron can be associated to ferrous cationic species, whereas ferric oxyhydroxides would be associated to iron precipitated.

All these reasons explain the no significant differences observed with the rate constants for both current densities (k_{PEC} = 0.02 s^{-1}) tested during PEC trials.

These results demonstrate that, in the studied conditions, the PEC is more recommendable than the EF process in terms of both mineralization rate and electricity consumption. After applying 0.32 and 0.64 A h L^{-1}, the electricity consumptions for BDD electrodes (EF) were around 3.7 kWh m^{-3} and 7.9 kWh m^{-3}, whereas they were reduced to 0.3 kWh m^{-3} and 0.6 kWh m^{-3} when Fe was used as anode (PEC), respectively. This means a reduction of almost 12-fold in the electricity costs by using Fe anodes.

Therefore, the PEC treatment was selected to study the influence of oxygen flowrate injected into the cathode.

3.2. Effect of the Oxygen Flowrate

The influence of two different oxygen flowrates (0.8 L min^{-1} O$_2$ and 2.0 L min^{-1} O$_2$) to the mineralization efficiency was studied by carrying out the PEC treatment applying two different current densities (5 and 10 mA cm^{-2}), at pH 2.8. Results are plotted in Figure 3.

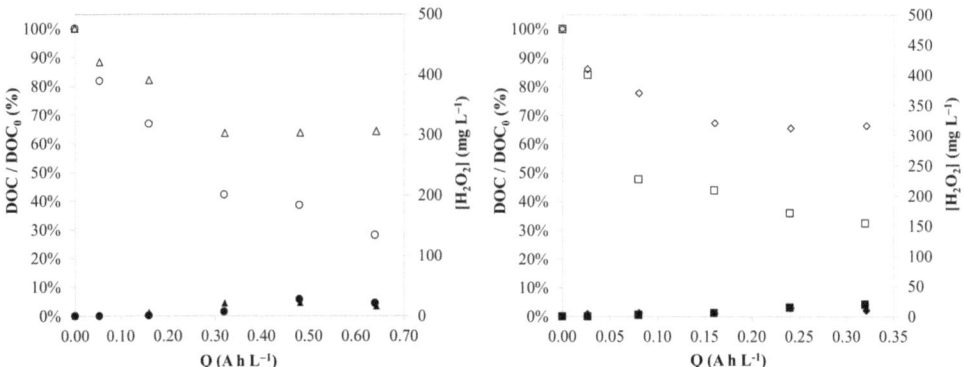

Figure 3. Evolution of DOC (○, △, □, ◇) and [H$_2$O$_2$] remained in water (●, ▲, ■, ◆) through the applied charge in 1 L of 10 mg L^{-1} erythromycin and 0.05M Na$_2$SO$_4$ (ultrapure water) treated by the PEC process at 25 °C, pH = 2.8, under the following conditions: (○, ●) 10 mA cm^{-2} at 2 L min^{-1} O$_2$ (left graphic); (△, ▲) 10 mA cm^{-2} at 0.8 L min^{-1} O$_2$ (left graphic); (□, ■) 5 mA cm^{-2} at 2 L min^{-1} O$_2$ (right graphic); and (◇, ◆) 5 mA cm^{-2} at 0.8 L min^{-1} O$_2$ (right graphic).

When injecting 0.8 L min^{-1} O$_2$, the mineralization kinetics followed a pseudo-first order constant during the first 30 min of treatment, being the constant rate of 0.013 s^{-1} for 5 and 10 mA/cm^{-2}. The kinetics mineralization decreased at 0.8 L min^{-1} O$_2$ (0.013 s^{-1}) compared with the value obtained at 2.0 L min^{-1} O$_2$ (0.020 s^{-1}). This fact can be associated to the concentration of dissolved oxygen. With 2.0 L min^{-1} O$_2$ injected into the cathode, continuous oxygen saturation in water can be obtained, resulting in more dissolved oxygen available to produce H$_2$O$_2$.

No significant differences on mineralization rates between the studied current densities (5 and 10 mA/cm^{-2}) were found in any of the tested oxygen flowrates (0.8 and 2.0 L min^{-1} O$_2$). The electricity consumption of the electrochemical cell also followed a similar pattern for both oxygen flowrates.

Therefore, the PEC process at 5 mA cm^{-2} applying 2.0 L min^{-1} O$_2$ was chosen to study the influence of the pH on the process efficiency.

3.3. Effect of the pH

The influence of the pH on the efficiency of the PEC treatment was also studied. The results obtained are depicted in Figure 4.

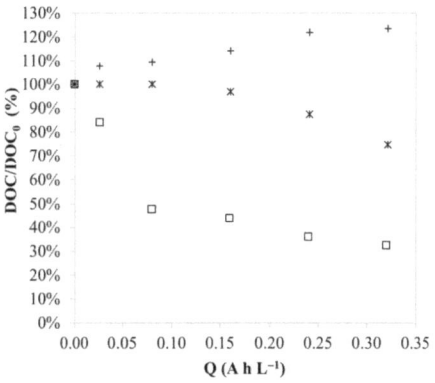

Figure 4. Evolution of DOC through the applied charge in 1 L of 10 mg L^{-1} erythromycin and 0.05 M Na$_2$SO$_4$ (ultrapure water) treated by the PEC process at 5 mA cm^{-2}, 25 °C, 2 L min^{-1} O$_2$, under the following pHs: (□) 2.8; (∗) 5.0; (+) 7.0.

The optimal pH for the PEC process was 2.8, the same as the most favourable pH found for Fenton process [42,43]. At pH values close to 2.8, the predominant ions of iron, ferrous and ferric, are dissolved in the form of aqua-complexes, highlighting as main compounds the $[Fe^{II}(H_2O)_6]^{2+}$, $[Fe^{III}(OH)(H_2O)_5]^{2+}$, $[Fe^{III}(OH)_2(H_2O)_4]^{+}$ and $[Fe^{III}(H_2O)_6]^{3+}$ and, to lesser extent, $[Fe_2(H_2O)_8(OH)_2]^{4+}$ [47,48]. However, we determined that at pH 5.0, around 0.2–0.6% iron was dissolved (0.2–1.1 mg L^{-1}), whereas at pH 7.0 all iron was precipitated. As observed by Serra-Clusellas et al. [49], minute concentrations of Fe(III) (0.6 mg L^{-1}) were enough to promote the Fenton-like reaction and the mineralization of organic pollutants at pH 3.6. This fact was then confirmed in the current work at pH 5.0, where around 25% of DOC was removed after applying 0.32 A h L^{-1}. Unlike pH 5.0, no mineralization was reached at pH 7.0; on the contrary, an increase of DOC throughout the treatment was observed. This rise can be associated to organic leaches from silicone tubs used for the peristaltic pump for recirculating the solution, which can be partially oxidized by hydrogen peroxide and other reactive oxygen species. This fact evidenced that mineralization percentage obtained in this article could be higher if other tubbing materials were used.

Despite this, the most suitable operating conditions found in this work are as follows: PEC treatment at 5 mA cm^{-2}, injecting 2.0 L min^{-1} O$_2$ at pH 2.8.

3.4. Identification of Aromatic Byproducts of Erythromycin and Degradation Pathway Proposed

Aromatic byproducts generated during the early stage of the erythromycin mineralization pathway were identified for the PEC process at the most suitable operational conditions, specified above.

Most of transformation products for erythromycin and its degradation pathways have already discussed in bibliography [16,35–39] under different conditions but it has not been investigated under PEC water remediation process. Pérez et al. [31] identified the intermediates formed at 30 and 120 min of the solar photo electro-Fenton process of a 165 mg L^{-1} erythromycin solution at pH 3 and $j_{cathode}$ = 0.16 mA cm^{-2}, determined by GC-MS. However, no information was found about the first intermediates generated during the early stages of the erythromycin mineralization pathway by the electrochemical-Fenton based processes.

In this paper we used chromatographic separation behaviour and high-resolution tandem mass spectrometry HRMS^2 for the identification and quantification of erythromycin transformation products. By using this approach, 23 erythromycin intermediates were putative identified by their theoretical exact mass, logical retention time (compared to bibliography) and diagnostic tandem mass ions. In addition, these compounds were quantified by using an erythromycin IS calibration curve and their concentration (erythromycin + 23 intermediates) throughout the PEC treatment time are presented on Table 3.

Table 3. Erythromycin and aromatic intermediates detected during PEC of 10 mg L^{-1} erythromycin and 0.05M Na$_2$SO$_4$ (ultrapure water) at 5 mA cm^{-2}, 25 °C, 2 L min^{-1} O$_2$, pH = 2.8.

N°	Compound	Molecular Formula	Pseudo-Molecular Ion (m/z)	Retention Time (min)	Compound Concentration (µg L^{-1})					
					0 min pH 2.8	5 min	15 min	30 min	45 min	60 min
1	Erythromycin (Erythromycin A)	C$_{37}$H$_{67}$NO$_{13}$	734.4685	4.0	457.4	17.9	0.8	Non detected (n.d.)	n.d.	n.d.
2	Erythromycin B	C$_{37}$H$_{67}$NO$_{12}$	718.4736	4.6	204.1	47.0	0.5	n.d.	n.d.	n.d.
3	Erythromycin C	C$_{36}$H$_{65}$NO$_{13}$	720.4529	3.5	5.2	n.d.	n.d.	n.d.	n.d.	n.d.
4	Erythromycin C anhydrous	C$_{36}$H$_{63}$NO$_{12}$	702.4423	4.2	61.1	19.2	0.2	n.d.	n.d.	n.d.
5	Erythromycin A desosamine N-demethylation	C$_{36}$H$_{65}$NO$_{13}$	720.4529	3.9	13.5	2.3	1.5	n.d.	n.d.	n.d.
6	Erythromycin E	C$_{37}$H$_{65}$NO$_{14}$	748.4478	3.9	301.3	71.0	2.9	n.d.	n.d.	n.d.

Table 3. Cont.

N°	Compound	Molecular Formula	Pseudo-Molecular Ion (m/z)	Retention Time (min)	Compound Concentration (µg L^{-1})					
					0 min pH 2.8	5 min	15 min	30 min	45 min	60 min
7	Erythromycin E dehydration + dehydroxylation	$C_{37}H_{63}NO_{12}$	714.4423	5.2	22.1	99.3	1.2	n.d.	n.d.	n.d.
8	Erythromycin E dehydration + dehydroxylation—cladinose	$C_{29}H_{49}NO_9$	556.348	4.3	n.d.	46.7	5.0	n.d.	n.d.	n.d.
9	Erythromycin F	$C_{37}H_{67}NO_{14}$	750.4634	3.7	241.7	14.5	n.d.	n.d.	n.d.	n.d.
10	Erythromycin F—cladinose TP-1	$C_{29}H_{53}NO_{11}$	592.3691	2.5	99.1	43.6	2.3	n.d.	n.d.	n.d.
11	Erythromycin F—cladinose TP-2	$C_{29}H_{53}NO_{11}$	592.3691	2.6	1.2	64.3	3.2	n.d.	n.d.	n.d.
12	Erythromycin A—cladinose	$C_{29}H_{53}NO_{10}$	576.3742	3.0	68.1	149.0	9.1	n.d.	n.d.	n.d.
13	Erythromycin F enol ether or anhydrous	$C_{37}H_{65}NO_{13}$	732.4529	3.6	3.5	20.8	0.8	n.d.	n.d.	n.d.
14	Erythromycin F dehydrogenation TP-1	$C_{37}H_{65}NO_{14}$	748.4478	3.5	0.7	31.4	0.8	n.d.	n.d.	n.d.
15	Erythromycin F dehydrogenation TP-2	$C_{37}H_{65}NO_{14}$	748.4478	3.9	12.8	61.0	3.1	n.d.	n.d.	n.d.
16	Erythromycin F enol ether or anhydrous + lactone dehydration TP-1	$C_{37}H_{63}NO_{12}$	714.4423	4.4	40.6	21.1	0.2	n.d.	n.d.	n.d.
17	Erythromycin F enol ether or anhydrous + lactone dehydration TP-2	$C_{37}H_{63}NO_{12}$	714.4423	4.8	20.9	27.4	0.2	n.d.	n.d.	n.d.
18	Erythromycin F enol ether or anhydrous + lactone dehydration—cladinose	$C_{29}H_{49}NO_9$	556.348	3.0	n.d.	5.6	0.4	n.d.	n.d.	n.d.
19	Erythromycin A enol ether TP-1	$C_{37}H_{65}NO_{12}$	716.458	4.8	5.1	4.4	n.d.	n.d.	n.d.	n.d.
20	Erythromycin A enol ether TP-2	$C_{37}H_{65}NO_{12}$	716.458	5.2	1.8	8.6	0.1	n.d.	n.d.	n.d.
21	Erythromycin A enol ether—cladinose	$C_{29}H_{51}NO_9$	558.3637	4.2	20.5	7.2	n.d.	n.d.	n.d.	n.d.
22	Erythromycin A anhydrous	$C_{37}H_{65}NO_{12}$	716.458	4.5	1533.5	432.0	2.6	n.d.	n.d.	n.d.
23	Erythromycin A anhydrous desosamine N-demethylation	$C_{36}H_{63}NO_{12}$	702.4423	4.4	56.2	3.5	n.d.	n.d.	n.d.	n.d.
24	Erythromycin A anhydrous—cladinose	$C_{29}H_{51}NO_9$	558.3637	3.5	46.7	79.4	4.7	n.d.	n.d.	n.d.

The erythromycin degradation pathway proposed for the early stage of the PEC process is represented in Figure 5.

As described in the literature, the decomposition of erythromycin A proceeds via the known intermediate erythromycin A-6,9-hemiketal. The monodehydroxylation or didehydroxylation of this intermediate produced the main compounds described in the literature, i.e., erythromycin A enol ether isomers (716.458 m/z) (compounds 19 and 20 of Figure 5) and anhydrous erythromycin A (716.458 m/z) (compound 22), respectively. Sequentially, the respective transformation products of the compounds 19, 20 and 22 from the cleavage of cladinose group were assigned (compounds 21 and 23) by their exact mass and the observation of unmodified desosamine fragment (158.1176 m/z) and cladinose neutral loss ([M+H-158.0937]$^+$ m/z). These results indicated that the delta mass observed was produced on the lactone part of erythromycin A. The last two compounds were assigned on the basis of their similar fragmentation pattern in respect of their precursor compound (480–580 m/z range), where neutral losses of water from lactone part are observed. Thus, anhydrous erythromycin A (compound 22) either with or without a cladinose moiety had two consecutive losses of water from 558.3637 m/z ion, whereas erythromycin A enol ether (compound 19) either with or without cladinose moiety had only one losses of water from 558.3637 m/z ion. In addition, an isomer of erythromycin A enol ether was identified (compound 20) since their fragmentation pattern was equal between the two chromatographic peaks.

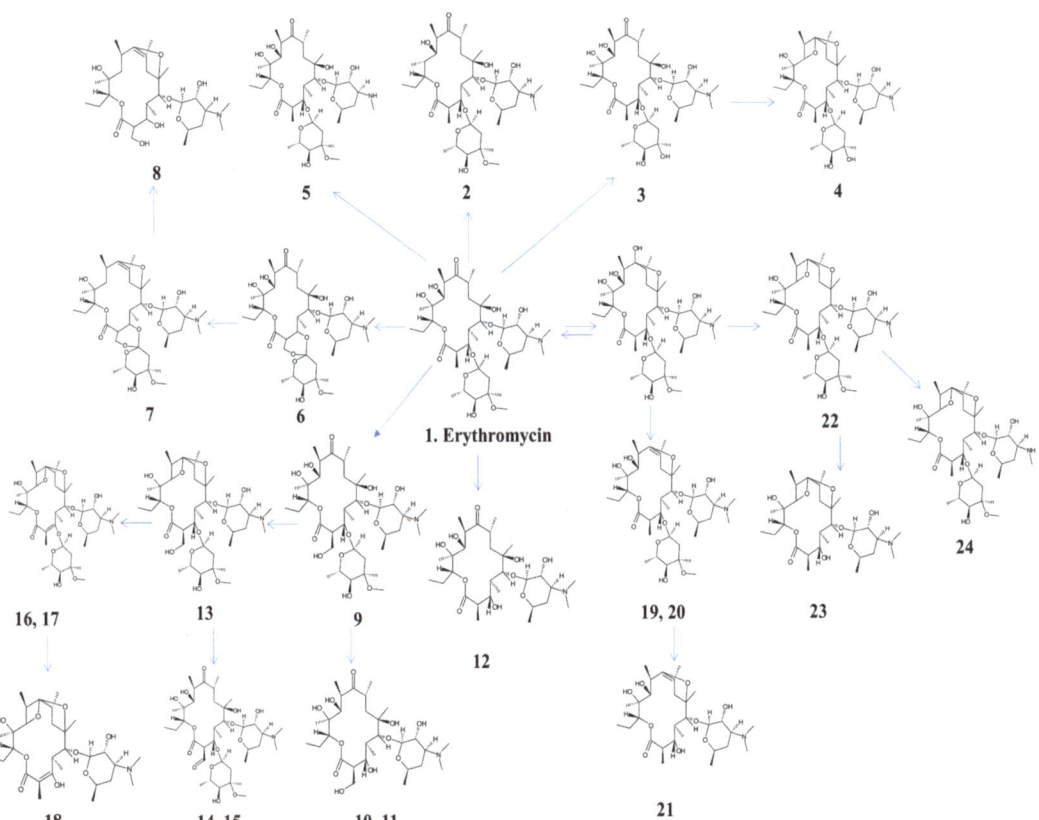

Figure 5. Proposed early-stage mineralization pathway for the removal of 10 mg L^{-1} erythromycin (ultrapure water) by the PEC process.

Moreover, erythromycin A without a cladinose sugar (compound 12) was identified based on the no observation of the neutral loos of cladinose and the observation of desosamine fragment, as well as other erythromycin isomers such as erythromycin B (compound 2), erythromycin C (compound 3), erythromycin E (compound 6) and erythromycin F (compound 9) and some of their transformation products concerning the cleavage of cladinose moiety (compound 8, 10, 11). Erythromycin isomers A, C, E and F dehydrate produced the compounds 22, 4, 13 and 7, respectively. For example, tandem mass spectrum of erythromycin C and erythromycin C anhydrous had the neutral loss of the modified cladinose (-OH instead of -OCH$_3$). All this transformation products have already been described for different remediation processes and follow the proposed degradation pathway indicated in Figure 5. Nonetheless, we found other interesting signals (732, 714, and 555 m/z) that had a delta mass of -2 m/z of the main erythromycin A transformation products explained before. Thus, their tandem mass spectra were also carefully inspected and we hypothesize that a degradation of erythromycin-A is promoted by the formation of erythromycin F (750.4634 m/z) intermediate (compound 9) and its degradation to erythromycin F enol ether or anhydrous (732.4529 m/z) (compound 13) and its lactone dehydration (714.4423 m/z) (isomers 16 and 17) and successive cladinose cleavage (556.348 m/z) (compound 18), or compound 13 dehydrogenation (748.4478 m/z) producing two isomers (compounds 14 and 15). Finally, an N-demethylation of erythromycin A generated the intermediate 5.

The results of the Table 3 show that none of these compounds were detected after 30 min of treatment, all of them thus being early-stage intermediates.

Another aspect to highlight is that after adjusting the pH to 2.8, and before starting the PEC process (0 min), only 5% of the initial erythromycin (erythromycin A, compound 1) remained in solution. 28% of this substance was degraded to the 23 intermediates detected in this work, being the erythromycin A anhydrous (compound 22) the most detected by-product produced (approximately 15%). This means that 67% of the initial erythromycin was degraded to other byproducts not determined in this article just due to the acidification. This acid-catalysed water molecule loss from erythromycin occurs rapidly at pH < 4 [50,51]. Therefore, this is an important aspect to be considered in future works, especially when working or analysing the erythromycin in acid pH solutions.

3.5. Extrapolation of the PEC Process to a Real Effluent

Finally, the PEC process under the most appropriate conditions (5 mA cm^{-2}, injecting 2.0 L min^{-1} O$_2$ into the cathode and pH 2.8) was validated by treating a real effluent obtained from the tertiary treatment (coagulation/flocculation and sand filtration) of an urban WWTP. Its composition is shown in Table 4.

Table 4. Composition of the effluent from the tertiary treatment of an urban WWTP.

Parameter	Units	Value
pH	-	7.6
Electrical conductivity	ms cm^{-1}	1.5
Total Inorganic Carbon	mg L^{-1}	72.8
DOC	mg L^{-1}	9.9
Total N	mg L^{-1}	33.1

No electrolyte was added to the water. The concentrations of the emerging compounds included in the Watch List 2018 in the influent and effluent of the electrochemical treatment are summarized in Table 5.

Table 5. Concentration of the compounds listed in the Watch List 2018 in the analysed effluents.

Compound	Effluent from the Tertiary Treatment of an Urban WWTP (ng L^{-1})	Treated Effluent from the PEC Treatment (ng L^{-1})	Removals (%)
E1	858	5	99%
E2	127	<MLOQ	>99%
EE2	117	1	99%
Amoxicillin trihydrate	<MLOQ	<MLOQ	-
Thiamethoxam	7	<MLOQ	>86%
Ciprofloxacin	1	<MLOQ	-
Clothianidin	6	<MLOQ	>83%
Imidacloprid	60	<MLOQ	>83%
Acetamiprid	30	<MLOQ	>97%
Thiacloprid	21	<MLOQ	>95%
Azythromycin	365	<MLOQ	100%
Erythromycin	23	<MLOQ	>74%
Methiocarb	30	<MLOQ	>33%
Clarythromycin	31	<MLOQ	>97%
Metaflumizone	17	<MLOQ	>65%

In general terms, the compounds found in the WWTP tertiary effluent (PEC influent) were not detected in the treated PEC effluent, except for E1 and EE2, although high degradation percentages were achieved (99% removal). As commented before, the determination of erythromycin, but also the amoxicillin, is significantly affected by the low pH in the treated water, since the internal standard is not stable in the acidic conditions. The initial DOC of the water was 9.9 mg L^{-1} and its evolution was plotted in Figure 6.

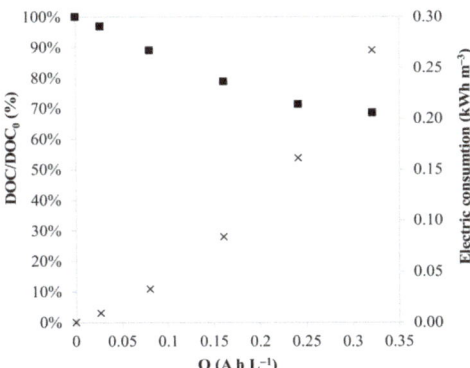

Figure 6. Evolution of the DOC (■) and the electric consumption (x) through the applied charge in 1 L of an effluent from an urban WWTP treated by the PEC process at 5 mA cm^{-2}, 25 °C, 2 L min^{-1} O$_2$ at pH 2.8.

Approximately 30% DOC was removed at the end of the PEC treatment, applying 0.32 A h L^{-1}. The electricity consumption associated to the electrochemical cell was around 0.4 kWh m^{-3}, that means a cost of around 0.09 € m^{-3}, considering the Spanish electricity price in first half 2020 was 0.2239 € kWh^{-1} [52]. This low electricity consumption with the high degradation efficiencies for all the compounds of the Watch List 2018, appoint the PEC as a potential technology to be applied for removing these types of emerging compounds from waters, preventing their arrival to the aquatic systems, and thus improving the water quality of the environment. Future studies should also include toxicity assays of the transformation byproducts generated during the PEC treatment to ensure a suitable final quality of water.

These results indeed suggested PEC as a promising technology to be implemented for water and wastewater remediation. For example, PEC can be applied to treat complex industrial wastewaters such as pharmaceutical or leachate effluents coupled before a biological treatment to increase water biodegradability. However, a neutralization step must be applied after the PEC process to send the effluent to a biological treatment. Also, this post-treatment would be required if the effluent was discharged to a municipal sewer in order to reduce the iron concentration below the limit sets by the legislation (10 mg L^{-1} Fe, in the case of Catalunya [44]).

4. Conclusions

The results of this investigation demonstrate the technical and economic feasibility of the peroxyelectrocoagulation (PEC) process to degrade contaminants of emerging concern present in water such as those included in the EU Decision 2018/840 (Watch List 2018). The article proves that PEC can degrade most of the compounds included in the Watch List 2018 below their quantification limit and mineralize the 25% DOC of a real WWTP tertiary effluent, applying an electrical charge of 0.32 A h L^{-1}, a current density j_{anode} = 5 mA cm^{-2}, injecting 2.0 L min^{-1} O$_2$ into the cathode and keeping the pH at 2.8. The electricity consumption of the cell was 0.27 kWh m^{-3}. These operational conditions were selected by previously studying the removal erythromycin from ultrapure water, obtaining 70% mineralization with an electricity consumption of 1.7 kWh m^{-3}. The research also revealed the early-stage transformation byproducts generated during the erythromycin mineralization, by identifying and quantifying 23 erythromycin intermediates.

Author Contributions: EF and PEC experiments: A.S.-C., L.S., M.d.L.C.; erythromycin byproducts determination: P.H., A.D.-R.; determination of the emerging compounds of the Watch List 2018: M.R.; project proposal: A.S.-C. and A.C.; project management: A.S.-C., A.C., N.C., X.M.-L. All authors have read and agreed to the published version of the manuscript.

Funding: This work was financially supported by the Catalan Government through the funding grant ACCIÓ-Eurecat (Project PRIV2019-AVANÇ-H$_2$O).

Institutional Review Board Statement: Not applicable.

Informed Consent Statement: Not applicable

Data Availability Statement: Data sharing not applicable. No new data were created or analyzed in this study. Data sharing is not applicable to this article.

Acknowledgments: The authors would like to acknowledge the Sustainability Laboratory of EURECAT—Manresa facilities for the analysis of DOC and iron of the generated samples.

Conflicts of Interest: The authors declare no conflict of interest.

References

1. Carlsson, C.; Johansson, A.K.; Alvan, G.; Bergman, K.; Kühler, T. Are pharmaceuticals potent environmental pollutants? Part I: Environmental risk assessments of selected active pharmaceutical ingredients. *Sci. Total Environ.* **2006**, *364*, 67–87. [CrossRef]
2. Sousa, J.C.G.; Ribeiro, A.R.; Barbosa, M.O.; Ribeiro, C.; Tiritan, M.E.; Pereira, M.F.R.; Silva, A.M.T. Monitoring of the 17 EU Watch List contaminants of emerging concern in the Ave and the Sousa Rivers. *Sci. Total Environ.* **2019**, *649*, 1083–1095. [CrossRef]
3. Cunningham, V.L.; Binks, S.P.; Olson, M.J. Human health risk assessment from the presence of human pharmaceuticals in the aquatic environment. *Regul. Toxicol. Pharmacol.* **2009**, *53*, 39–45. [CrossRef]
4. European Commission. *Commission Implementing Decision (EU) 2015/495 of 20 March 2015 Establishing a Watch List of Substances for Union-Wide Monitoring in the Field of Water Policy Pursuant to Directive 2008/105/EC of the European Parliament and of the Council*; Office for Official Publications of the European Communities: Luxembourg, 2015. Available online: https://eur-lex.europa.eu/legal-content/EN/TXT/PDF/?uri=CELEX:32015D0495&from=EN (accessed on 18 April 2021).
5. European Commission. *Commission Implementing Decision (EU) 2018/840 of 5 June 2018 Establishing a Watch List of Substances for Union-Wide Monitoring in the Field of Water Policy Pursuant to Directive 2008/105/EC of the European Parliament and of the Council and Repealing Commission Implementing Decision (EU) 2015/495*; Office for Official Publications of the European Communities: Luxembourg, 2018. Available online: https://eur-lex.europa.eu/legal-content/EN/TXT/PDF/?uri=CELEX:32018D0840&from=EN (accessed on 18 April 2021).
6. European Commission. *Commission Implementing Decision (EU) 2020/1161 of 4 August 2020 Establishing a Watch List of Substances for Union-Wide Monitoring in the Field of Water Policy Pursuant to Directive 2008/105/EC of the European Parliament and of the Council*; Office for Official Publications of the European Communities: Luxembourg, 2020. Available online: https://eur-lex.europa.eu/legal-content/EN/TXT/PDF/?uri=CELEX:32020D1161&from=EN (accessed on 18 April 2021).
7. Rodriguez-Mozaz, S.; Vaz-Moreira, I.; Varela Della Giustina, S.; Llorca, M.; Barceló, D.; Schubert, S.; Berendonk, T.U.; Michael-Kordatou, I.; Fatta-Kassinos, D.; Martinez, J.L.; et al. Antibiotic residues in final effluents of European wastewater treatment plants and their impact on the aquatic environment. *Environ. Int.* **2020**, *140*, 105733. [CrossRef]
8. Polianciuc, S.I.; Gurzău, A.E.; Kiss, B.; Georgia Ştefan, M.; Loghin, F. Antibiotics in the environment: Causes and consequences. *Med. Pharm. Rep.* **2020**, *93*, 231–240. [CrossRef]
9. Hanna, N.; Sun, P.; Sun, Q.; Li, X.; Yang, X.; Ji, X.; Zou, H.; Ottoson, J.; Nilsson, L.E.; Berglund, B.; et al. Presence of antibiotic residues in various environmental compartments of Shandong province in eastern China: Its potential for resistance development and ecological and human risk. *Environ. Int.* **2018**, *114*, 131–142. [CrossRef] [PubMed]
10. World Health Organization. Antibiotic Resistance. 21 July 2020. Available online: https://www.who.int/news-room/fact-sheets/detail/antibiotic-resistance (accessed on 18 April 2021).
11. Veseli, A.; Mullallari, F.; Balidemaj, F.; Berisha, L.; Švorc, L.; Arbneshi, T. Electrochemical determination of erythromycin in drinking water resources by surface modified screen-printed carbon electrodes. *Microchem. J.* **2019**, *148*, 412–418. [CrossRef]
12. Carraro, E.; Bonetta, S.; Bertino, C.; Lorenzi, E.; Bonetta, S.; Gilli, G. Hospital effluents management: Chemical, physical, microbiological risks and legislation in different countries. *J. Environ. Manag.* **2016**, *168*, 185–199. [CrossRef] [PubMed]
13. Kovalova, L.; Siegrist, H.; von Gunten, U.; Eugster, J.; Hagenbuch, M.; Wittmer, A.; Moser, R.; McArdell, C.S. Elimination of micropollutants during post-treatment of hospital wastewater with powdered activated carbon, ozone, and UV. *Environ. Sci. Technol.* **2013**, *47*, 7899–7908. [CrossRef] [PubMed]
14. Ohlsen, K.; Ternes, T.; Werner, G.; Wallner, U.; Löffler, D.; Ziebuhr, W.; Witte, W.; Hacker, J. Impact of antibiotics on conjugational resistance gene transfer in *Staphylococcus aureus* in sewage. *Environ. Microbiol.* **2003**, *5*, 711–716. [CrossRef]

15. Rueda-Márquez, J.J.; Palacios-Villarreal, C.; Manzano, M.; Blanco, E.; Ramírez del Solar, M.; Levchuk, I. Photocatalytic degradation of pharmaceutically active compounds (PhACs) in urban wastewater treatment plants effluents under controlled and natural solar irradiation using immobilized TiO_2. *Sol. Energy* **2020**, *208*, 480–492. [CrossRef]
16. Llorca, M.; Rodríguez-Mozaz, S.; Couillerot, O.; Panigoni, K.; de Gunzburg, J.; Bayer, B.; Czaja, R.; Barceló, D. Identification of new transformation products during enzymatic treatment of tetracycline and erythromycin antibiotics at laboratory scale by an on-line turbulent flow liquid-chromatography coupled to a high resolution mass spectrometer LTQ-Orbitrap. *Chemosphere* **2015**, *119*, 90–98. [CrossRef] [PubMed]
17. Moslah, B.; Hapeshi, E.; Jrad, A.; Fatta-Kassinos, D.; Hedhili., A. Simultaneous Decontamination of Seven Residual Antibiotics in Secondary Treated Effluents by Solar Photo-Fenton and Solar TiO_2 Catalytic Processes. In *Recent Advances in Environmental Science from the Euro-Mediterranean and Surrounding Regions. EMCEI 2017. Advances in Science, Technology & Innovation (IEREK Interdisciplinary Series for Sustainable Development)*; Kallel, A., Ksibi, M., Ben Dhia, H., Khélifi, N., Eds.; Springer: Cham, Switzerland, 2017. [CrossRef]
18. Klavarioti, M.; Mantzavinos, D.; Kassinos, D. Removal of residual pharmaceuticals from aqueous systems by advanced oxidation processes. *Environ. Int.* **2009**, *35*, 402–417. [CrossRef] [PubMed]
19. Brillas, E.; Martínez-Huitle, C.A. Decontamination of wastewaters containing synthetic organic dyes by electrochemical methods. An updated review. *Appl. Catal. B Environ.* **2015**, *166–167*, 603–643. [CrossRef]
20. Lima, V.B.; Goulart, L.A.; Rocha, R.S.; Steter, J.R.; Lanza, M.R.V. Degradation of antibiotic ciprofloxacin by different AOP systems using electrochemically generated hydrogen peroxide. *Chemosphere* **2020**, *247*, 125807. [CrossRef]
21. Liu, X.; Zhou, Y.; Zhang, J.; Luo, L.; Yang, Y.; Huang, H.; Peng, H.; Tang, L.; Mu, Y. Insight into electro-Fenton and photo-Fenton for the degradation of antibiotics: Mechanism study and research gaps. *Chem. Eng. J.* **2018**, *347*, 379–397. [CrossRef]
22. Fenton, H.J.H. Oxidation of tartaric acid in the presence of iron. *J. Chem. Soc.* **1894**, *65*, 899–910. [CrossRef]
23. Peres Ribeiro, J.; Nunes, M.I. Recent trends and developments in Fenton processes for industrial wastewater treatment—A critical review. *Environ. Res.* **2021**, *197*, 110957. [CrossRef] [PubMed]
24. Sychev, A.Y.; Isak, V.G. Iron compounds and the mechanisms of the homogeneous catalysis of the activation of O_2 and H_2O_2 and of the oxidation of organic substrates. *Russ. Chem. Rev.* **1995**, *64*, 1105–1129. [CrossRef]
25. Vasudevan, S. An efficient removal of phenol from water by peroxi-electrocoagulation processes. *J. Water Process Eng.* **2014**, *2*, 53–57. [CrossRef]
26. Sandhwar, V.K.; Prasad, B. Comparison of electrocoagulation, peroxi-electrocoagulation and peroxi-coagulation processes for treatment of simulated purified terephthalic acid wastewater: Optimization, sludge and kinetic analysis. *Korean J. Chem. Eng.* **2018**, *35*, 909–921. [CrossRef]
27. Farhadi, S.; Aminzadeh, B.; Torabian, A.; Khatibikamal, V.; Alizadeh Fard, M. Comparison of COD removal from pharmaceutical wastewater by electrocoagulation, photoelectrocoagulation, peroxi-electrocoagulation and peroxi-photoelectrocoagulation processes. *J. Hazard. Mater.* **2012**, *219–220*, 35–42. [CrossRef] [PubMed]
28. Marselli, B.; Garcia-Gomez, J.; Michaud, P.-A.; Rodrigo, M.A.; Comninellis, C. Electrogeneration of hydroxyl radical on boron-doped diamond electrodes. *J. Environ. Sci.* **2003**, *150*, D79–D83. [CrossRef]
29. Oturan, N.; Oturan, M.A. Chapter 8—Electro-Fenton Process: Background, New Developments, and Applications. In *Electrochemical Water and Wastewater Treatment*; Martínez-Huitle, C.A., Rodrigo, M.A., Scialdone, O., Eds.; Butterworth-Heinemann: Oxford, UK, 2018; pp. 193–221. ISBN 9780128131602. [CrossRef]
30. Li, S.; Liu, Y.; Ge, R.; Yang, S.; Zhai, Y.; Hua, T.; Ondon, B.S.; Zhou, Q.; Li, F. Microbial electro-Fenton: A promising system for antibiotics resistance genes degradation and energy generation. *Sci. Total Environ.* **2020**, *699*, 134160. [CrossRef] [PubMed]
31. Pérez, T.; Sirés, I.; Brillas, E.; Nava, J.L. Solar photoelectro-Fenton flow plant modeling for the degradation of the antibiotic erythromycin in sulfate medium. *Electrochim. Acta* **2017**, *228*, 45–56. [CrossRef]
32. Sánchez Ruíz, C. Fenton Reactions (FS-TER-003). Inditex. 2015. Available online: https://www.wateractionplan.com/documents/177327/558166/Fenton+reactions.pdf/087c01a6-7f9c-2f33-95e7-b8a5945b9162 (accessed on 12 April 2021).
33. APHA. Method 3500-Fe B. ASTMD 1068-77, Iron in Water, Test Method. In *Standard Methods Standard Methods for the Examination of Water and Wastewater*, 22nd ed.; American Public Health Association: Washington, DC, USA, 2012.
34. Welcher, F.J. *Standard Methods of Chemical Analysis*, 6th ed.; Huntington 2B; R.E. Krieger Publishing Company: Florida, FL, USA, 1975; pp. 1827–1828.
35. Hassanzadeh, A.; Barber, J.; Morris, G.A.; Gorry, P.A. Mechanism for the degradation of erythromycin A and erythromycin A 2′-Ethyl succinate in acidic aqueous solution. *J. Phys. Chem. A* **2007**, *111*, 10098–10104. [CrossRef]
36. Volmer, D.A.; Hui, J.P. Study of erythromycin A decomposition products in aqueous solution by solid-phase microextraction/liquid chromatography/tandem mass spectrometry. *Rapid Commun. Mass Spectrom.* **1998**, *12*, 123–129. [CrossRef]
37. Chitneni, S.K.; Govaerts, C.; Adams, E.; Van Schepdael, A.; Hoogmartens, J. Identification of impurities in erythromycin by liquid chromatography–mass spectrometric detection. *J. Chromatogr. A* **2004**, *1056*, 111–120. [CrossRef]
38. Deubel, A.; Fandino, A.S.; Sörgel, F.; Holzgrabe, U. Determination of erythromycin and related substances in commercial samples using liquid chromatography/ion trap mass spectrometry. *J. Chromatogr. A* **2006**, *1136*, 39–47. [CrossRef]
39. Luiz, D.B.; Genena, A.K.; Virmond, E.; José, H.J.; Moreira, R.F.; Gebhardt, W.; Schröder, H.F. Identification of degradation products of erythromycin A arising from ozone and advanced oxidation process treatment. *Water Environ. Res.* **2010**, *82*, 797–805. [CrossRef]

40. Malvar, J.L.; Abril, C.; Martín, J.; Santos, J.L.; Aparicio, I.; Escot, C.; Basanta, A.; Alonso, E. Development of an analytical method for the simultaneous determination of the 17 EU Watch List compounds in surface waters: A Spanish case study. *Environ. Chem.* **2018**, *15*, 493–505. [CrossRef]
41. Ganzenko, O.; Trellu, C.; Oturan, N.; Huguenot, D.; Péchaud, Y.; van Hullebusch, E.D.; Oturan, M.A. Electro-Fenton treatment of a complex pharmaceutical mixture: Mineralization efficiency and biodegradability enhancement. *Chemosphere* **2020**, *253*, 126659. [CrossRef] [PubMed]
42. Pignatello, J.J.; Liu, D.; Huston, P. Evidence for an additional oxidant in the photoassisted Fenton reaction. *Environ. Sci. Technol.* **1999**, *33*, 1832–1839. [CrossRef]
43. Wu, K.; Xie, Y.; Zhao, J.; Hidaka, H. Photo-Fenton degradation of a dye under visible light irradiation. *J. Mol. Catal. A Chem.* **1999**, *144*, 77–84. [CrossRef]
44. Diari Oficial de la Generalitat de Catalunya. DECRET 130/2003, de 13 de maig, pel qual s'aprova el Reglament dels serveis públics de sanejament. *Diari Oficial de la Generalitat de Catalunya* **2003**, *3894*, 11143–11158.
45. Buxton, G.V.; Greenstock, C.L.; Helman, W.P.; Ross, A.B. Critical-Review of Rate Constants for Reactions of Hydrated Electrons, Hydrogen-Atoms and Hydroxyl Radicals($^{\bullet}$OH/$^{\bullet}$O^{-}) in Aqueous-Solution. *J. Phys. Chem. Ref. Data* **1988**, *17*, 513–886. [CrossRef]
46. Grundl, T.; Delwiche, J. Kinetics of ferric oxyhydroxide precipitation. *J. Contam. Hydrol.* **1993**, *4*, 71–87. [CrossRef]
47. Faust, B.C.; Hoigné, J. Photolysis of Fe(III)-hydroxy complexes as source of OH radical in clouds, fog and rain. *Atmos. Environ.* **1990**, *24*, 79–89. [CrossRef]
48. Gallard, H.; De Laat, J.; Legube, B. Spectrophotometric study of the formation of iron(III)-hydroperoxy complexes in homogeneous aqueous solutions. *Water Res.* **1999**, *33*, 2929–2936. [CrossRef]
49. Serra-Clusellas, A.; De Angelis, L.; Lin, C.-H.; Vo, P.; Bayati, M.; Sumner, L.; Lei, Z.; Amaral, N.B.; Bertini, L.M.; Mazza, J.; et al. Abatement of 2,4-D by H2O2 solar photolysis and solar photo-Fenton-like process with minute Fe(III) concentrations. *Water Res.* **2018**, *144*, 572–580. [CrossRef]
50. Hirsch, R.; Ternes, T.; Haberer, K.; Kratz, K.-L. Occurrence of antibiotics in the aquatic environment. *Sci. Total Environ.* **1999**, *225*, 109–118. [CrossRef]
51. Zhang, Y.; Duan, L.; Wang, B.; Liu, C.S.; Jia, Y.; Zhai, N.; Blaney, L.; Yu, G. Efficient multiresidue determination method for 168 pharmaceuticals and metabolites: Optimization and application to raw wastewater, wastewater effluent, and surface water in Beijing, China. *Environ. Pollut.* **2020**, *261*, 114113. [CrossRef] [PubMed]
52. Eurostat. Electricity Prices (Including Taxes) for Household Consumers, First Half 2020. Available online: https://ec.europa.eu/eurostat/statistics-explained/index.php/Electricity_price_statistics (accessed on 5 March 2021).

MDPI
St. Alban-Anlage 66
4052 Basel
Switzerland
Tel. +41 61 683 77 34
Fax +41 61 302 89 18
www.mdpi.com

Water Editorial Office
E-mail: water@mdpi.com
www.mdpi.com/journal/water

www.ingramcontent.com/pod-product-compliance
Lightning Source LLC
LaVergne TN
LVHW072346090526
838202LV00019B/2490